Radio Propagation

for

Modern Wireless Systems

ISBN 0-13-026373-7

9 780130 263735

Radio Propagation for Modern Wireless Systems

Henry L. Bertoni

PH PTR

Prentice Hall PTR
Upper Saddle River, NJ 07458
www.phptr.com

Editorial/production supervision: *Mary Sudul*
Cover designer: *Wee Design Group*
Cover design director: *Jerry Votta*
Manufacturing manager: *Alexis R. Heydt*
Marketing manager: *Lisa Konzelman*
Acquisitions editor: *Bernard Goodwin*
Editorial assistant: *Diane Spina*
Composition: *Aurelia Scharnhorst*

Prentice Hall, Inc.
Upper Saddle River, New Jersey 07458

Prentice Hall books are widely used by corporations and government agencies for training, marketing, and resale.
The publisher offers discounts on this book when ordered in bulk quantities. For more information, contact Corporate Sales Department. Phone: 800-382-3419; Fax: 201-236-7141; E-mail: corpsales@prenhall.com; Or write: Prentice Hall PTR, Corp. Sales Dept., One Lake Street, Upper Saddle River, NJ 07458

Printed in the United States of America
10 9 8 7 6 5 4 3 2 1

ISBN 0-13-026373-7

Prentice-Hall International (UK) Limited, *London*
Prentice-Hall of Australia Pty. Limited, *Sydney*
Prentice-Hall Canada Inc., *Toronto*
Prentice-Hall Hispanoamericana, S.A., *Mexico*
Prentice-Hall of India Private Limited, *New Delhi*
Prentice-Hall of Japan, Inc., *Tokyo*
Prentice-Hall (Singapore) Pte. Ltd., *Singapore*
Editora Prentice-Hall do Brasil, Ltda., *Rio de Janeiro*

This book is dedicated to the students who came to work with me so that we could learn together.

Contents

Preface

The commercial success of cellular mobile radio since its initial implementation in the early 1980s has led to an intense interest among wireless engineers in understanding and predicting radio propagation characteristics within cities, and even within buildings. In this book we discuss radio propagation with two goals in mind. The first is to provide practicing engineers having limited knowledge of propagation with an overview of the observed characteristics of the radio channel and an understanding of the process and factors that influence these characteristics. The second goal is to serve as text for a master's-level course for students intending to work in the wireless industry. Books on modern wireless applications typically survey the issues involved, devoting only one or two chapters to radio channel characteristics, or focus on how the characteristics influence system performance. Now that the wireless field has grown in scope and size, it is appropriate that books such as this one examine in greater depth the various underlying topics that govern the design and operation of wireless systems. The material for this book has grown out of tutorials given by the author to engineering professionals and a course on wireless propagation given by the author at Polytechnic University as part of a program in wireless networks. It also draws upon the 15 years of experience the author and his students have had in understanding and predicting propagation effects.

Cellular telephones gave the public an active role in the use of the radio spectrum as opposed to the previous role of passive listener. This social revolution in the use of the radio spectrum ultimately changed governmental views of its regulation. Driven by the requirement to allow many users to operate in the same band, cellular telephones also created a technical revolution through the concept of spectral reuse. Systems that do not employ spectral reuse avoid interference by operating in different frequency bands and are limited in performance primarily by noise. In these systems, lack of knowledge of the propagation conditions can be compensated for by increasing the transmitted power, up to regulatory limits. In contrast, the concept of spectral reuse acknowledges that in commercially successful systems, interference from other users will be the primary factor limiting performance. In designing these systems, it is necessary to balance the desired signal for each user against interference from signals intended for other users. Finding the balance requires knowledge of the radio channel characteristics. Chapter 1 is intended to introduce the student reader to the concept of spectrum reuse and in the process to give examples of how the propagation characteristics influence the balance between desired signal and interference, and thereby influence system design. As in all chapters, examples are discussed to illustrate the concepts, and problems are included at the end of the chapter to give the students experience in applying the concepts.

In modern systems, the radio links are about 20 kilometers or less, the antennas that create the links lie near to or among the buildings or even inside the buildings, and the wavelength is small compared to the building dimensions. As a result, the channel characteristics are strongly influenced by the buildings as well as by vegetation and terrain. In this environment, signals propagate from one antenna to the other over multiple paths that involve the processes of reflection and transmission at walls and by the ground and the process of diffraction at building edges and terrain obstacles. The multipath nature of the propagation makes itself felt in a variety of ways that have challenged the inventiveness of communication engineers. Although initially a strong limitation on channel capacity, engineers have begun to find ways to harness the multipath signals so as to achieve capacities that approach the theoretical limit. However, each new concept for dealing with multipath calls for an even deeper understanding of the statistical characteristics of the radio channel. In Chapter 2 we describe many of the propagation effects that have been observed in various types of measurements, ranging from path loss for narrowband signals, to angle of arrival and delay spread for wideband transmission. As in other chapters, an extensive list of references is cited to aid the professional seeking a detailed understanding of particular topics. For the student reader, this chapter serves as an introduction to the types of measurements that are made, the methods used to process the data, and some of the statistical approaches used to represent the results. Understanding the measurements, their processing, and their representation also serves to guide the theoretical modeling described in subsequent chapters.

The level of presentation assumes that the reader has had an undergraduate course in electromagnetics with exposure to wave concepts. The presentation does not attempt to derive the propagation characteristics from Maxwell's equations rigorously; rather, the goal is to avoid vector calculus. The reader's background is relied on for acceptance of some wave properties; other properties are motivated through heuristic arguments and from basic ideas, such as conservation of power. For example, in Chapter 3 we start with the fundamental properties of plane waves and call on the reader's background in transmission lines when discussing reflection and transmission at the ground and walls. Wherever possible in this and following chapters, the theoretical results are compared to measurements. Thus plane waves are used to model observed interference effects, which are referred to as fast fading, and to model Doppler spreading. Plane wave properties and conservation of power are used in Chapter 4 to justify the properties of spherical waves radiated by antennas and to motivate the ray description of reflection at material surfaces. By accounting for these reflections, propagation on line-of-sight paths in urban canyons is modeled. Circuit concepts are used to obtain the reciprocity of propagation between antennas, and to derive expressions for path gain or loss.

Diffraction at building edges is an important process in wireless communications. It allows signals to reach subscribers who would otherwise be shadowed by the buildings. Because the reader is not expected to be familiar with this process, Chapter 5 explores diffraction in some detail. For simplicity, the scalar form of the Huygens–Kirchhoff integral is use as a starting point. We first use it to give physical meaning to the Fresnel ellipsoid about a ray, which is widely employed in propagation studies to scale physical dimensions. The geometrical and uniform forms of the fields diffracted by an absorbing half screen are derived. In these expressions we identify a universal component that applies to diffraction by any straight building edge or corner and a diffraction coefficient whose specific form is dependent on the nature of the edge. Diffraction coefficients for several types of edges and corners are given without derivation. Using heuristic ray arguments, the results obtained for plane waves are generalized to spherical waves radiated by antennas and to multiple edges. These results are cast in terms of path gain or loss, which is convenient for wireless applications.

Chapter 6 formulates the problem of average path loss in residential environments in terms of multiple diffraction past rows of buildings. Relying on the Huygens–Kirchhoff formulation, the diffraction problem is solved for various ranges of base station and subscriber antenna height. These results show how the frequency, average building height, and row separation influence the range dependence and height gain of the signal. This approach to diffraction is used in Chapter 7 to investigate the effects of randomness in building construction on shadow fading. Chapter 7 also makes use of diffraction to examine the effects of terrain and vegetation on the average path loss.

Propagation predictions that make use of a geometrical description of individual buildings are discussed in Chapter 8. Various ray-based models that incorporated the processes of reflection and diffraction at buildings have been developed to make such site-specific predictions. Their accuracy has been evaluated primarily by comparing predictions against measurements of the small area average received signal. However, the ray models have started to be used to predict higher-order channel statistics, such as time delay and angle spread, through Monte Carlo simulations. This approach can generate values for the statistical descriptors of the radio channel that are employed in advanced communication systems and show how these values depend on the distribution of building size and shape in different cities.

Acknowledgment

I would like to acknowledge the people who contributed to this book. First I must thank my wife Helene Ebenstein for continued encouragement during is writing, and for proof reading the final manuscript. Byung-Chul Kim of Polytechnic University prepared many of the computed curves, as well as many of the drawings. Additional computed curves and drawings were prepared by Jeho Lee, Hyun-Kyu Chung and Cheolhang Cheon of Polytechnic, and George Liang of Site Ware Technologies. The programs appearing on the web site for this book were prepared by Byung-Chul Kim, Hyun-Kyu Chung, George Liang, Jeho Lee and Leonard Piazzi. I was fortunate to have worked with a number of students whose ideas and efforts are embodied in the book. Finally, I would like to thank Mohsen Gharabaghloo for introducing me to the problems of propagation for cellular mobile radio in the early 1980's.

CHAPTER 1

The Cellular Concept and the Need for Propagation Prediction

The introduction of cellular mobile radio (CMR) telephone systems in the early 1980s marked an important turning point in the application of radio technology in communications. Prior to that time, the commercial use of radio spectrum was dominated by broadcast radio and television, although other applications, such as fleet dispatch, walkie-talkies, and citizens band (CB) radio were of growing importance. Broadcast systems are intended to cover an entire metropolitan area from a single transmitter whose signal is in an assigned frequency channel. As a result, the system design goal is to achieve the largest possible coverage area in which the received power is sufficiently strong compared to background noise. This goal is achieved by locating the transmitting antenna on a tall building or tower and radiating the maximum allowed power. In fringe areas, the receiving antennas are also located atop buildings or masts whenever possible. To support the design of broadcast systems, experimental and theoretical studies were made of radio propagation over long distances of 100 km or more, accounting for the earth's curvature, refraction in the atmosphere, and large-scale terrain features.

In contrast, CMR telephone systems were designed to give mobile subscribers access to a communication system by using radio over short links at the end of an otherwise wired network. The short link is intended to cover only a fraction of the metropolitan area, so that the decrease of the radio signal with distance allows the spectrum to be reused elsewhere within the same metropolitan area. In this case the system design goal is to make the received power adequate to overcome background noise over each link, while minimizing interference to other more distant links operating at the same frequency. Achieving such a balance between coverage over the desired links, while avoiding undue interference to other links, greatly complicates the system design problem. Fueled by the enormous commercial success of CMR telephones, many additional wireless systems and applications have been introduced or proposed [1,2] that in one way or another make use of the cellular concept to accommodate many users through spatial reuse of the limited radio spectrum assigned. The purpose of this chapter is to acquaint the reader with the reuse concept and to show why the characteristics of radio propagation are a fundamental feature in determining system design.

The initial deployment of CMR in metropolitan areas employed radio links covering up to about 20 km from the subscriber to the base station (access point to the wired system). However, as more base stations have been added to accommodate the growing number of subscribers, the maximum propagation distance has been reduced significantly. For newer systems, especially those envisioned for indoor applications such as wireless local area networks (W-LANs) and wireless private branch exchanges (W-PBXs), the maximum distance may be no more than a few hundred meters. Over these short links, the buildings have a profound influence on the radio propagation. It is the influence of the buildings on the radio signal that is the principal subject of this book. We also consider the influence of terrain and vegetation, but atmospheric effects are not significant.

Cellular systems have operated in the frequency band from 450 to 900 MHz. Additional bands near 1.9 GHz are now being used throughout the world for second-generation cellular systems, while 3.9 GHz is used for wireless local loops. Unlicensed frequency bands near 900 MHz, 2.4 GHz, and 5.2 GHz are or will be used for wireless LANs, PBXs, and other applications. The wavelength is less than 1 m for these frequencies, making it significantly smaller than the typical dimensions of buildings but larger than the roughness of building materials. As a result, the propagation of radio waves can be understood and described mathematically in terms of the processes of reflection and transmission at walls and diffraction by building edges. To make use of such a description, several chapters are devoted to these processes. Understanding these processes is also important for other systems that have been proposed for operation at higher frequencies, ranging up to 30 GHz.

One cannot speak of the history of CMR, however briefly, without acknowledging the importance of integrated-circuit technology, which has allowed intelligence, control functions, and signal processing to be located in the fixed system and in the subscriber units. The steady progress in microminiaturization since the introduction of CMR is seen in the reduction in size of mobile telephone units from that of a large briefcase to today's pocket phones. Continued progress in miniaturization allows the use of more intelligence and signal processing capability to overcome the limitations imposed by the propagation characteristics of the radio channel (e.g., through the use of smart antenna systems) [3]. However, the design of systems that make use of the intelligence requires an even deeper knowledge of the radio channel.

1.1 Concept of spatial reuse

Multiple access methods allow the separation of simultaneous radio signals sent to or from an individual base station or access point and several mobile subscribers in the same local area. The first access method implemented for cellular telephones was frequency-division multiple access (FDMA), which is used in connection with analog frequency modulation (FM) of the transmitted signal. In FDMA systems for two-way transmission, each subscriber is assigned one frequency channel for the uplink from the subscriber to the base station, and a second channel for the downlink to the subscriber. In North America the currently operating FDMA system, known as AMPS (Advanced Mobile Phone System), has a total of $N_c = 395$ two-way channels, each one-way channel being 30 kHz wide [1, pp. 80–81]. In FDMA systems, spatial reuse amounts to reusing the same N_c frequency channels in different geographical subareas. In this case the frequency reuse plan must place the local areas far enough apart to limit the interference from the other cochannel signals.

The introduction of digital transmission of voice signals has led to new approaches to multiple access. One approach is time-division multiple access (TDMA), in which short segments of digitized voice, or other information, are compressed into shorter time intervals and transmitted at an assigned time slot in a recurring sequence. In one approach known as IS-54, which is intended to permit simple migration from the AMPS system, each 30 kHz AMPS channel is used to carry three digitized voice channels using three time slots [1]. A widely used system known as GSM transmits eight TDMA signals in frequency bands that are 200 kHz wide [1]. If the time slots of all subscribers and base stations are synchronized, multiple access is essentially that of FDMA during a particular time slot. As a result, spatial reuse is similar to that of FDMA.

The most recent cellular telephone systems employ digital transmission with code-division multiple access (CDMA) for distinguishing the signals to and from individual subscribers [4]. In this approach each information bit is transmitted as a coded sequence of shorter-duration bits called *chips*. Because of this encoding, a greater bandwidth must be used for an individual call than would otherwise be needed to send the information bits. However, all subscribers in the system use the same radio-frequency band, with each subscriber having a distinct code that can be distinguished upon reception. To any one subscriber, the interference from signals sent to all the other subscribers appear as background noise, whose level will be inversely proportional to the number of chips per bit (known as the *processing gain*). For one commonly employed CDMA system known as IS-95, each bit of information is transmitted using 128 chips, and the voice channel is spread over an entire 1.23 MHz band. Separate frequency bands are used for downlink and uplink transmission. In this case the spatial reuse plan must limit to an acceptable level the total interference generated by all other subscribers.

Because spatial reuse must balance the need to provide an adequate signal for each subscriber, and at the same time limit the cochannel interference due to signals for other subscribers, radio propagation characteristics play an important role in system design. In the remainder of this chapter we illustrate the concepts of spatial reuse through several examples, and through these examples show why the propagation characteristics are critical in system design. We start by describing spatial reuse in FDMA systems and then briefly discuss the interference limitations in CDMA systems.

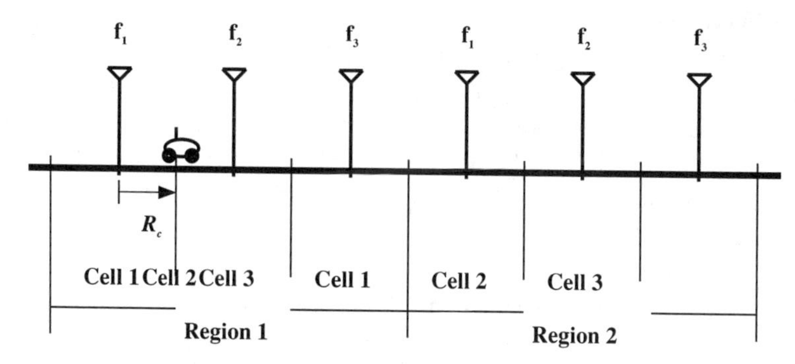

Figure 1-1 One-dimensional cellular FDMA system serving a highway and making use of three cells per frequency reuse region.

1.2 Linear cells as an example of FDMA spectrum reuse

To achieve greater capacity for FDMA systems within a single metropolitan or other large area, the area is divided into a number of regions N that reuse the same set of N_c radio-frequency channels. As a result, the total number of simultaneous phone calls is NN_c, which can be increased as demand requires by dividing the metropolitan area into a greater number N of regions. Each region is further subdivided into N_R cells, each being serviced by a base station to which are assigned a fraction N_c/N_R of the available radio-frequency channels. The base stations are connected through wired lines to switching centers that control the network and connect calls to the wired telephone network. The role of this subdivision is to separate those cells using the same frequency channels by a distance that is large enough to keep the interference small. For example, acceptable voice quality in the North American AMPS system requires that the received power P (watts) of the desired signal from the serving base station must be 50 or more times stronger than the total received interference I (watts) from all other cochannel base stations [5]. Thus the signal to interference ratio P/I must be greater than 50, which on a decibel scale is 10 log 50 = 17 dB.

Figure 1–1 shows a one-dimensional view of how regions might be divided into cells to give coverage along a highway. In this example, each region is divided into $N_R = 3$ cells of radius R_c. On the downlink, a mobile in cell 1 of region 1 will experience the lowest value of received signal P and the highest interference I from region 2 when it is near the right-hand edge of the cell, as shown. In this case the mobile is at a distance R_c from the serving base station and at a distance $(2N_R - 1)R_c$ from the nearest interfering base station using the same frequency, which is located in cell 1 of region 2 in Figure 1–1. To evaluate the ratio P/I at the cell radius, it is necessary to understand how signals propagate in the particular environment.

As discussed in Chapter 4, for propagation in free space the power P received by one antenna due to radiation of P_T watts by a distant antenna has the form $P = P_T\, A/R^n$, where A is a constant, R is the distance between the antennas, and the range index n is 2 [6]. We also show in Chapter 4 that for low base station antennas above a flat earth, a similar expression holds for large enough separation, but in this case the range index is $n = 4$ [6]. Assuming that all base stations radiate the same power, and accounting only for the interference from the nearest cochannel base station at the distance $(2N_R - 1)R_c$, the foregoing dependence for the power received by a subscriber gives the downlink signal-to-interference ratio at the cell boundary:

$$\frac{P}{I} = \frac{P_T A / R_c^{\,n}}{P_T A / [(2N_R - 1)R_c]^n} = (2N_R - 1)^n \tag{1–1}$$

If $N_R = 3$, as in Figure 1–2, the range index $n = 2$ for free space gives $P/I = 25$, which is too small, while the range index $n = 4$ for flat earth gives $P/I = 625$, which is much more than needed. To achieve signal to interference values close to 50 requires $N_R = 4$ if $n = 2$ and $N_R = 2$ if $n = 4$.

The implication of (1–1) can be seen from a simple example for coverage along a six-lane highway at rush hour. If the cars have a center-to-center spacing of 10 m, there are 100 cars per lane per kilometer, and a total of 600 cars/km. If each car has a cell phone that is used 2% of the time, there are 12 simultaneous calls per kilometer on average. If 200 channels of an AMPS system are devoted to cover these calls, then for $n = 2$ it was shown above that $N_R = 4$ and each cell will have 50 channels. In this case each cell has a diameter of $2R_c = 50/12 = 4.2$ km, which is also the spacing between base stations. However, if $n = 4$, then $N_R = 2$ and each cell has 100 channels, in which case the spacing between base stations is $2R_c = 8.4$ km. As seen from this example, the number of base stations needed to serve a given number of subscribers along a roadway is proportional to N_R if trunking efficiency is neglected. Thus the simple example of linear cells shows that the design of the fixed system is strongly dependent on the propagation characteristics of the operating environment. Since installation costs and real estate rental make base stations expensive, economic operation of cellular systems calls for the use of the least number to achieve an acceptable level of service. The foregoing analysis was for the downlink from the base station to the subscriber. A similar analysis applies to the signal-to-interference ratio on the uplink, except that the transmission power of individual mobiles is controlled by the communicating base station, so that the mobiles do not all transmit the same power.

The propagation characteristics encountered in actual metropolitan environments differ in important ways from the simple dependence indicated above. For high base stations, the range dependence of the received power is of the form $P = P_T\, A/R^{\,n}$, where n is typically between 3 and 4. For the low base stations of more advanced systems, A and n can depend on the direction, relative to the street grid, and n can be greater than 4. In addition to the range dependence, the signal is found to have significant random variations over two smaller scale lengths, which are referred to as *fast fading* and *shadow fading*. The effect of this fading is to greatly increase the minimum value of P/I in (1–1). For systems operating inside buildings, propagation is again different. The signal variations observed in different environments are discussed in Chapter 2.

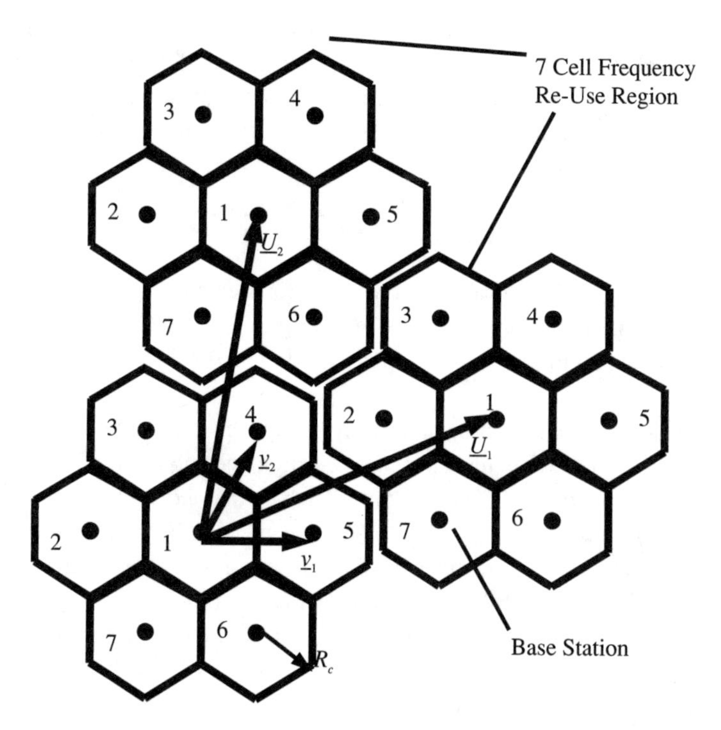

Figure 1-2 Hexagonal cells that are used to cover an area for the case of regions having a symmetric pattern with $N_R = 7$.

1.3 Hexagonal cells for area coverage

The ideas illustrated above for one-dimensional cells have been employed to achieve two-dimensional coverage over a metropolitan area using cells whose conceptual shape is in the form of a hexagon. The choice of the hexagon to represent cell shape is made because hexagons are the highest-degree regular polygons that can tile a plane and because they approximate the circular contours of equal received signal strength when the propagation is isotropic in the horizontal plane. While hexagonal cells are widely used to understand and evaluate system concepts, in actual system planning, terrain and other effects result in cells that are far less regular, even for elevated base station antennas. Also, locating the base stations is strongly influenced by the practical problem of finding acceptable sites and may not follow the regular hexagonal grid.

Hexagonal cells are shown in Figure 1–2 for the case of $N_R = 7$ cells per region. The displacements between any two cells can be expressed as a linear combination of the two basis vectors $\mathbf{v}_1$ and $\mathbf{v}_2$ having an included angle of 60°, as shown in Figure 1–3. If the cell radius R_c is defined to be the distance from the center to any vertex of the hexagon, then $|\mathbf{v}_1| = |\mathbf{v}_2| = \sqrt{3}R_c$. The area of the parallelogram defined by $\mathbf{v}_1$ and $\mathbf{v}_2$ has the same transitional periodicity as the hexagons, hence both have the same area. Thus the area of a hexagon is given by $|\mathbf{v}_1 \times \mathbf{v}_2| = 3R_c^2 \sin 60°$.

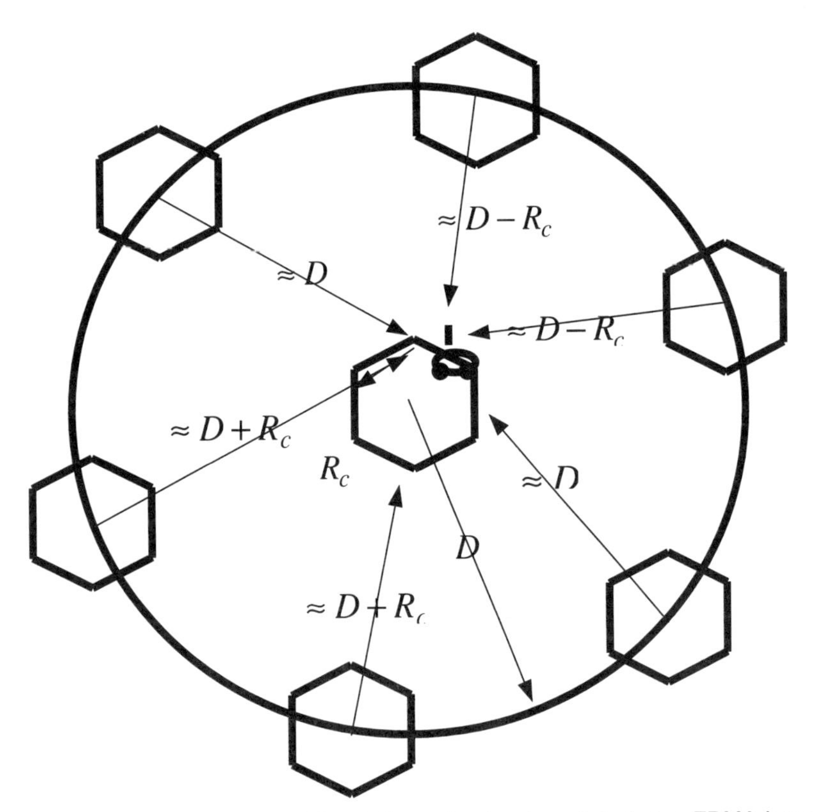

Figure 1-3 Distances to a mobile at the cell edge from interfering FDMA base stations located in the first tier of the serving base station.

The regions of frequency reuse can be composed of any integer number N_R of contiguous cells, and after the region shape is defined, all other regions are obtained by translation of the defining region through a linear combination of the frequency reuse vectors $\mathbf{U}_1$ and $\mathbf{U}_2$ [7], as indicated in Figure 1–2. The displacement between any two cells using the same frequencies can also be expressed as a linear combination of the two reuse vectors. Because the regions have the same transitional periodicity as the parallelogram defined by $\mathbf{U}_1$ and $\mathbf{U}_2$, the area of the region is given by $|\mathbf{U}_1 \times \mathbf{U}_2|$. This area is also equal to N_R times the area of an individual cell. The displacement vector can be expressed in terms of the basis vectors as

$$\begin{aligned} \mathbf{U}_1 &= k_1\mathbf{v}_1 + m_1\mathbf{v}_2 \\ \mathbf{U}_2 &= k_2\mathbf{v}_1 + m_2\mathbf{v}_2 \end{aligned} \tag{1–2}$$

where the constants $k_{1,2}$ and $m_{1,2}$ are integers. In terms of these constants, the area covered by a region is

$$|\mathbf{U}_1 \times \mathbf{U}_2| = |k_1m_2 - k_2m_1||\mathbf{v}_1 \times \mathbf{v}_2| \tag{1–3}$$

Since $|\mathbf{U}_1 \times \mathbf{U}_2|$ is equal to N_R times the cell area $|\mathbf{v}_1 \times \mathbf{v}_2|$, it is seen that

$$N_R = |k_1 m_2 - k_2 m_1| \tag{1–4}$$

1.3a Symmetric reuse patterns

Regions may take various shapes for different choices of the integers $k_{1,2}$ and $m_{1,2}$, which can be selected to give any integer value of N_R. Different choices of $k_{1,2}$ and $m_{1,2}$ can result in regions having different shapes but the same value of N_R. For some choices of integers $k_{1,2}$ and $m_{1,2}$, the reuse vectors $\mathbf{U}_1$ and $\mathbf{U}_2$ will be of equal magnitude and the angle between them is 60°, as shown in Figure 1–2. This choice results in a symmetric arrangement of the cochannel cells using the same frequencies into circular tiers about any reference cell. The center-to-center distance from the reference cell to the cochannel cells in a tier are all equal. In the first tier, there are six cochannel cells, as indicated in Figure 1–3, whose distance from the reference cell is $D = |\mathbf{U}_{1,2}|$. Nonsymmetric arrangements having the same value of N_R have some co-channel cells that are closer to the reference cell than the value D for the symmetric region shape, leading to higher values of interference.

For symmetric reuse patterns, only certain values of the reuse factor N_R are possible. To find these values, the coefficients k_2 and m_2 are expressed in terms of k_1 and m_1 by requiring that $\mathbf{U}_2$ have magnitude equal to that of $\mathbf{U}_1$, and be rotated 60° counterclockwise from $\mathbf{U}_1$. This can be achieved by noting that $\mathbf{v}_2$ is rotated 60° counterclockwise from $\mathbf{v}_1$ and that the vector $\mathbf{v}_2 - \mathbf{v}_1$ has the same length as $\mathbf{v}_1$ and $\mathbf{v}_2$, and is rotated 60° counterclockwise from $\mathbf{v}_2$. Thus the rotated vector $\mathbf{U}_2$ can be found in terms of the rotated basis vectors using the same coefficients as in (1–2) for $\mathbf{U}_1$, or

$$\mathbf{U}_1 = k_1\mathbf{v}_2 + m_1(\mathbf{v}_2 - \mathbf{v}_1) = -m_1\mathbf{v}_1 + (m_1 + k_1)\mathbf{v}_2 \tag{1–5}$$

Comparing (1–2) and (1–5) it is seen that $m_2 = -m_1$ and $k_2 = m_1 + k_1$. It is easily verified that $|\mathbf{U}_2|$ is the same as $|\mathbf{U}_1|$ and that the angle between them is 60°. Substituting these expressions for m_2 and k_2 into expression (1–4) gives

$$N_R = m_1{}^2 + m_1 k_1 + k_1{}^2 \tag{1–6}$$

Substituting integer values for m_1 and k_1 into (1–6) gives the values of N_R for which the cochannel cells are symmetrically located on circles about any reference cell. These values for N_R are 1, 3, 4, 7, 9, 12, 13, and so on.

1.3b Interference for symmetric reuse patterns

While (1–6) relates N_R to the reuse pattern, an alternative formulation is more useful for examining the relation between N_R and the signal-to-interference ratio for the symmetric patterns. As noted previously, the area of a region can be expressed in terms of D via $|\mathbf{U}_1 \times \mathbf{U}_2| = D^2 \sin 60^\circ$, while the area of a cell is $|\mathbf{v}_1 \times \mathbf{v}_2| = 3(R_c)^2 \sin 60°$. Because the area of a region is N_R times the area of a cell,

$$N_R = \frac{|\mathbf{U}_1 \times \mathbf{U}_2|}{|\mathbf{v}_1 \times \mathbf{v}_2|} = \frac{1}{3}\left(\frac{D}{R_c}\right)^2 \tag{1–7}$$

To evaluate the downlink interference from cochannel cells, consider a mobile at the cell radius, as shown in Figure 1–3, and for simplicity assume that the signal received from the controlling base station is $P = P_T A/R^n$. The distance to the nearest two cochannel base stations is approximately $D - R_c$, so that the interference signal from these two base stations is $I = 2P_T A/(D - R_c)^n$. The requirement on P/I for analog FM therefore becomes

$$50 \le \frac{P}{I} = \frac{1}{2}\left(\frac{D}{R_c} - 1\right)^n \tag{1–8}$$

or conversely,

$$\frac{D}{R_c} \ge 1 + \sqrt[n]{100} \tag{1–9}$$

For free-space propagation, the range index is $n = 2$, so that from (1–9) we see that $D/R_c = 11$. When this inequality is substituted into (1–7) we obtain $N_R = 40.3$, which is satisfied for a symmetric region having $k_1 = 6$, $m_1 = 1$, so that from (1–6), $N_R = 43$. Accounting for other cochannel base stations in the first tier will require even higher values of N_R. By way of comparison, for propagation over flat earth, $n = 4$, so that from (1–9), $D/Rc = 4.2$. When substituted into (1–7) this inequality requires that $N_R = 5.9$, which is well satisfied by a symmetric region when $N_R = 7$. The significance of this result is that providing adequate service to a frequency reuse region requires at least 43 base stations if $n = 2$ but only 7 base stations if $n = 4$ (assuming that the propagation is of the form $P = P_T A/R^n$, and accounting for only two interferers). Since the AMPS system has approximately 400 channels, each base station would handle about 10 calls for $n = 2$, but nearly 60 calls for $n = 4$, giving great savings in infrastructure costs. Even greater savings are achieved for $n = 4$ when the statistical nature of teletraffic is taken into account [2, pp. 44–54].

1.4 Sectored cells

The foregoing examples show how important the propagation characteristics are for an interference-limited system. For high base station antennas covering large cells, the range dependence is of the form $P = P_T A/R^n$, usually with n somewhat less than 4, about which there are additional variations that have a random nature. Accounting for the actual range index n and the additional variations, the $N_R = 7$ reuse factor is not adequate to limit the interference from all cochannel base stations for good performance of an AMPS system. For example, if the signal P from the base station is reduced by 10 dB due to a fade, P/I will be a factor of 10 smaller than found from (1–8). Note that if $n = 4$ and $D/R_c = 4.6$, as found from (1–7) for $N_R = 7$, P/I obtained from (1–8) is 82.1 not counting the fading, but only 8.21 when the fade is taken into account.

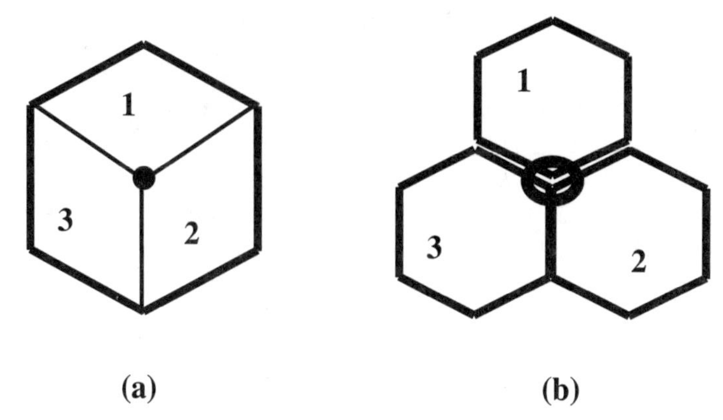

Figure 1-4 Directional antennas at the base station are used to improve signal/interference ratio: a) the conventional approach using 120° beam antennas to illuminate individual sectors in a cell; b) an $N_R = 21$ pattern with three cells served by a single base station.

To improve the *P*/*I* ratio without increasing the number of base stations, the cellular concept is modified by the use of directional antennas at the base station that introduce anisotropy in the coverage. Typically, three antennas having 120° beam width are used to divide each hexagon in an $N_R = 7$ pattern into three sectors, as indicated in Figure 1–4a. Sectorization has the effect of reducing the number of base stations that cause interference and increasing the distance from mobiles at a cell boundary to the nearest interfering base stations, at the cost of increasing the actual reuse pattern to $N_R = 21$. Alternatively, one can think of the cells as being organized in a symmetric reuse pattern having $N_R = 21$ but with the base station located at the cell vertex rather than at the center of the cell. Such an approach is suggested in Figure 1–4b, where it is seen that each base station serves three cells. Such cells can be served by antennas having 60° beam width. Because the antenna beam width is smaller for this approach than for the sectorization of Figure 1–4a, this approach will result in even lower interference from other base stations using the same frequency [8].

1.5 Spatial reuse for CDMA

To understand how the propagation characteristics influence CDMA system design, it is simplest to consider the downlink from base station to mobile. Communication to all subscribers takes place in the same band. Thus the total interference to any one subscriber will come from the signals being sent to the other subscribers in the cell and from the signals being sent to mobiles in other cells from their respective base stations. A mobile will receive all the signals being sent by its base station with equal amplitude since all signals travel over the same path. However, the relative strength of the signals received from other base stations will depend on the propagation characteristics.

Consider a mobile located at the junction of three cells labeled A, B, and C, as shown in Figure 1–5. Through a design feature in CDMA called *soft hand-off*, the subscriber can simulta-

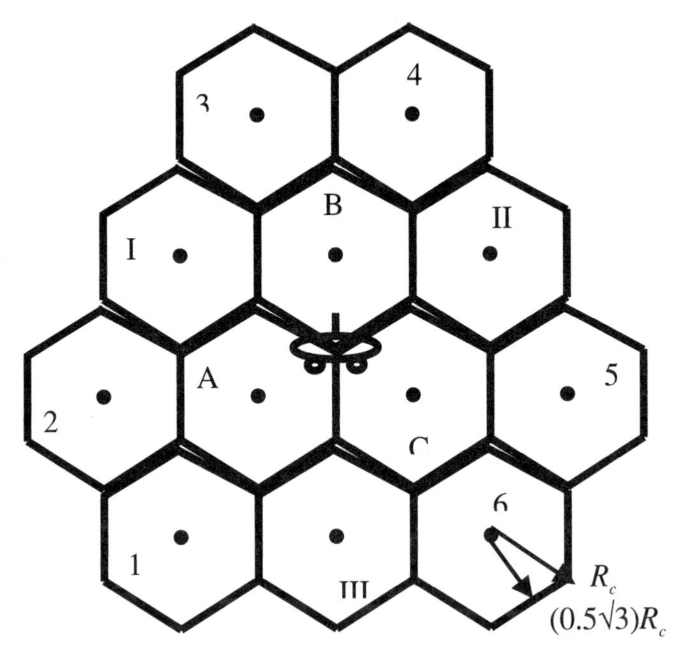

Figure 1-5 Distances from interfering CDMA cells that are located about a mobile at the intersection of three cells.

neously receive and detect the desired signal transmitted independently from the three base stations. If we again assume that the power received from a base station has the dependence $P_T A/R^n$, the sum of the received signals from three base stations is equivalent to a received power

$$P = \frac{3P_T A}{R_c^n} \tag{1–10}$$

If there are N_s subscribers in each cell, the interference power due to transmission to the other $N_s - 1$ subscribers in the three cells A, B, C is $3(N_s - 1)P_T A/R_c^n$. Interference also comes from transmission to the N_s subscribers in the surrounding cells. For cells labeled I, II, and III in Figure 1–5, the distance from the base station to the subscriber in question is $2R_c$, while the distance to the six cells labeled 1, 2, ..., 6 is $\sqrt{7}R_c$. Thus the total interference signal I received by the subscriber from the cells shown is

$$I = 3(N_s - 1)\frac{P_T A}{R_c^n} + 3N_s\frac{P_T A}{(2R_c)^n} + 6N_s\frac{P_T A}{(\sqrt{7}R_c)^n} \tag{1–11}$$

For adequate detection, the interference I must be less than some multiple $F > 1$ of the desired signal P. The value of F depends on the processing gain, the fraction of the time that a signal is being sent to other subscribers (voice activity factor), and so on. [4]. For this discussion the value of F is not important. After some manipulation using (1–10) and (1–11), the condition $I < FP$ can be written

$$N_s < \frac{F+1}{1 + 1/2^n + 2/(\sqrt{7})^n} \tag{1–12}$$

From (1–12) it is seen that the number of subscribers that can be accommodated by each base station is directly dependent on the value of n. For example, if $n = 4$, then from (1–12), $N_s < 0.91(F + 1)$, while for $n = 2$ we find from (1–12) that $N_s < 0.65(F + 1)$. Thus when the signals decrease rapidly with distance ($n = 4$), the interfering signals received by a particular mobile from other base stations will be small compared to those received from the base station of the mobile's cell, and the capacity N_s is close to the capacity that would be obtained for a single cell. However, if the signals decrease more slowly with distance ($n = 2$), the contribution to the total interference from other base stations will be larger, and hence the capacity N_s will be reduced. In other words, to cover an area with a given number of users will take more base stations for $n = 2$ than for $n = 4$. Similar interference considerations apply on the uplink, but the analysis is more complex due to the base station's control of the power radiated by the subscribers. When random fading of the received signal is taken into account, it is found desirable to sector the cells, just as in the case of FDMA systems.

1.6 Summary

The foregoing discussions have demonstrated in some detail the importance of the range dependence of radio propagation for system design. Many properties of the radio channel in addition to range dependence affect the capacity and performance of radio communications. Subsequent chapters are devoted to understanding the physical properties of the radio channel and using this understanding to predict channel characteristics. At many points throughout the theoretical development, we compare the theoretical predictions with measurements that have been made.

During the development of the AMPS and related European systems, many measurements were made of the propagation characteristics using signals having a narrow bandwidth of 30 kHz or less. For systems using digital transmission, the signals generated by individual users will have a bandwidth that may be 5 MHz or more, in which case it is often convenient to think in terms of the pulse response in the time domain rather than the frequency domain. In Chapter 2 we review the results of narrowband measurements made for outdoor propagation in cities, as well as wideband pulse measurements that have been made. We also introduce the various statistical measures that have been used to account for the random nature of the channel. In Chapters 3, 4, and 5 we introduce the essential features of plane wave reflection from surfaces, spherical waves radiated from antennas, and diffraction. In Chapter 6 we discuss diffraction models that have been developed to predict the range dependence for portions of cities where the buildings have nearly uniform height. The mechanisms leading to shadow fading are discussed in Chapter 7, along with the effects of terrain and foliage. Ray models that have been developed to predict propagation in environments of high-rise or mixed-height buildings, accounting for the individual buildings in a particular region, are treated in Chapter 8.

Problems

1.1 Consider the arrangement of linear cells in Figure 1–1 and suppose that the received power varies as $P_T A/R^n$ with range index $n = 3$. Find the value of N_R to achieve $P/I \geq 50$, and find the base station separation for the teletraffic of a six-lane highway, as discussed in the paragraph following (1–1).

1.2 Show that $\mathbf{U}_2$ of (1–5) has magnitude equal to that of $\mathbf{U}_1$ and that the angle between $\mathbf{U}_1$ and $\mathbf{U}_2$ is equal to 60°.

1.3 Make up a table showing the values of N_R obtained from (1–6) for symmetric frequency reuse patterns for $1 \leq k_1 \leq 6$ and $0 \leq m_1 \leq 3$. Sketch a frequency reuse pattern for $N_R = 9, 21$.

1.4 For a subscriber at the cell edge, as in Figure 1–3, find the expression corresponding to (1–8) for P/I on the downlink that accounts for interference from all six cochannel base stations in Figure 1–3 (assume that the received power varies as $P_T A/R^n$). Solving (1–7) for D/R_c, show that when all six interferences are accounted for, the $N_R = 7$ reuse pattern satisfies the condition $P/I \geq 50$ when $n = 4$. When $n = 2$, find the value of N_R for symmetric reuse patterns that satisfies $P/I \geq 50$ accounting for the six interferences.

1.5 Consider a set of CDMA cells along a highway, as in Figure 1–1. Assume that a subscriber can receive signals from two base stations, so that a reference subscriber at the boundary of two cells receives the effective signal power $P = 2P_T A/R_c^n$. If there are N_s subscribers per cell, find the interference at the reference subscriber due to a) transmission from the base stations to the other subscribers in the two cells neighboring the reference subscriber; and b) transmission from base stations located at $3R_c$ and $5R_c$ from the reference subscriber. Derive an inequality for N_s from the condition $I < FP$. Evaluate the percent change in N_s when the range index changes from $n = 2$ to $n = 4$.

References

1. D. J. Goodman, Wireless Personal Communications Systems, Addison-Wesley, Reading, Mass., 1997.
2. T. S. Rappaport, Wireless Communications; Principles and Practice, Prentice Hall, Upper Saddle River, N. J., 1996.
3. A. F. Naguiv, A. Paulraj, and T. Kailath, Capacity Improvement with Base-Station Antenna Array Receiver in Cellular CDMA, *IEEE Trans. on Veh. Technol.*, vol. 43, no. 3, pp. 691–698, 1994.
4. A. J. Viterbi, CDMA: Principles of Spread Spectrum Communication, Addison-Wesley, Reading, Mass., 1995.
5. W. C. Y. Lee, Mobile Communications Engineering, McGraw-Hill, New York, 1982.
6. E. C. Jordan and K. G. Balmain, Electromagnetic Waves and Radiating Systems, 2nd Ed., Prentice Hall, Upper Saddle River, N.J., 1968, Chap. 10.
7. M. Mouly, Regular Cellular Reuse Patterns, Proc. 1991 Vehicular Technology Conference, pp. 681–688, 1991.
8. L. C. Wang, K. Chawla, and L. J. Greenstein, Performance Studies of Narrow-Beam Trisector Cellular Systems, Int. J. of Wireless Inf. Networks, vol. 5, no. 2, 1998.

CHAPTER 2

Survey of Observed Characteristics of the Propagation Channel

The original development of cellular mobile radio (CMR) required knowledge of 450-to-900 MHz radio propagation between elevated base station antennas and subscribers located at ground or street level. In cities, the base station antennas are placed above the surrounding buildings but not usually on the tallest buildings. The subscribers are located among the buildings, which shadow them from the base station, except on streets aligned with the base station or when no buildings are present for a few blocks in front of the subscriber. Extensive measurements were made using continuous-wave (CW) or narrowband signals to determine the propagation characteristics to shadowed subscribers [1–7].

Since then, many additional measurements have been made in support of new applications and more advanced system concepts. Narrowband measurements were made at 1800 MHz in connection with the development of personal communication services (PCS), envisioned to use base station antennas near or below the surrounding buildings [8–16]. To evaluate the influence of the radio channel on digital transmission, the time dependence of the received signal has been measured when the channel is excited by a pulsed signal having finite time duration [17–23]. To support indoor voice and data applications, CW and pulsed measurements have been made for transmitter and receiver inside buildings at frequencies of 900 MHz and above [24–37].

In this chapter we survey the measurements so that the reader is aware of the types of signal variations that have been observed as well as the measurement procedures and the attendant data reduction processes. Data reduction yields various statistical quantities that are used by system designers to characterize the radio channel. We discuss how these statistical quantities depend on the propagation environment and on parameters such as the radio frequency. Simple qualitative description of the wave phenomena that are thought to give rise to the observed signal characteristics are also presented in this chapter. In later chapters predictions based on quantitative descriptions of the wave phenomena are presented and compared to the observations.

The qualitative discussions in this chapter make use of ray and wave concepts at the level taught in introductory physics and electromagnetics courses. For frequencies in the ultrahigh frequency (UHF; 300 MHz $\leq f \leq$ 3 GHz) and microwave bands, the wavelength is less than 1 m and hence is small compared to building dimensions. As a result, it is meaningful to describe the propagation from transmitter to receiver in terms of individual ray paths. In an urban environment many such paths exist and involve the phenomena of reflection, diffraction, and scattering from buildings, vehicles, people, and so on. These paths make it possible to receive signals even when the transmitter and receiver are not within line of sight (LOS) of each other. Note that the wavelengths of UHF radio waves are the same as those of acoustic waves in air having frequencies above about 300 Hz, and it is the same phenomena of reflection, diffraction, and scattering that make it possible to hear the sound produced by sources that are not visible. When there is a LOS path to the transmitter, the direct ray usually gives the dominant component of the received signal, although signals arrive via many other paths as well. For subscribers not visible to the base station, many multipath arrivals can have nearly equal amplitudes, which results in a variety of observed effects that are discussed here.

2.1 Narrowband signal measurements

For outdoor radio links, typical procedures used for measuring narrowband or CW signals involved radiating a signal from a base station antenna mounted on a building, tower, or retractable mast carried by a small truck or van. In most measurements the signal is received by a simple vertically polarized antenna having isotropic reception in the horizontal plane, which is mounted on a van or automobile that also contains position-locating equipment and recording equipment. In this way a file or record can be generated of the mobile's position and the received signal at that position for subsequent processing. As discussed in Chapter 4, the radio link is reciprocal, so that the same characteristics would be obtained if the transmitter and receiver were interchanged. The received signal for narrowband excitation is found to exhibit three scales of spatial variation (fast fading, shadow fading, and range dependence), as well as temporal variation and polarization mixing.

A typical scenario of a subscriber traveling along a street and shadowed by the nearby buildings is depicted in Figure 2–1, where we have shown a few of the many possible rays arriving at the subscriber. Individual ray arrivals have been observed for CW excitation using directive antennas [38] and for pulsed excitation using high-resolution space-time processing [23]. As the subscriber moves along the street, rapid variations of the signal are found to occur over distances of about one-half the wavelength $\lambda = c/f$, where c is the speed of light and f is the frequency in hertz. Figure 2–2 shows an example of this spatial variation in the received voltage amplitude measured at 910 MHz [39]. It is seen that over a distance of a few meters the signal can vary by 30 dB, which on a voltage scale is a factor of 31.6. Even over distances as small as $\lambda/2$, the signal is seen to vary by 20 dB. This small-scale variation results from the arrival of the signal at the subscriber along multiple ray paths due to reflection, diffraction, and scattering by buildings, vehicles, and other objects in the vicinity of the subscriber, as suggested in Figure 2–1.

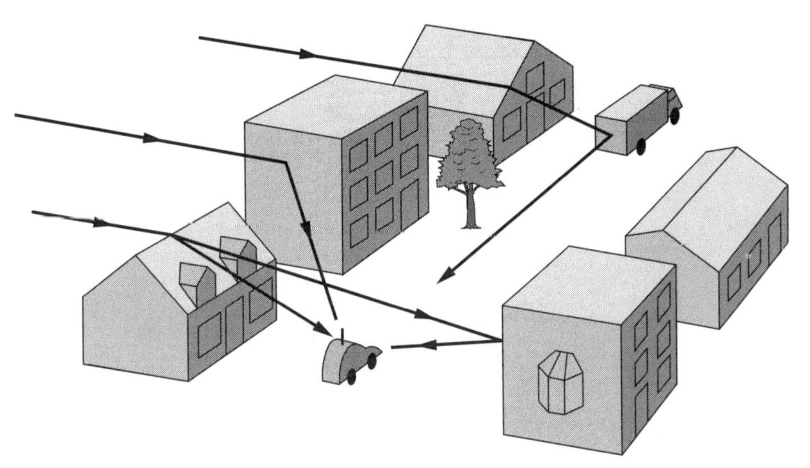

Figure 2-1 Scattering from objects in the vicinity of the mobile result in signals arriving at the subscriber from many directions.

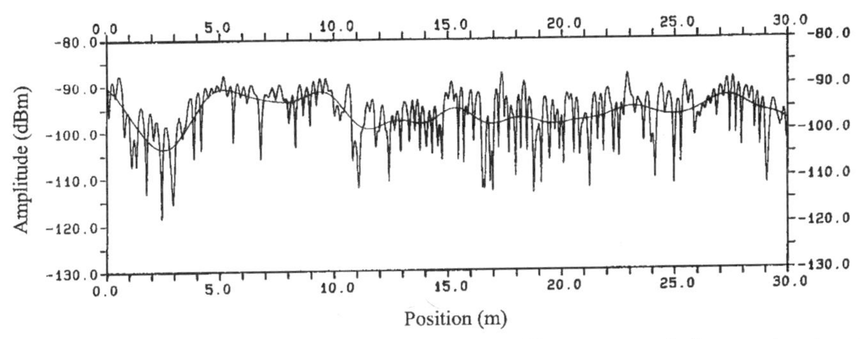

Figure 2-2 Narrowband (CW) signal variation in dBm measured along a street and the sliding average (shadow fading) as a function of position for non-LOS conditions [39] (©1988 IEEE).

As shown in Chapter 3, the multiple rays set up an interference pattern in space through which the subscriber moves. Since the signals arrive from all directions in the plane, fast fading will be observed for all directions of motion (e.g., a pedestrian crossing the street with a portable phone, or walking down the street).

When moving farther down the street, the subscriber will pass buildings of different height, vacant lots, street intersections, and so on. In response to the variation in the nearby buildings, there will be a change in the average about which the rapid fluctuations take place, as shown in Figure 2–2. This middle scale over which the signal varies, which is on the order of the building dimensions, is referred to by various authors as *shadow fading*, *slow fading*, or *lognormal fading*. Finally, the overall average of the signal decreases systematically with increasing distance or range R from the base station.

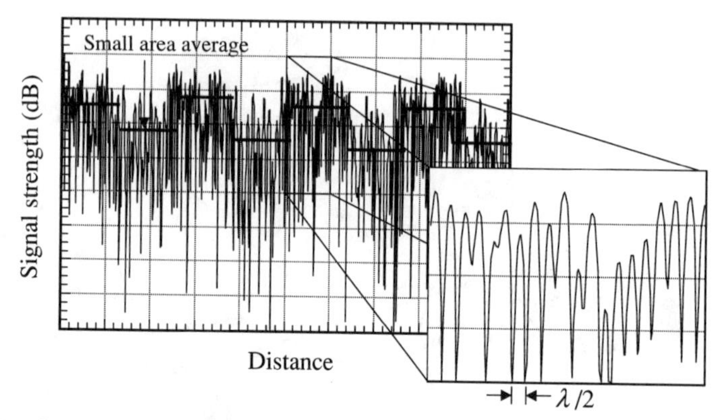

Figure 2-3 Small-area or sector averages obtained by averaging over discrete intervals.

2.1a Signal variation over small areas: fast fading

To separate out fast fading from shadow fading, the envelope or magnitude of the received voltage $V(x)$ is averaged over a distance on the order of 10 m, and the result is referred to as the *small-area average* or the *sector average*. The averaging is often done during data acquisition by taking a 1 s average of the voltage or received power so as to reduce the amount of data that must be stored. In 1 s a vehicle traveling 40 km/h (25 mph) covers approximately 11 m, corresponding to 33λ at 900 MHz. In this way the average in a discrete small area is obtained, as suggested in Figure 2–3. Alternatively, if a complete record of the received voltage $V(x)$ is available, as in Figure 2–2, the shadow fading can be separated from the fast fading using a sliding average over a window of length $2W$ that is on the order of 10 m [39,40]. The sliding average $\overline{V(x)}$ shown in Figure 2–2 is defined by averaging over such a window centered at x, or

$$\overline{V(x)} = \frac{1}{2W}\int_{-W}^{W} V(x+s)ds \tag{2–1}$$

Whereas $V(x)$ represents fast fading in combination with shadow fading, the normalized voltage $V(x)/\overline{V(x)}$ gives fast fading alone.

Because fast fading has an irregular appearance, its statistical properties have been used to anticipate system performance. Let $r_i = V(x_i)/\overline{V(x_i)}$ be the sampled value of the normalized voltage at a point x_i of the fast fading. Using 200 or more sampled values of the random variable r_i, a probability distribution function (PDF) can be constructed. For locations that are heavily shadowed by surrounding buildings, it is typically found that the PDF, denoted by $p(r)$, approximates a Rayleigh distribution [39]–[42]. The PDF of a Rayleigh distribution is defined only for $r \geq 0$ and is given by

$$p(r) = \frac{r}{\rho^2}\exp(-r^2/2\rho^2) \tag{2–2}$$

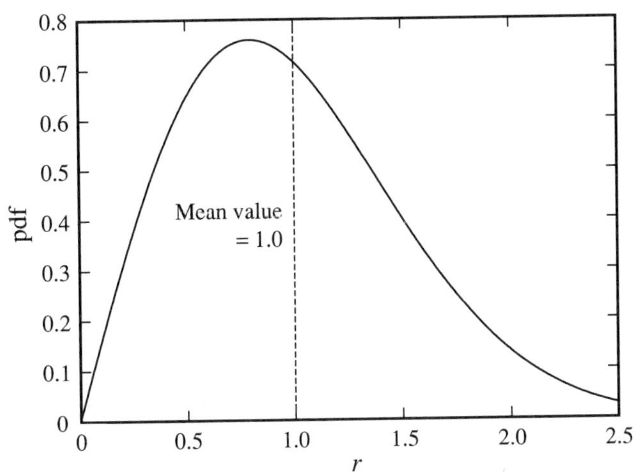

Figure 2-4 Probability distribution function $p(r)$ for a Rayleigh distribution having $\rho = \sqrt{2/\pi}$.

The variation of $p(r)$ is shown in Figure 2–4 for the case when $\rho = \sqrt{2/\pi}$. The peak of the distribution occurs at $r = \rho$, and the mean value of r can be shown to be $<r> = \rho\sqrt{\pi/2} \approx 1.25\rho$. Because the mean value of $V(x)$ over a window of length $2W$ is $\overline{V(x)}$, the mean value $\langle r\rangle = \langle V(x)/\overline{V(x)}\rangle = 1$ and therefore $\rho = \sqrt{2/\pi} \approx 0.80$. Substituting this value of ρ into (2–2), it is seen that the Rayleigh distribution for the normalized received voltage is completely determined. Thus in heavily shadowed environments, knowing the small-area average voltage $\overline{V(x_i)}$ is sufficient to completely characterize the statistical properties of the signal.

For locations where there is one path making a dominant contribution to the received voltage, such as when the base station is visible to the subscriber, the distribution function is typically found to be that of a Rician distribution [43], whose probability density function $p(r)$ is again defined only for $r \geq 0$ and is given by

$$p(r) = \frac{r}{\rho^2}\exp\left(-\frac{r^2 + r_0^2}{2\rho^2}\right)I_0\left(\frac{rr_0}{\rho^2}\right) \tag{2–3}$$

Here $I_0(\cdot)$ is a modified Bessel function of the first kind and zero order, $r_0^2/2$ is proportional to the power of the dominant wave signal, and ρ^2 is proportional to the net power of all the other waves [43]. The relative amplitude of the dominant signal is often measured by means of the parameter $K = r_0^2/2\rho^2$. When r_0 vanishes, $I_0(0) = 1$ and the distribution reduces to the Rayleigh distribution (2–2). When r_0 is large, the distribution approaches a Gaussian distribution centered at r_0.

The mean of the random variable r for a Rician distribution can be found using the integral properties of the incomplete Bessel function [44] and is given by the expression

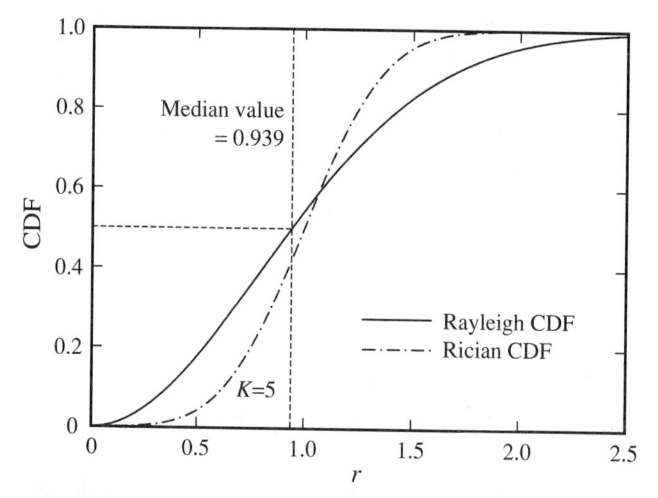

Figure 2-5 Cumulative distribution functions $P(r)$ for a Rayleigh distribution and a Rician distribution with $K = 5$. Both distributions have mean value $<r> = 1$.

$$<r> = \rho\sqrt{\pi/2}[(1+K)I_0(K/2)+KI_1(K/2)]e^{-K/2} \tag{2–4}$$

where $I_1(\cdot)$ is a modified Bessel function of the first kind and order 1. If the random variable r represents the fast fading $V(x)/\overline{V(x)}$, then $<r> = 1$ and (2–4) may be used to solve for ρ in terms of K. If this value of ρ is used in (2–3) and r_0 is expressed in terms of K, the Rician distribution is seen to have K as its only parameter. In the limit as K vanishes, this solution gives $\rho = \sqrt{2/\pi}$, as in the case of the Rayleigh distribution, whereas for K large, $\rho = 1/\sqrt{2K}$.

In comparing distribution functions obtained from measurements with each other and with theoretical distributions, it is sometimes more convenient to use the cumulative distribution function (CDF), denoted by $P(r)$. The CDF is defined by the integral

$$P(r) = \int_{-\infty}^{r} p(u)du \tag{2–5}$$

and is a monotonically increasing function of r, such as that shown in Figure 2–5 for a Rayleigh distribution and a Rician distribution with $K = 5$. For both distributions in Figure 2–5 we have chosen the value for ρ that makes $<r> = 1$. A plot of the CDF can be interpreted as giving the fraction of the locations at which the random variable r is less than some value. For example, at the median value $r = r_m$, the CDF $P(r_m) = 0.5$, indicating that at half the measurement points the value of r is less than r_m. In the case of a Rayleigh distribution $r_m = \rho\sqrt{2\ln 2} \approx 1.18\rho$. Thus the median value of r for a Rayleigh distribution is slightly less than its mean value of $<r> = 1.25\rho$. The Rician distribution has the property that the distribution of r is more concentrated about the mean value, and hence the CDF switches more rapidly from 0 to 1.

In addition to the distribution of the signal about its mean value, system designers are concerned with other statistical parameters, such as the correlation distance of the fading pattern.

When the mobile is moving at road or train speeds, the signals arriving from various directions exhibit a range of Doppler frequency shifts that also affect receiver performance. Multipath fading is also observed at elevated base station antenna locations, although the scale length over which it occurs is much greater than λ. For communication links inside buildings, or for mobile-to-mobile communications, multipath fading effects will appear equally at both ends of the link. We discuss some of these issues in subsequent sections and chapters.

2.1b Variations of the small-area average: shadow fading

In the preceding section we have seen how the signal variation within a small area can be characterized by its fast fading about an average. As a result of shadowing by buildings and other objects, the average within individual small areas also varies from one small area to the next in an apparently random manner, referred to as the shadow fading. The variation of the average is frequently described in terms of the average power in decibels, rather than in terms of the average voltage. The mean or average power in a small area will be proportional to the spatial average $\langle V^2(x)\rangle$ of the voltage amplitude squared. *If* $V(x)$ is Rayleigh distributed, then $\langle V^2(x)\rangle = (4/\pi)[\langle V(x)\rangle]^2$.

Using the subscript i to denote different small areas, we now define the new random variable

$$U_i = 10\log\langle V^2(x_i)\rangle \tag{2–6}$$

where x_i is the location of the ith small area. For small areas at approximately the same distance from the base station, the distribution observed for U_i about its mean value $\langle U\rangle$ is found to be close to the Gaussian distribution [1, 4, 5, 17, 45]:

$$p(U_i - \langle U\rangle) = \frac{1}{\sigma_{SF}\sqrt{2\pi}}\exp\left[\frac{-(U_i - \langle U\rangle)^2}{2\sigma_{SF}^2}\right] \tag{2–7}$$

In cities, the standard deviation σ_{SF} of the shadow fading is usually in the range 6 to 10 dB. In place of $\langle V^2(x)\rangle$ in the definition of U_i in (2–6), some authors have used the median of $V^2(x)$ in a small area, which is equal to the square of the median of $V(x)$. For a Rayleigh distribution the median value of $V^2(x)$ is proportional to $\langle V^2(x)\rangle$ so that both definitions lead to the same shadow fading distribution.

Because the log of the power displays a Gaussian or normal distribution, shadow fading is often referred to as *lognormal fading*. It is also known as *slow fading*, since in a moving vehicle it is observed over a longer time scale than the fast fading. Because the lognormal distribution of the small-area averages has been observed under many different circumstances, it is thought that it arises as a result of a cascade of events along the propagation path, each of which multiplies the signal by a randomly varying amount. When expressed on a logarithmic scale, the product of random events becomes a sum of random variables, which by the mean value theorem tends to a Gaussian or normal distribution. The statistical properties of the shadow fading show only small

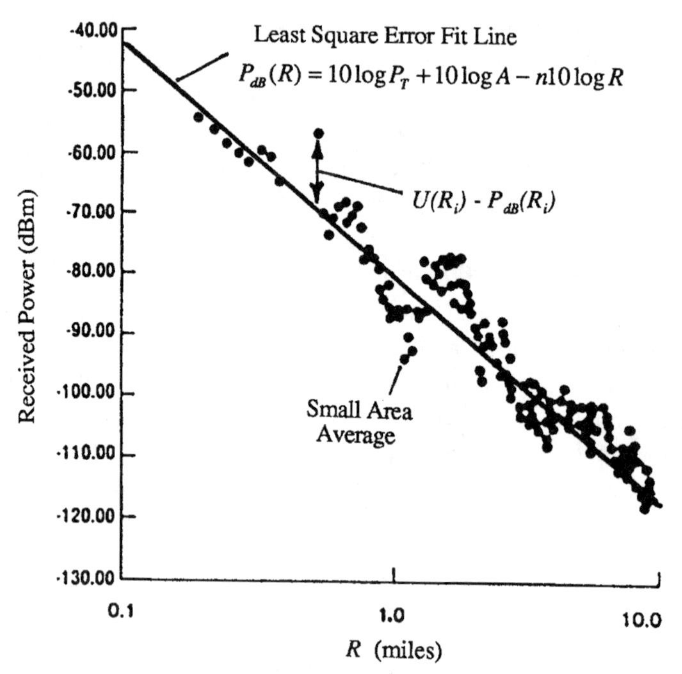

Figure 2-6 For measurements on flat terrain, the least-squares fit line to the small-area average is the range dependence, while the deviations from the least-squares fit line give the shadow fading. Small-area averages are based on [5] (©1978 IEEE).

dependence on frequency. In a later chapter we investigate mechanisms that explain these properties of shadow fading.

2.1c Separating shadow fading from range dependence

When making measurements of received signal, the mobile seldom remains at a fixed distance from the base station. It is more common for the drive path to meander from within 100 m from the base station out to distances of 20 km or more for macrocellular systems. For such drive paths, the methodology used to separate the shadow fading from the range dependence involves plotting the small-area average power P_{dB} on a decibel scale versus the radial distance R on a logarithmic scale. Such a plot is shown in Figure 2–6 for measurements made in Philadelphia [5], where each dot represents one small-area average. A straight line is then fit to the data using the method of least square errors. The equation of the straight line is of the form

$$P_{\mathrm{dB}}(R) = 10\log P_T + 10\log A - n10\log R \tag{2–8}$$

or in watts $P = P_T A/R^n$, where P_T is the transmitted power, n the slope index, and A the intercept at $R = 1$ unit. The straight line represents the average range dependence of the signal, and the shadow loss corresponds to the deviation $U(R_i) - P_{\mathrm{dB}}(R_i)$ of the small-area average from the average range dependence. The distribution of the shadow fading defined in this way will

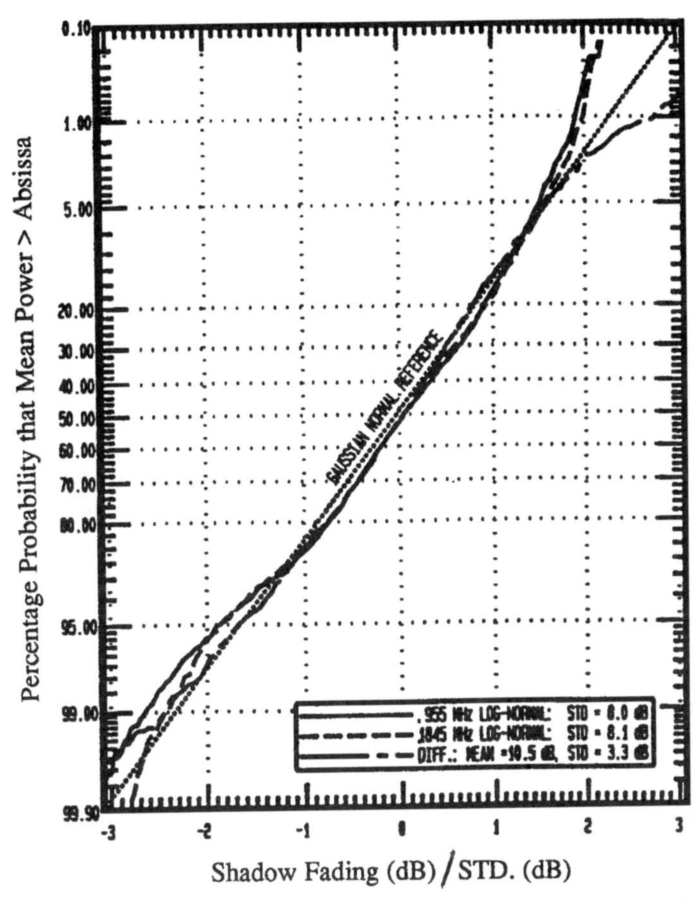

Figure 2-7 Cumulative distribution functions plotted on a Gaussian scale of slow fading at 955 and 1845 MHz obtained from simultaneous measurements taken at the two frequencies on a drive route in Copenhagen by Mogensen et al. [13]. Also shown is the ideal Gaussian distribution and the distribution of the difference in slow fading at the two frequencies (©1991 IEEE).

approximate the Gaussian distribution [13,45].

Examples of the cumulative distribution functions for 955 and 1845 MHz obtained from simultaneous measurements made at the two frequencies on a drive route in Copenhagen by Mogensen et al. [13] are shown in Figure 2–7, where a Gaussian scale has been used for the vertical axis. Using this distorted scale, the CDF for a Gaussian distribution (2–7) plots as a straight line, as shown. In this plot the random variable U for each frequency has been normalized to its standard deviation σ_{SF}, which is 8.0 at 955 MHz and 8.1 at 1845 MHz. The vertical scale gives the percent of points at which the received power was greater than or equal to the value indicated on the horizontal axis. For example, at 1845 MHz about 20% of the measurement points had received power that was greater than the mean power in decibels plus one standard deviation

σ_{SF} = 8.1 dB. It is seen from Figure 2–7 that distribution function for the shadow fading follows a Gaussian distribution, except at high values of received power. The difference of 0.1 between the standard deviations at the two frequencies is somewhat less than the difference of 0.8 found by Okumura et al. [1]. In the measurements leading to Figure 2–7, at each measurement point the difference between the received power at 955 MHz and at 1845 MHz was obtained for the same transmitted power. On average the power at 1845 MHz was 10.5 dB lower than the received power at 955 MHz. The cumulative distribution of this difference is also shown in Figure 2–7, where it has been normalized to its standard deviation of 3.3 dB. This low value of standard deviation of the difference in received power indicates that shadow fading at the two frequencies is highly correlated.

As in the case of fast fading, other statistical parameters of shadow fading are also of interest to the system designer. One such parameter is the degree of correlation between shadow fading for the links from the subscriber to two different base stations, which influences the interference experiences from other cells. A subscriber moving into an open region will experience a stronger signal from both base stations, whereas a subscriber located in midblock may be shadowed to one base station by large buildings on one side of the street, but not to a base station in the other direction. It is observed that strong correlation occurs when the base stations lie in the same direction from the subscriber, and that correlation is small when the base stations are in opposite directions [46]. The correlation length of shadow fading, which a moving subscriber would experience as an average fade duration, has been found to range from 5 m in urban environments to 300 m in the suburbs [47]. This correlation is of interest in carrying out system simulation, and may be of interest in selecting the time constant of power control features.

2.2 Slope-intercept models for macrocell range dependence

In connection with the design of the original CMR systems, extensive measurements of the sector average power versus R were made by various groups around the world [1–7]. Because the number of subscribers was expected to be small, at least in the startup phase, measurements were made for R in the range from about 1 km out to 20 km or more to support the use of large cells called *macrocells*. An extensive and influential set of measurements was made in and around Tokyo by Okumura et al. [1]. Their work discussed the effects of base station antenna height, frequency, building environment, terrain roughness, and so on, on the range dependence of the received signal, as well as the fast fading and shadow fading. The range dependence was presented as curves of median received field strength (proportional to voltage) versus R for various parameters. Subsequently, Hata [48] expressed these results in terms of path loss L between isotropic antennas. When expressed on a decibel scale, path loss L is given by $L = 10\log(P_T/P)$, where P_T is the transmitted power in watts and P is the received power in watts. Isotropic antennas are an idealization that makes L independent of the antennas' characteristics; the use of specific antennas can be accounted for by subtracting their gain in decibels from L (see Chapter 4). Hata also fitted the resulting path loss curves with simple formulas based on the slope-intercept form $L = -10\log A + 10n\log R$, and his expressions have become an important tool in the design of cellular systems.

The Hata formulas give the median path loss L, which for a normal distribution of shadow fading is equal to the mean or average, as a function of the following parameters: frequency f_M in megahertz, distance from the base station R_k in kilometers, height of the base station antenna h_{BS} in meters, and height of the subscriber antenna h_m in meters. The formulas were made to fit the measurements over the following range of parameters: $150 \le f_M \le 1500$ MHz, $1 \le R_k \le 20$ km, $30 \le h_{BS} \le 200$ m, and $1 \le h_m \le 10$ m. Over this range of parameters the standard formula for the path loss in decibels in an urban area is

$$L = 69.55 + 26.16 \log f_M - 13.82 \log h_{BS} - a(h_m) + (44.9 - 6.55 \log h_{BS}) \log R_k \qquad (2\text{–}9)$$

The term $a(h_m)$ gives the dependence of path loss on subscriber antenna height and is defined such that $a(1.5) = 0$. For regions classified as a "large city,"

$$a(h_m) = \begin{cases} 8.29(\log 1.54 h_m)^2 - 1.10 & \text{for } f_M \le 200 \text{ MHz} \\ 3.2(\log 11.75 h_m)^2 - 4.97 & \text{for } f_M \ge 400 \text{ MHz} \end{cases} \qquad (2\text{–}10)$$

In a small-to-medium-sized city,

$$a(h_m) = (1.1 \log f_M - 0.7) h_m - (1.56 \log f_M - 0.8) \qquad (2\text{–}11)$$

Hata also gives correction factors for suburban and open areas for subscriber antennas of height $h_m = 1.5$ m but does not give the dependence on h_m. The correction factor K_r to be subtracted from (2–9) for suburban areas is

$$K_r = 2\left[\log\left(\frac{f_M}{28}\right)\right]^2 + 5.4 \qquad (2\text{–}12)$$

In open rural areas the correction factor Q_r to be subtracted from (2–8) is given by

$$Q_r = 4.78(\log f_M)^2 - 18.33 \log f_M + 40.94 \qquad (2\text{–}13)$$

The first four terms in (2–9), together with the correction factors (2–10) – (2–13), represent the intercept obtained at $R_k = 1$ km, while the range index n in (2–8) is one tenth the coefficient of the term $\log R_k$, or

$$n = \frac{1}{10}(44.9 - 6.55 \log h_{BS}) \qquad (2\text{–}14)$$

Expression (2–14) gives $n = 3.52$ for a 30-m-high base station antenna, which corresponds to 10 stories and was the lowest height for which measurements leading to (2–14) were made. This antenna height is greater than commonly used in urban environments. For example, for six of seven base station locations used in making measurements in Philadelphia [5], the antenna height was under 22 m. When averaged together, the Philadelphia measurements at 820 MHz yielded the value $n = 3.68$. If we use $h_{BS} = 20$ m in (2–14), it gives a value of $n = 3.64$, which suggests that the Hata model may be extrapolated to base station antenna heights that are lower than those used for the Tokyo measurements. The definition of building environments used by Okumura and Hata is subjective, and it is not clear how well the results for Tokyo apply to other cities. It is therefore common practice in the cellular industry to make drive tests to obtain the correct values of the intercept $10\log A$ and the slope index n for the actual system operating environment. In a later chapter we model some of the physical processes that lead to A and n and show how they are influenced by the building environment.

Some of the ways in which the average range dependence and shadow fading play a role in system design can be seen from the following two examples. First we consider the power that must be radiated in order that the received signal be strong enough to overcome noise. At room temperature, the received thermal noise is $kT = 4 \times 10^{-21}$ W per hertz of system bandwidth. For a 30 kHz AMPS channel, the received thermal noise is 1.2×10^{-13} mW, or –129.2 dBm (decibels above 1 milliwatt). Human-made noise adds about 3 dB at 900 MHz, and the receiver adds its own noise (≈ 3 dB). For adequate voice quality in an analog FM system, the received signal must be at least 10 dB greater than the noise, so that the minimum detectable signal is –129.2 dB, plus the other three factors, or –113.2 dB. To achieve this level of received power, the transmitted power must be set high enough to overcome the path loss. The Hata standard formula (2–9) for $f_M = 900$ MHz, $h_{BS} = 30$ m, $h_m = 1.5$ m, and $R_k = 5$ km gives a path loss of $L = 151.0$ dB. If dipole antennas with gain of 2.2 dB (see Chapter 4) are used at both ends of the link, the net path loss is reduced to 146.6 dB. The minimum transmitter power needed to achieve the minimum detectable signal of –113.2 dB is therefore 33.4 dBm, which is equivalent to 2.2 W. Fading that causes the path loss to vary about its average value will alter the minimum transmission power required for detection.

The second example examines the influence of the shadow fading on the frequency reuse pattern. A subscriber can readily experience shadow fading that reduces the signal it receives from the communicating base station by a standard deviation σ_{SF} decibels, but increases the interference from the nearest interfering base station by the same amount. If $\sigma_{SF} = 8$ dB, this effect corresponds to dividing or multiplying by a factor of 6.3. For a subscriber at the cell boundary $R = R_c$, as in Figure 1–3, the received signal is then $P = P_T A/(6.3R_c{}^n)$, while the interference from the strongest interfering base station is $I = 6.3A/(D - R_c)^n$. From (2–14) it is seen that $n \approx 3.5$ for $h_{BS} = 30$ m, so that $P/I = (D - R_c)^{3.5}/(6.3^2 R_c{}^{3.5})$. For a symmetric reuse pattern with $N_R = 7$, we have from (1–7) that $D/R_c = \sqrt{21}$. Under these circumstances it is seen that $P/I = (\sqrt{21} - 1)^{3.5}/6.3^2$, or $P/I = 2.2$, which is far less than the minimum value of 50 needed for proper operation in an AMPS system. Because of the fading, it is necessary to sector the cells, as discussed in Chapter 1.

2.3 Range dependence for microcells: influence of street geometry

As noted above, the measurements made in connection with macrocellular systems were carried out for high antennas to provide coverage over radial distances as great as 20 km. Over such a large scale, the character of the buildings, land use, terrain, and even the street grid can show considerable variations. Because of these variations, propagation over several kilometers in different radial directions from the base station will not be dependent on the street grid. However, as the number of cellular phone subscribers has increased, it has been necessary to use smaller cells to increase system capacity. Currently in high-density areas, and in the future as systems

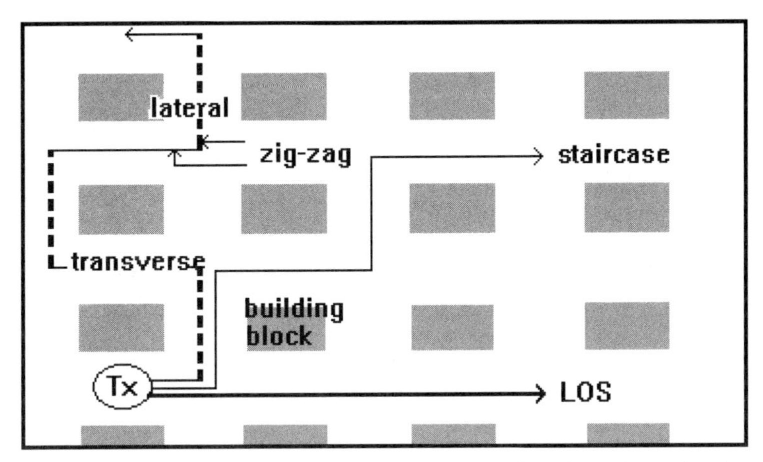

Figure 2-8 Drive routes used for measurements in San Francisco showing LOS, staircase, and zigzag routes, the last of which is divided into transverse and lateral segments [49] (©1999 IEEE).

mature, the cell radius may be 1 km or less. The reduction in cell size is accomplished by lowering the base station antennas so that they are near, or even below, that of the surrounding buildings. Over distances of 1 or 2 km in an urban area, the street grid can be very regular, the buildings quite uniform in size, and the terrain variation small. With the use of low antennas, propagation in such an environment will depend strongly on the direction from the base station to the subscriber relative to the street grid.

To study the influence of the street grid for low base station antennas, measurements have been made over kilometer-length paths by various groups in the United States and Europe [8–16]. An extensive series of measurements was made by a group working for what is now AirTouch Communication [15,16] in the high-rise core of San Francisco and in two residential districts, Mission and Sunset, that have buildings of fairly uniform height. In all regions the street grid is rectangular and the terrain is nearly that of a slightly tilted plane, although the terrain under the high-rise core shows some undulation and is bordered by hills. In each region measurements were made for LOS paths by parking the van with the transmitter on a street and driving the car-mounted receiver down the street. The transmitter van was also moved to the middle of a block and the receiver driven on zigzag and staircase routes, shown in Figure 2–8. For each route, measurements were made at frequencies near 900 and 1900 MHz, and for three transmitting antenna heights h_{BS}, as measured from the street, of 3.2, 8.7, and 13.4 m. In all cases, the mobile antenna height was $h_m = 1.6$ m.

2.3a LOS paths

The received signal on the LOS path in the Mission district for a frequency of 1937 MHz and a transmitter height of h_{BS} = 8.7 m is shown in Figure 2–9, together with a two-segment least-

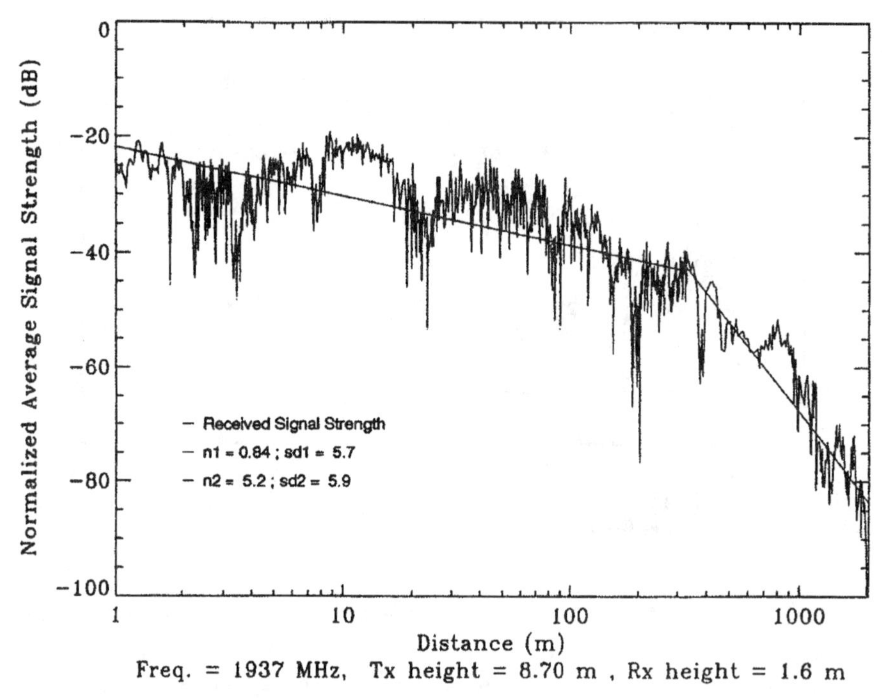

Figure 2-9 Received signal variation measured along a LOS path in the Mission district of San Francisco at 1937 MHz and a transmitter antenna height of h_{BS} = 8.7 m [15] (©1993 IEEE).

squares fit line. It is seen that the signal has variations of a random nature about a range dependence that has two distinct slopes, one before and one after the breakpoint. This characteristic of LOS paths on flat terrain was observed in all regions in the San Francisco area and has been observed by others in various locations, although some have observed a higher slope index after the breakpoint. In Chapter 4 the propagation mechanisms that underlie the observed variations are discussed, and it is shown that the breakpoint distance R_b, expressed in kilometers, is located at

$$R_b = 4\frac{h_{BS}h_m}{\lambda} \times 10^{-3} \tag{2–15}$$

Slope-intercept models were fit to the measurements for LOS paths on flat terrain by Har et al. [49], who obtained the following expressions for the path loss in decibels for isotropic antennas. Before the breakpoint ($R_k < R_b$), the path loss L is given by

$$L = 81.1 + 39.4 \log f_G - 0.1 \log h_{\mathrm{BS}} + (15.8 - 5.7 \log h_{\mathrm{BS}}) \log R_k \qquad (2\text{–}16)$$

where f_G is the frequency in gigahertz and R_k is the distance in kilometers, while beyond the breakpoint ($R_k > R_b$),

$$L = 48.4 + 47.5 \log f_G + 25.3 \log h_{\mathrm{BS}} + (32.1 + 13.9 \log h_{\mathrm{BS}})(\log R_k - \log R_b) \qquad (2\text{–}17)$$

The measurements used as a basis for (2–16) and (2–17) were carried out to distances R_k approaching 2 km. The path loss given by (2–16) and (2–17) is not continuous at $R_k = R_b$, but the discontinuity is less than about 3 dB for the range of values of f_G and h_{BS} over which the measurements were made.

Expression (2–17) depends implicitly on f_G and h_{BS} through R_b. If R_b is replaced by (2–15), and λ is expressed in terms of f_G, the coefficient of $\log f_G$ becomes $(13.6 - 13.9 \log h_{BS})$. Over the range of values of h_{BS} used in the measurements, the coefficient ranges between 6.6 and –2.1, indicating that the frequency dependence of the path loss L is very weak for $R_k > R_b$. In Chapter 4 we examine the theoretical basis for this observation.

2.3b Zigzag and staircase paths in Sunset and Mission districts

For the zigzag paths in the Sunset and Mission districts, values of the small-area average were separated into two sets for (1) mobile locations in the lateral or side street running perpendicular to the main street on which the transmitter was placed, and (2) mobile locations along the streets parallel to the main street on which the transmitter was placed, so that the propagation path was transverse to the intervening rows of buildings. For each set, the small-area average signal was plotted as a function of the radial distance R between the transmitter and the mobile on a logarithmic scale, and a least-squares error fit made to the result. The least-squares fit lines described above are shown in Figure 2–10 for the transverse and lateral paths at f_G = 1.937 GHz, along with the two-segment fit for the LOS path. The strongest signal is obtained on the LOS path, and the lateral path is seen to have a slope that is significantly less than that for the transverse path.

As the mobile drove along the staircase route, propagation was diagonal to the street grid at an angle of about 30° to the main street on which the transmitter was located. For this route there was no difference between the signal received when the mobile was on the side (perpendicular) streets and on the parallel streets, so that all measurements were plotted versus radial distance on a logarithmic scale, and a least-squares fit obtained. The fit line is also shown in Figure 2–10, and is seen to lie close to the transverse path measurements.

Results similar to those shown in Figure 2–10 were obtained at frequencies f_G of 0.9 and 1.937 GHz for all base station antenna heights and for the Mission and Sunset districts. These two districts are composed of attached residential buildings whose average building height H_B is about 11 and 8 m, respectively. When the base station antenna is near or above the surrounding

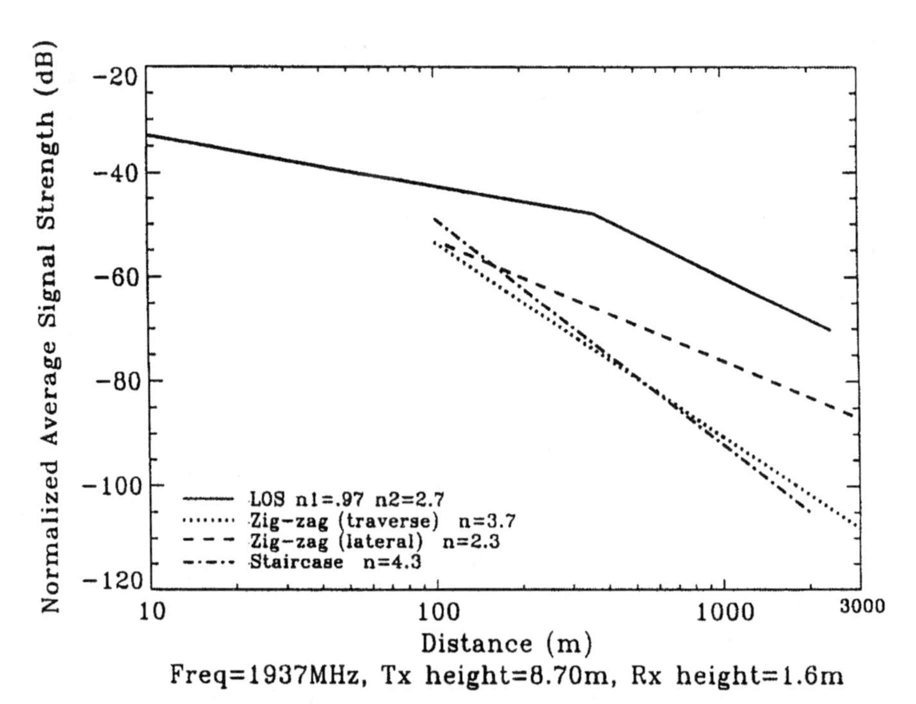

Figure 2-10 Regression fits to the small-area average received signal at 1937 MHz for LOS, lateral, transverse and staircase propagation paths in the Sunset district for a transmitter antenna height of 8.7 m [16] (©1994 IEEE).

buildings, propagation takes place over the buildings, so that Har et al. [49] have constructed slope-intercept formulas by fitting the measurements using the base station antenna height relative to the buildings as a variable parameter. The relative antenna height

$$y_0 = h_{BS} - H_B \tag{2–18}$$

for the measurements was in the range $-7.8 \text{ m} = y_0 = 5.4 \text{ m}$.

Because the measurements of path loss on paths lying transversely over the roofs or obliquely over them are close to each other, they can be combined into a single staircase-transverse (S-T) formula, given by

$$\begin{aligned} L = {} & (138.3 + 38.9 \log f_G) - (13.7 + 4.6 \log f_G) \operatorname{sgn}(y_0) \log(1 + |y_0|) \\ & + [40.1 - 4.4 \operatorname{sgn}(y_0) \log(1 + |y_0|)] \log R_k \end{aligned} \tag{2–19}$$

The S-T formula is expected to apply to all subscriber locations except for those on LOS streets, or on the nearest street oriented perpendicular or parallel to that of the base station. It is seen that the slope index n depends on the base station antenna height, being less than 4 when the antenna is above the buildings, close to 4 when $h_{BS} = H_B$, and greater than 4 for antennas below the buildings. If (2–19) is extrapolated to an antenna height of $y_0 = 9$ m above the buildings, corresponding to a curb height of 20 m in the Mission district, then from (2–19) $n = 3.57$, which is close to the value of 3.64 found by extrapolating the Hata formulas down to 20 m. The fact that measurements in different cities produce consistent values of n suggests that there is an underlying physical mechanism controlling the propagation, as we discuss in Chapter 6.

The formula obtained from fitting the measurements on lateral paths applies to locations on the first street parallel to that on which the base station is located, as well as to the first perpendicular street. This lateral formula is

$$L = (127.4 + 31.6 \log f_G) - (13.1 + 4.4 \log f_G) \operatorname{sgn}(y_0) \log(1 + |y_0|) + [29.2 - 6.7 \operatorname{sgn}(y_0) \log(1 + |y_0|)] \log R_k \quad (2\text{–}20)$$

In this case the slope index n is significantly less than that found from the S-T formula. For example, when $y_0 = 0$, from (2–19), $n = 2.92$ instead of 4. This difference will be shown to result from the mechanism by which waves propagating over the buildings are diffracted down to ground level.

2.3c Non-LOS paths in the high-rise core of San Francisco

Measurements were made in downtown San Francisco as the mobile was driven over the staircase and zigzag routes shown in Figure 2–8 for the same frequencies and base station antenna heights. Although the plots of path loss versus radial distance R are similar to those of Figure 2–10, the dependence on h_{BS} and R is different. In this region most of the buildings are tall and hence the base station antenna is well below the tops of almost all buildings, so that propagation is thought to go around the buildings instead of over them. For such propagation paths, the height of the antenna above the ground is the significant parameter.

Because the measured path loss on the staircase and transverse routes were close to each other, Har et al [49] combined them into a single staircase-transverse (S-T) formula, given by

$$L = 143.2 + 29.7 \log f_G - 1.0 \log h_{\text{BS}} + (47.2 + 3.7 \log h_{\text{BS}}) \log R_k \quad (2\text{–}21)$$

From measurements on the first streets perpendicular to that on which the base station was located, the lateral formula was found to be

$$L = 135.4 + 12.5 \log f_G - 5.0 \log h_{\text{BS}} + (46.8 - 2.3 \log h_{\text{BS}}) \log R_k \quad (2\text{–}22)$$

It is seen from (2–21) and (2–22) that the base station antenna height has little effect on the path loss, and that the slope index n is significantly larger than 4. The difference in slope index for LOS and non-LOS paths in high-rise building environments has been noted by others [14]. The frequency dependence in (2–21) is substantially stronger then in (2–22). In a later chapter we discuss this difference in terms of the frequency dependence of diffraction at building corners.

2.4 Multipath model for fast fading and other narrowband effects

The total signal arriving at a subscriber from a base station is a superposition of many contributions that travel along different paths, which we call *rays*, from the base station. For a subscriber located at position x, the contribution to the received voltage from each ray has amplitude $V_i(x)$ and phase $\phi_i(x)$. The contributions combine to given the total received voltage, whose magnitude $V(x)$ is given by

$$V(x) = \left|\sum_i V_i(x) e^{j\phi_i(x)}\right| \tag{2–23}$$

The ray amplitudes $V_i(x)$ will vary relatively slowly over distances on the order of a wavelength, but the phases will show significant variation. One component of $\phi_i(x)$ is given by $-kL_i$, where $k = 2\pi/\lambda = 2\pi f/c$ is the wavenumber and L_i is the path length of the individual ray. The rays arrive at the subscriber from different directions so that as x varies, the path lengths L_i of the various rays will undergo different variations, some increasing and some decreasing, and all at different rates, as discussed in greater detail in Chapter 3. These variations result in relative phase changes of the different rays, leading to rapid amplitude variations of $V(x)$. The variation of $V(x)$ with x is what we have called *fast fading*.

By averaging $V^2(x)$ over a distance of 20λ or more, and assuming that $V_i(x)$ has little variation over this distance, the small-area average is obtained as

$$\langle V^2(x)\rangle = \sum_i \sum_j V_i(x) V_j(x) \langle e^{j[\phi_i(x) - \phi_j(x)]}\rangle \tag{2–24}$$

For rays arriving at the subscriber from different directions, the phase difference $\phi_i(x) - \phi_j(x)$ will also undergo 2π variations, so that the spatial averages of the individual exponential terms in (2–24) will be small for $i \neq j$. For $i = j$ the average is unity and (2–24) reduces approximately to

$$\langle V^2(x)\rangle = \sum_i [V_i(x)]^2 \tag{2–25}$$

Expression (2–25) indicates that the small-area average power is the sum of the individual ray powers. As seen in Figure 2–2, this average does not vary significantly over distances of a few wavelengths, from which we may conclude that the individual amplitudes $V_i(x)$ are nearly constant over distances on the order of a wavelength, as we had assumed.

2.4a Frequency fading

Because the phase of the individual ray paths in (2–23) depends on frequency through the term kL_i, the relative phases of ray paths with different lengths will vary with frequency. As a result, the amplitude of the received voltage will also vary with frequency. An example of the frequency dependence of the radio channel about the center frequency of 910 MHz is shown in Figure

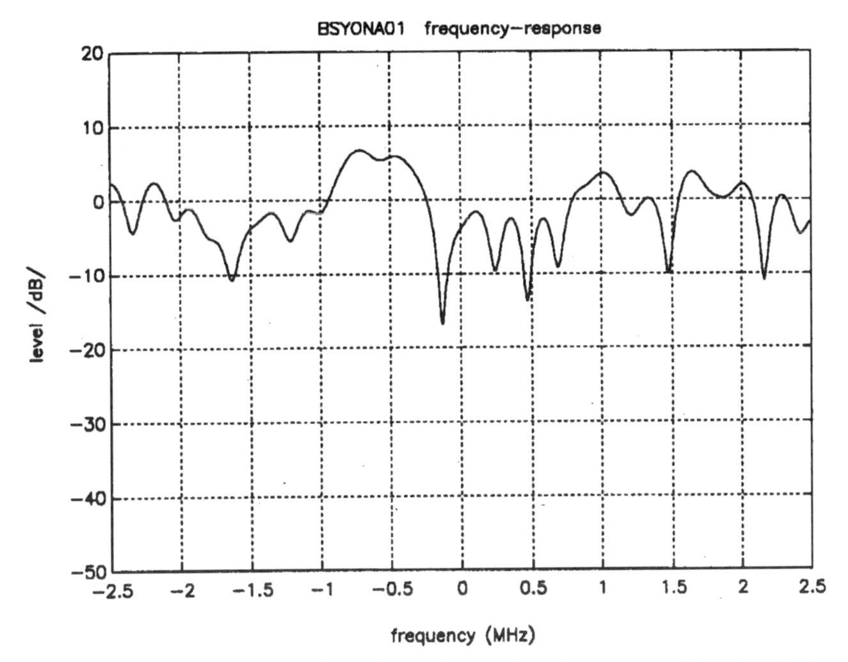

Figure 2-11 Frequency variation about the center frequency of 910 MHz for a 1.2-km radio link in downtown Toronto [21] (©1994 IEEE).

2–11 for a 1.2-km link in downtown Toronto [21]. Accounting for the difference in length L_i and L_j for two ray paths, the difference in phase is $\phi_i - \phi_j = (L_j - L_i)\ 2\pi f/c$. For outdoor paths it is easy to conceive of scatterers giving rise to path length differences on the order of 1 km, so that $\phi_i - \phi_j$ will undergo 2π phase when f changes by 0.3 MHz. In other words, the two ray contributions can go from constructive to destructive and back to constructive addition, or vice versa, in 0.3 MHz, which is consistent with the spacing between minima in Figure 2–11. For indoor paths the differences in path length will be much smaller, so that a greater change in frequency is required to produce the same cycle in the interference [35]. The frequency variation will be discussed further in connection with the channel impulse response.

2.4b Time-dependent fading

The path length L_i is the total distance along a ray from the base station to the various reflecting points and/or scatterer, and then to the subscriber. The motion of scattering objects, such as vehicles, pedestrians, and trees, causes a change in L_i even when the subscriber is stationary. The resulting change in the phases of the individual rays will therefore change with time, causing the amplitude of the received signal to change. Figure 2–12 shows time variation observed in the amplitude of the received voltage for a 900 MHz signal measured at a vehicle parked in a suburban environment [50]. It is seen that rapid variations in the voltage amplitude occur over a time scale of ½ seconds or less. Pedestrian walking speed is around 4 km/h (2.4 mph) or 1.1 m/s, so

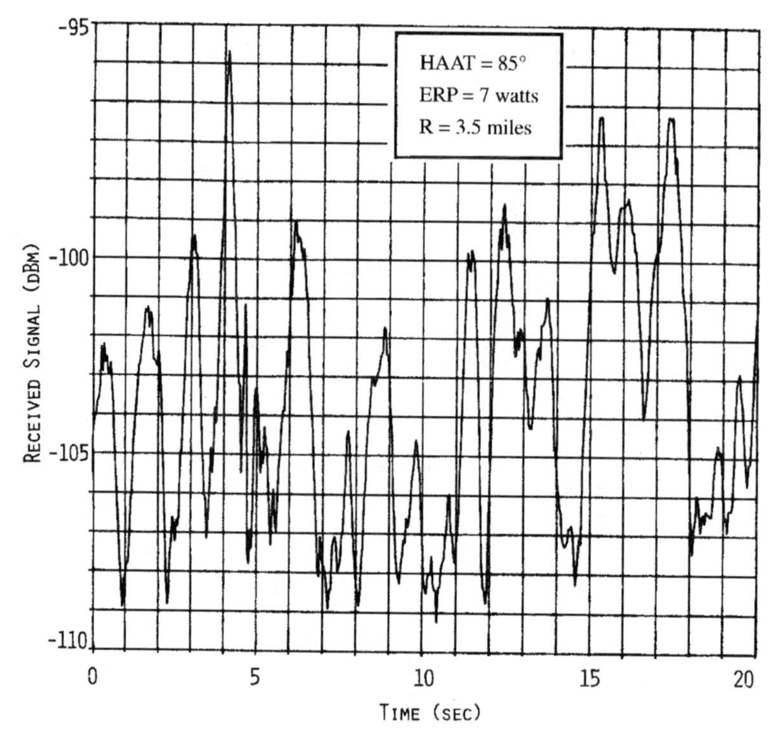

Figure 2-12 Time variation of a 900-MHz CW signal received at a vehicle parked on a suburban street [50] (©1998 IEEE).

that in ½ s a person walks about ½ m, which is slightly larger than λ at 900 MHz. Scattering from more rapidly moving vehicles can be expected to produce more rapid variations of the signal.

2.4c Doppler spread

When the subscriber is in motion, there will be a constant change in the phase ϕ_i, whose rate of change is given by $-k(dL_i/dt)$. The time variation of ϕ_i corresponds to a Doppler frequency shift $\Delta f = -(f/c)(dL_i/dt)$. If the subscriber is traveling directly opposite to the ray direction with velocity v, the shift will be $+vf/c$, while for rays in the direction of travel the shift will be $-vf/c$. For rays making an angle θ to the velocity of the subscriber, the Doppler shift is $\Delta f = (vf/c)\cos\theta$. The effect of the Doppler shifts for the individual multipath ray contributions will be to spread the received signal over a band of frequencies. An example of the measured Doppler spectrum of the received signal is shown in Figure 2–13 [51]. In Chapter 3 a simple theory is presented for the Doppler spectrum that assumes a continuous and uniform distribution of arrival directions in the horizontal plane, and the results are shown in the smooth curve labeled classical spectrum in Figure 2–13.

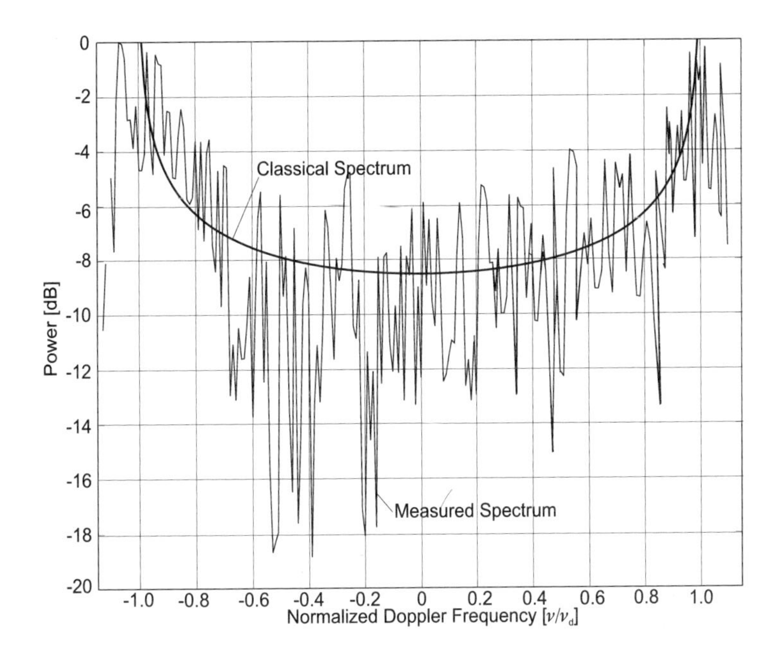

Figure 2-13 Comparison of the received power spectral density due to Doppler spread measured at 1800 MHz with a simple theoretical curve [51]. The maximum Doppler shift is v/λ for a subscriber velocity v.

2.4d Depolarization

The diffraction and scattering processes that are responsible for some of the multipath arrivals will also result in depolarization of the waves. Waves propagating down to the subscriber from the surrounding rooftops will have both vertical and horizontal components of electric field. Scattering by irregular metal objects such as vehicles, lampposts, stoplights, and signs can also produce cross-polarized fields. Measurements have been made of the copolarized (vertical) and cross-polarized (horizontal) electric fields received at base stations for vertically polarized radiation from a mobile [52–55]. It is found that both polarizations exhibit fast fading with fading patterns that are uncorrelated. The ratio of the average power carried by the cross-polarized field to that of the copolarized fields is around –7 dB in cities and –13 dB in suburban areas. In the case of hand-held terminals, where the inclination of the antenna is significant and for vehicle antennas mounted on the trunk or rear window, one can expect more nearly equal average power to be received at the base station in the vertical and horizontal polarizations.

2.5 Narrowband indoor signal propagation

Measurements of narrowband and pulse propagation characteristics inside buildings have been made in support of applications such as wireless LAN and PBX [24–37]. The fast fading characteristics as a portable unit is moved over distances of a few wavelengths are similar to those observed for outdoor portables, although the range dependence is different. When the base station is not within LOS of the portable, the fast fading exhibits Rayleigh statistics [29,30]. If a

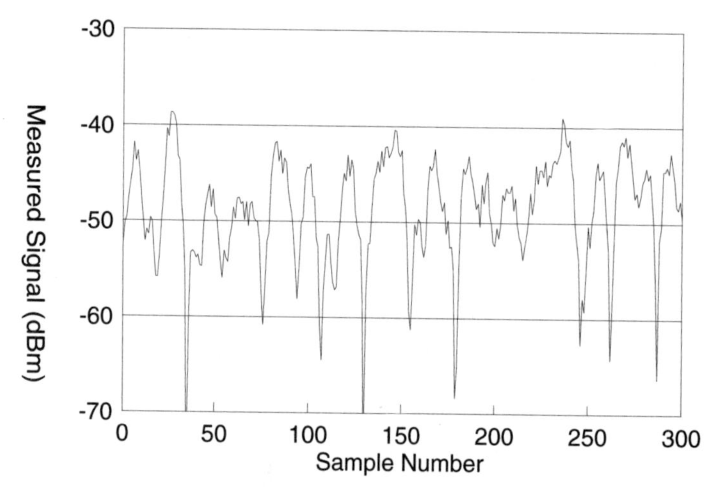

Figure 2-14 Fast fading observed on an indoor link at 900 MHz as one end of the link is moved around a circle having a radius of 3 ft [56] (©1995 IEEE).

LOS path such as that down a hallway exists between the base station and the portable, the fast-fading statistics are Rician.

2.5a Fast fading for indoor links

Following the tradition established for the macrocells of CMR, a small-area average signal is usually defined by averaging the signal measured as the portable end of the radio link is moved over a path whose length is about 20λ. Because of the limited space indoors, this is accomplished by moving the portable around a circle of radius about 1 m [56], or on a raster path [32]. Figure 2–14 shows an example of the received power as the portable is moved over a circle of radius 3 ft when the portable and base station were located in different rooms in a large office building [56]. The power was sampled at 67-ms intervals and is plotted versus sample number over one revolution. Averaging over a revolution gives the small-area average.

For the case when both the base station and portable are located inside a building, as suggested in Figure 2–15, the link is symmetric in the sense that both ends are located in a strongly scattering environment, and fast fading is experienced when either end of the link is moved. This behavior is in contrast to the case of high base stations used for outdoor cellular mobile radio, which are located above the scatterers. The high base station antennas receive multipath arising from a small region about the subscriber, so that there is much less spatial variation at the base

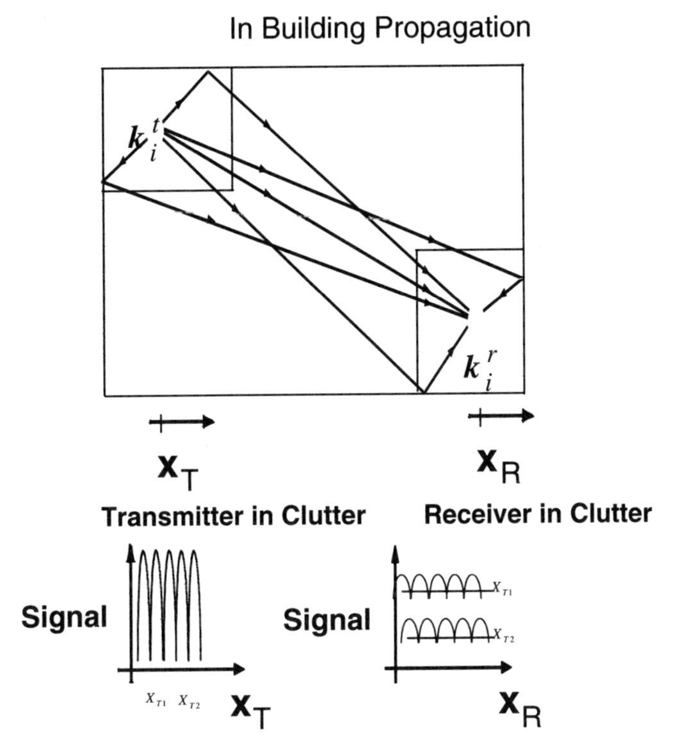

Figure 2-15 Multipath at both ends of an indoor link result in fast fading of the received signal when either end of the link is moved [56] (©1995 IEEE).

station, and a unique number representing the path loss to a small area can be obtained by averaging only over the position of the subscriber, as indicated in going from (2–23) to (2–25). Because both ends of the link experience fading for in-building propagation, it is found that the average obtained by moving one end of the link over a circular path will change if the other end is moved a distance on the order of a wavelength [56,57].

The cause of the change in the sector average as a result of the displacement of the other end of the link is indicated in Figure 2–15. Two rays that are nearly parallel when they arrive at the receiver may leave the transmitter at widely different directions. When the receiver is moved to obtain the small-area average, the sum of the contributions from these two rays will be nearly constant. However, the sum of the two ray contributions will depend strongly on the exact position of the transmitter. Thus the average obtained by moving the receiver will depend on the position of the transmitter, as indicated for positions x_{T1} and x_{T2} in Figure 2–15. Conversely, rays that are nearly parallel when they leave the transmitter may arrive at very different angles at the receiver. To obtain a true small area-to-small area path loss, it is necessary to average over both ends of the link.

Actual measurements of the variation suggested in Figure 2–15 are shown in Figure 2–16

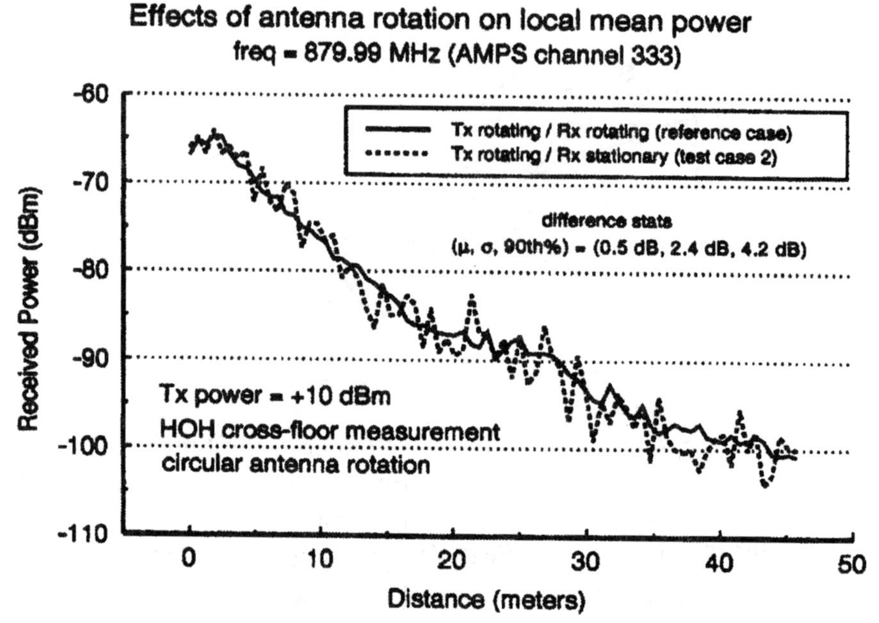

Figure 2-16 Comparison of the small-area averages obtained on indoor links when averaging is made around circles at both ends of the link, or only on a circle at one end of the link [57] (©1997 IEEE). Transmitter and receiver were on different floors of a two-story building, and the horizontal axis represents the center position of the receiver circle as it was moved down a corridor.

for a transmitter and receiver on different floors of a two-story building [57]. Here the average obtained by moving one end of the link around a circular path several wavelengths in diameter is compared to the average obtained when both ends are moved around circular paths. In Figure 2–16 the horizontal axis gives the position down a corridor of the center of the receiver circle. The curve obtained by averaging over both ends of the link gives a smoother distance dependence than is obtained when averaging only at the transmitter end. The difference between the two curves has a random appearance, with peak differences of ±5 dB. One implication of the measurements shown in Figure 2–16 is that some of the scatter in the data that have been reported for in-building links is due to use of single-end averaging. A more significant implication for in-building radio system design is that the performance can be enhanced by using antenna diversity at both ends of the link. This effect will also be of significance for improving outdoor peer-to-peer or mobile-to-mobile communications, such as are employed in small unit military communications, where both ends of the link are located among scatterers.

2.5b Distance dependence of small-area average

The small-area average will vary with location within a building as a result of the signal passing through walls and floors. The exact nature of this variation will depend not only on the shape and construction of the building and the geometry of the radio link, but may also depend on the presence of other buildings. To depict this variation, the small-area average signal in decibels is typically plotted versus the straight-line distance R between the base station and the portable on a logarithmic scale. For propagation down hallways, the measurements indicated a signal dependence of the form $1/R^n$, with n less than the value 2 for free space [27,28,32]. For propagation

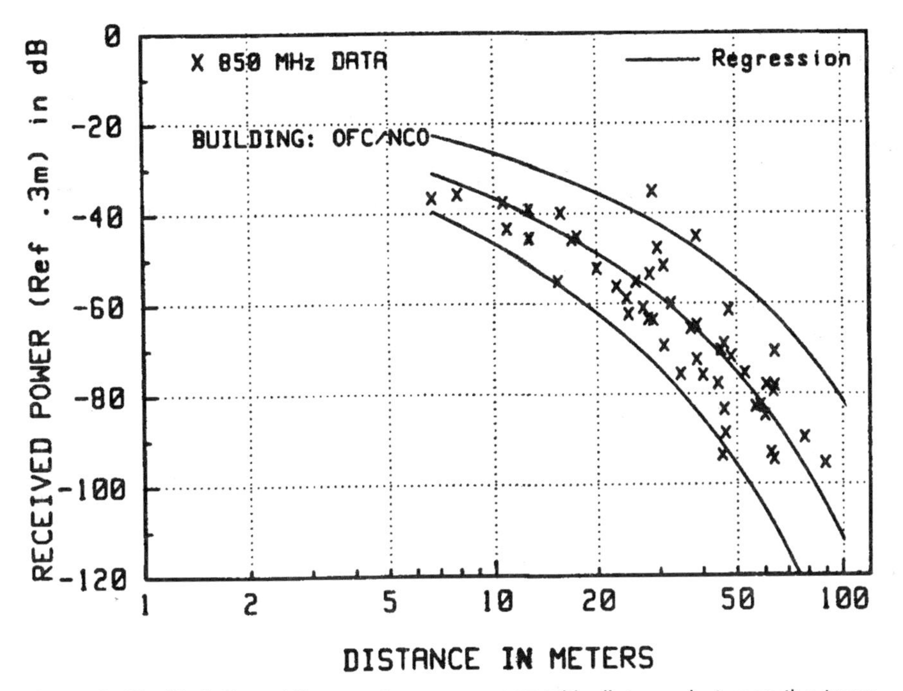

Figure 2-17 Variation of the small-area average with distance between the transmitter and receiver when they are located on one floor of a large office building for narrowband 850-MHz transmission [28] (©1990 IEEE).

through rooms and walls, the sector averages are distributed about a curve whose slope index n is larger than 2 and can even increase with R [27,28,32], as seen, for example, in Figure 2–17, which shows the distance dependence of the small-area averages measured at 850 MHz over one floor of a large office building [28]. For propagation between floors, a different dependence has been obtained [58]. These variations are dependent on the building geometry, and considerable effort has gone into modeling them for specific buildings using ray optics. The ray models and the observed variations are discussed in greater detail in Chapter 8.

For base stations located outdoors, the mechanisms for penetration of the signal into a building can be different for freestanding buildings than it is when the building is surrounded by others. Measurements in freestanding office buildings has shown a variation of up to 20 dB on different sides of the buildings but little variation from floor to floor above the ground floor [59]. For freestanding houses, there was no systematic variation as the portable was moved about, except for an additional 10-dB loss when propagating into the basement [8]. When located among other buildings that shadow the base station, the paths by which the signal reach the building in question can involve diffraction over or around the shadowed buildings, so that the penetration loss shows greater variation from floor to floor [60].

2.6 Channel response for pulsed excitation

Whereas the initial deployment of cellular mobile radio made use of analog voice transmission, subsequent systems have all employed digital transmission of both voice and data. The effect of the radio channel on transmission of digital signals can be determined from the channel response to pulsed excitation. The multipath character of the channel makes itself felt through the echoes that spread the received pulse over time. Existing and proposed telecommunication systems employ a limited bandwidth for each channel, which is typically 1 to 5 MHz wide, although some concepts have discussed far wider bandwidths. For center frequency in the UHF band, such radiated pulses have a small fractional bandwidth, and hence each pulse contains many radio-frequency cycles. As a result, it is the finite-bandwidth channel response that is of importance rather than the infinite-bandwidth impulse response.

Let $w(t)e^{j\omega t}$ be the voltage, normalized to the free-space path loss, that would be received when the transmitter and receiver are in free space with a small separation to avoid time delays. When placed in a multipath environment, the voltage received will be $V(\mathbf{r},t)e^{j\omega t}$, where $V(\mathbf{r},t)$ is the complex amplitude. Neglecting distortion of the individual pulses as a result of reflection, diffraction, or scattering, the complex received voltage amplitude is found by summing the contributions from the individual ray paths as

$$V(\mathbf{r}, t) = \sum_i w[t - \tau_i(\mathbf{r})]V_i(\mathbf{r})e^{j\phi_i(\mathbf{r})} \tag{2–26}$$

Here $\tau_i(\mathbf{r})$ is ray time delay given by L_i/c (where L_i is the ray path length and c is the speed of light), $V_i(\mathbf{r})$ is the ray amplitude, and the ray phase $\phi_i(\mathbf{r})$ includes the phase delay given by $-\omega\tau_i$.

2.6a Power delay profile

At any fixed point $\mathbf{r}$, the received voltage envelope is $|V(\mathbf{r}, t)|$, while the envelope power $P(\mathbf{r}, t)$ is proportional to $|V(\mathbf{r}, t)|^2$. When plotted as a function of time, the envelope power is referred to as the *power delay profile*. Figure 2–18 shows an example of a power delay profile that was made using a 850-MHz channel sounder having the equivalent of a 50-ns pulse duration on an obstructed path in Red Bank, New Jersey [61]. Since it is difficult to have an absolute reference time at the receiver, the time origin in Figure 2–18 has been taken to coincide with the onset of the received pulse. Significant power is received even for $t > 2$ μs, which is much larger than the duration of the original pulse. Overall, the received power decreases with time until it

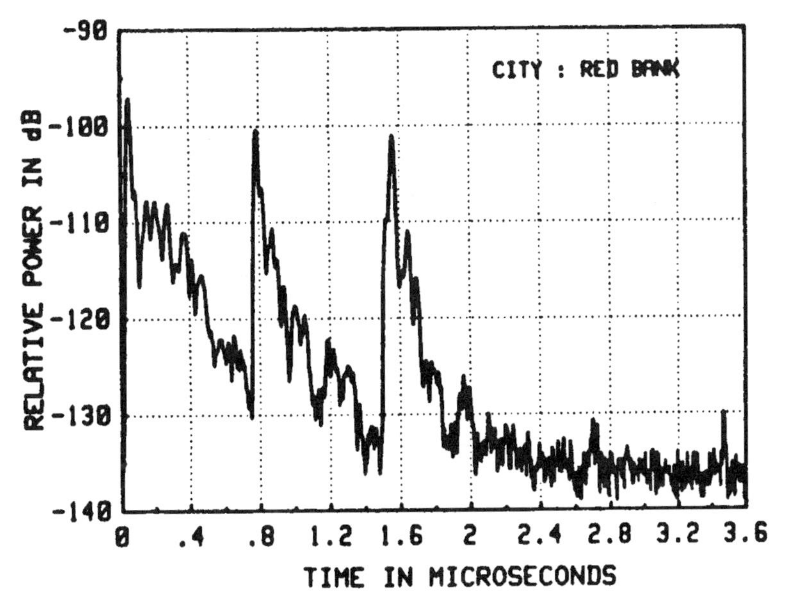

Figure 2-18 Power delay profile measured in Red Bank, New Jersey using a 850-MHz channel sounder with the equivalent of a 50-ns pulse width [61] (©1988 IEEE).

cannot be distinguished from the noise.

In an urban environment, vehicles, buildings, trees, and other objects near the subscriber represent one class of scatterers producing the multipath arrivals, as indicated in Figure 2–19. Since the dimensions of a city block are on the order of 100 m, these nearby objects can produce signals delayed by around 0.3 μs. Large buildings and structures at greater distances can cause stronger scattered signals that arrive with delays greater than 1 μs. In mountainous regions, scattering from steep slopes can result in significant signals having delays greater than 10 μs. Signals scattered from large buildings and terrain features will be rescattered by objects nearer the subscriber, resulting in a cascade of scattered arrivals at the subscriber. This observation is embodied in the cluster model proposed by Saleh and Valenzuela [27] to explain observed time-delay characteristics. Each cluster of arriving rays exhibits decreasing signal amplitude and lasts a few tenths of a microsecond, as shown in Figure 2–18, while the individual clusters arrive with a longer time delay, and the cluster amplitudes decrease with increasing time delay.

2.6b Fading characteristics of individual pulses

When the pulse $w(t)$ is of finite duration, many of the individual received pulses will partially

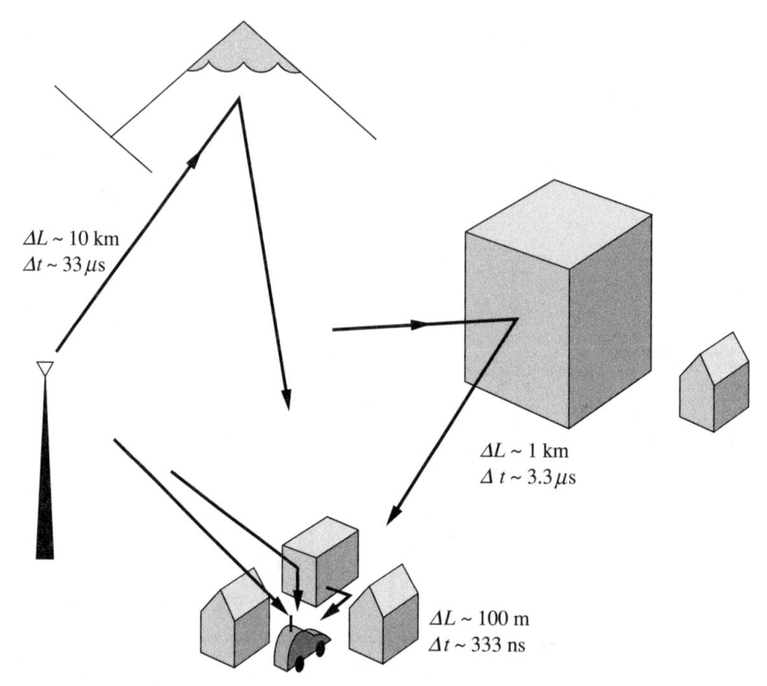

Figure 2-19 Scattering by local and more distant objects leads to clusters of multipath arrivals.

overlap in time. A group of overlapping pulses can appear as a single broadened pulse in the time-delay profile, whose amplitude will depend on the relative phases of the individual ray contributions. The amplitude of the broadened pulse will exhibit fading when the subscriber antenna is moved over distances that are a fraction of a wavelength, as in the case of fast fading observed for CW signals. This effect is demonstrated by the measured pulse response shown in Figure 2–20 observed on indoor LOS and obstructed links using a 5-ns-wide pulse at 2.4 GHz [36,63]. For each path condition there are 16 power delay profiles $|V(x_k,t)|^2$ measured at locations x_k (k = 1, 2, ..., 16) that were separated by $\lambda/4$. For the LOS case of Figure 2–20a , there appears to be a dominant first arrival. However, the amplitude of this pulse shows dependence on the subscriber location, which is thought to result from interference between the direct ray and the ray reflected from the floor. For the obstructed paths, the fading of individual peaks in the power delay profiles of Figure 2–20b is much more severe. Analysis of the measurements has shown that the statistical properties for the fading of individual peaks are nearly Rayleigh for obstructed paths, and Rician for the LOS case [36].

2.6c Measures of time-delay spread

Two commonly used measures of the spreading of the pulse in time are the mean excess delay T_0 and the RMS delay spread τ_{RMS} [18,25,27]. Recognizing from Figure 2–20 that the received

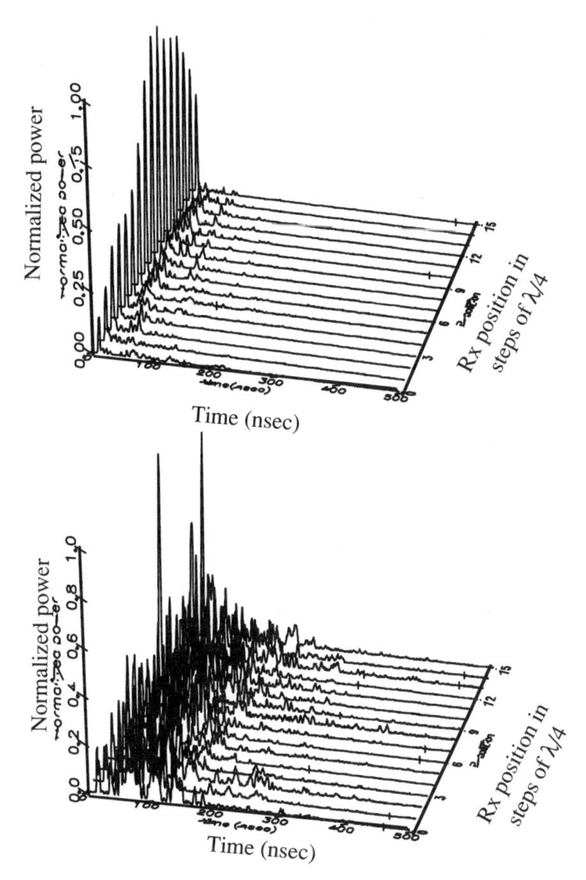

Figure 2-20 Power delay profiles measured at 16 locations separated by λ/4 on indoor links using a 5-ns-wide pulse at 2.4 GHz: (a) LOS path; (b) obstructed path [62] (©1996 IEEE).

voltage envelope depends on the exact position of the subscriber, these quantities are functions of position defined as

$$T_0(\mathbf{r}) = \frac{\int_0^{\infty} t|V(\mathbf{r}, t)|^2 dt}{\int_0^{\infty} |V(\mathbf{r}, t)|^2 dt} \tag{2–27}$$

$$\tau_{\mathrm{RMS}}^2(\mathbf{r}) = \frac{\int_0^{\infty} (t - T_0)^2 |V(\mathbf{r}, t)|^2 dt}{\int_0^{\infty} |V(\mathbf{r}, t)|^2 dt} \tag{2–28}$$

In the case of LOS paths, where the fading of the main peak greatly influences the integrals in (2–27), $T_0(\mathbf{r})$ and $\tau_{RMS}(\mathbf{r})$ can have considerable variation with position (by a factor of about 3 for the power delay profiles in Figure 2–20a). For heavily obstructed paths, where no single peak dominates, the variations of $T_0(\mathbf{r})$ and $\tau_{RMS}(\mathbf{r})$ are smaller (by a factor less than 2 for the profiles in Figure 2–19b), but their mean values are greater than for LOS paths.

For outdoor systems, the mean delay spread τ_{RMS} on LOS paths increases with distance to about 0.2 μs at R = 1 km [20]. On non-LOS paths in cities, the mean value of τ_{RMS} is larger but generally less than 1 μs. High-rise areas have a somewhat larger mean τ_{RMS} than that in suburban areas [21], and in very hilly areas even larger delay spreads are observed [63]. For links inside office buildings, the mean delay spreads τ_{RMS} are typically less than 50 ns, which is an order of magnitude smaller than found for outdoor links. The small delay spread on indoor links is due to the small differences in the path lengths of the individual rays, as compared to outdoor links. Aside from the general concepts cited above, there has been little work devoted to providing theoretical models that can predict the delay spread observed in various circumstances [64]. Monte Carlo simulations based on site-specific ray tracing codes that have recently been developed (see Chapter 8) may allow such predictions.

To define a measure of the pulse spread that is independent of the exact position within a small area, Devasirvatham [25,26] has averaged the power delay profiles over the small area, and from this average profile has defined T_0 and τ_{RMS}. The significance of the average can be understood by expanding the received voltage into a series of individual ray arrivals, as done in (2–24). The power delay profile is proportional to $|V(x_k, t)|^2$, which from (2–23) can be written as

$$|V(\mathbf{r}, t)|^2 = \sum_i \sum_j V_i(\mathbf{r}) V_j(\mathbf{r}) w(t-\tau_i) w^*(t-\tau_j) e^{j[\phi_i(\mathbf{r}) - \phi_j(\mathbf{r})]} \tag{2–29}$$

For pulses of such extremely short duration that they do not overlap in time, the double sum in (2–28) reduces to a single sum of the individual ray amplitudes multiplying $|w(t-\tau_i)|^2$. However, for pulses of more realistic width, the individual ray arrivals overlap in time, with the result that the phase terms $[\phi_i(\mathbf{r}) - \phi_j(\mathbf{r})]$ for $i \neq j$ give rise to the spatial fading evident in Figure 2–20.

Now consider the spatial average of (2–28) over distances of about 20λ. For small displacements, $V_i(\mathbf{r})$ will not vary significantly, and the change in the pulse delay t_i will be small compared to the pulse width. However, $[\phi_i(\mathbf{r}) - \phi_j(\mathbf{r})]$ for $i \neq j$ will vary by 2π or more, so that the spatial average of the corresponding exponential terms in (2–28) will be small. As a result, spatial averaging reduces the double summation in (2–28) to the single sum

$$\langle |V(\mathbf{r}, t)|^2 \rangle = \sum_i |V(\mathbf{r}, t)|^2 |w(t-\tau_i)|^2 \tag{2–30}$$

Thus spatial averaging removes the spatial fading generated by overlapping pulses and gives an average power delay profile that is the sum of the individual ray powers. This average profile is representative of the local environment of the subscriber, as will be the values of T_0 and τ_{RMS} obtained from it. However, it must be remembered that individual delay profiles can give significantly different values.

If the spatial average time-delay profile is integrated over time to find the total energy W, it is seen from (2–29) that

$$W(\mathbf{r}) = \int_0^{\infty} |V(\mathbf{r}, t)|^2 dt = \left[\int_0^{\infty} |w(t)|^2 dt\right] \sum |V_i(\mathbf{r})|^2 \tag{2–31}$$

The spatial average total energy is seen from (30) to be equal to the individual pulse energy multiplied by the spatial average received power for narrowband (CW) excitation, as given by (2–25). Thus the small-sector average signal obtained for narrowband excitation is a direct measure of the received energy for pulsed sources.

2.6d Coherence bandwidth

In our earlier discussion of narrowband transmission, it was pointed out that the phase of the individual multipath rays is dependent on the frequency, so that the interference will depend on frequency, leading to frequency fading that has characteristics similar to those of spatial fading. Since the pulse response contains channel information over a band of frequencies, the Fourier transform of the measured pulse response can be used to characterize the frequency dependence. For indoor paths that are 100 m or less, a network analyzer can be used to find the pulse response by taking the Fourier transform of the measured frequency response. For long paths it is more difficult to accurately measure the phase of the received signal, as well as its amplitude, over a band of frequencies, which is required for the Fourier approach [21].

The coherence function $R_c(\omega)$ is an alternative way of studying the frequency dependence of a channel that does not require measuring the phase of the received signal. The coherence function is obtained by taking the Fourier transform of the average power delay profile [40, Chap. 1; 65], or

$$R_c(\omega) = \int_{-\infty}^{\infty} \langle |V(\mathbf{r}, t)| \rangle e^{-j\omega t} dt \tag{2–32}$$

As an example, Figure 2–21 shows the normalized coherence functions for indoor LOS and obscured paths obtained from the average power delay profiles measured using a 5-ns-wide pulse at 2.4 Ghz [36]. For the LOS path the coherence function is seen to be greater than 0.5 over a bandwidth of about 50 MHz, while for the obscured path it remains above 0.5 only over the bandwidth of about 5 MHz.

The coherence bandwidth is often taken to be that for which the coherence function, normalized to its peak value, is above 0.5. To understand its meaning, recognize that for free-space propagation, where there is no multipath interference, the received pulse will show no spreading in time, and hence the channel is independent of frequency. In this case the coherence bandwidth is infinite. For LOS paths, where the direct ray gives the dominant contribution to the signal, the

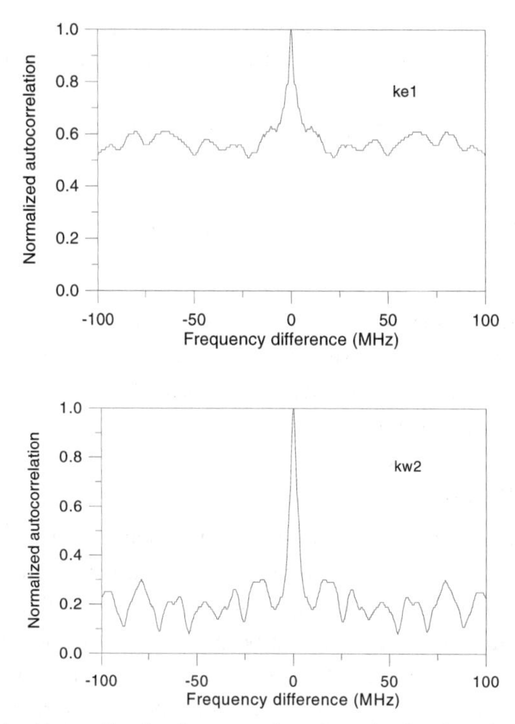

Figure 2-21 Normalized coherence function obtained on indoor links using a 5-ns-wide pulse at 2.4 GHz for (a) LOS path; (b) obstructed path [36] (©1996 IEEE).

pulse will show only small spreading. In this case there will be only a weak frequency dependence and coherence bandwidth is large. On obstructed paths the pulse is expanded in time due to many multipath arrivals having roughly equal amplitudes. The CW response in this case will show rapid variation with frequency and hence a small coherence bandwidth. Thus the coherence bandwidth is inversely related to the RMS delay spread.

2.7 Multipath observed at elevated base station antennas

Multipath can occur even at elevated base stations, as can be visualized by reversing the arrows on the rays in Figure 2–19 to represent transmission from the subscriber to the base station. Even scattering in the vicinity of the subscriber will result in a spread of arrival directions at the base station. The multipath is most clearly seen when directive antennas are used at the base station to receive pulsed transmissions from the subscriber, so that the arriving rays can be resolved in both time and direction. Individual rays arriving at the antenna in direction and time can be extracted from the measurements using high-resolution techniques [23,51] and lead to a display such as that shown in Figure 2–22 for measurements made in Aalborg, Denmark using a 0.923 μs wide pulse in the 1800 MHz band [51]. It is seen that the earliest arrivals are spread over a few degrees as a result of scattering near the subscriber. A second cluster of arrivals occur at a

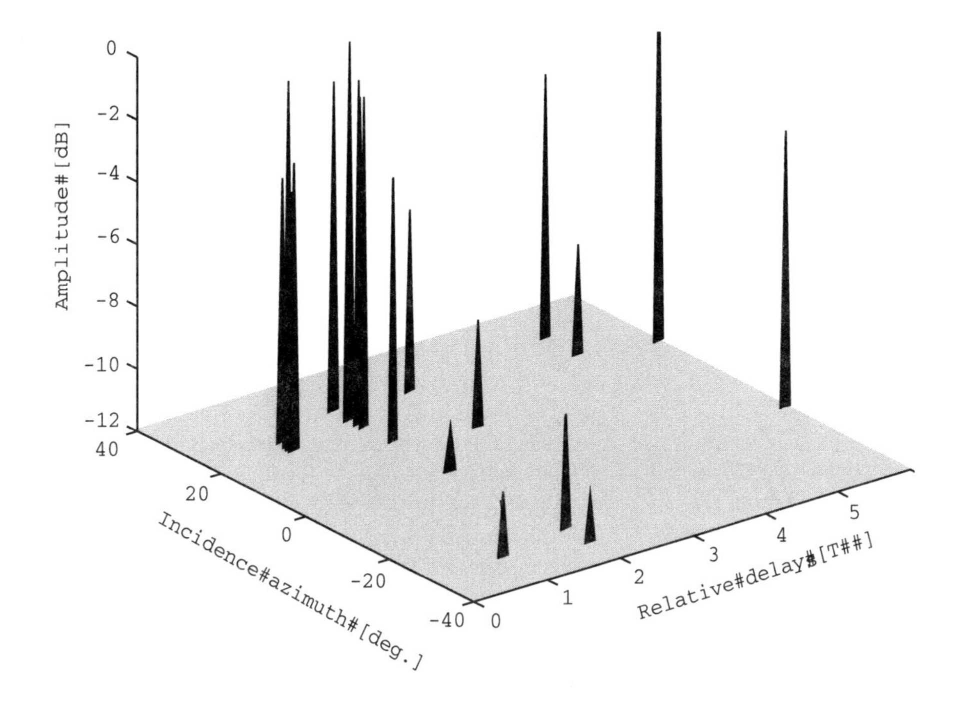

Figure 2-22 Estimated angle of arrival and time delay of individual rays for pulsed signals received at an elevated base station antenna in Aalborg, Denmark [51].

delay of about 0.9 μs with angular spread of about 8°, and other arrivals come from very different directions and much larger delays.

The multipath seen in Figure 2–21 is also significant in narrowband systems, where it gives rise to spatial fading at the base station. To overcome this fading, two or more antennas separated by six or seven wavelengths or more are used for diversity reception [66]. Measurements of the spatial correlation at base station antenna sites in an urban environment indicate that the multipath arrivals come from a region about the subscriber that extends along the street for a distance of about 100 m from the subscriber [67]. At a distance R from the base station of 1 km, such a region appears from the base station to have angular half width ±3°, which is consistent with the spread seen in Figure 2–22.

2.8 Summary

The measurements cited above show that the buildings have a major influence on the characteristics of the radio channel. Reflection and scattering from the buildings produce many paths by which the signal can propagate from the transmitter to the receiver. Depending on the type of transmitting and receiving system used, the multipath can be experienced simply as a spatial fading pattern with a scale length of λ, as a spreading out in time of received pulses, or as the directionally dependent arrival of individual pulses. On the larger scale of building dimensions, the average received signal shows variations due to shadowing by the buildings, while on an even larger scale the signal shows a systematic decrease with distance that is enhanced by the buildings. In subsequent chapters we make use of quantitative descriptions of wave phenomena to understand some of the ways in which buildings can influence the received signal. We show the relation between these processes and relevant measurements and use the processes as a basis for prediction.

Problems

2.1 For the Gaussian distribution given in (2–7), find the probability that U_i lies within the range $\pm\sigma_{SF}$ about the mean value $\langle U \rangle$. Repeat for the range $\pm 2\sigma_{SF}$ about $\langle U \rangle$. What is the probability that $U_i \leq \langle U \rangle + \sigma_{SF}$?

2.2 To compare the slope index n of the Hata and Har ST models, plot n given by (2–14) for h_{BS} ranging from 20 to 50 m. On the same graph, plot n taken from (2–19) for $h_{BS} = 10 + y_0$ with y_0 in the range –5 to +10 m.

2.3 For a frequency of 900 MHz, plot the mobile antenna height factor $a(h_m)$ for a large city using (2–10) and for a small-to-medium-sized city using (2–11) for h_m ranging from 1 to 10 m.

2.4 In the Hata model assume that $f_M = 900$ MHz, $h_{BS} = 30$ m and $h_m = 1.5$ m.

(a) On semilog paper plot the mean path loss L in decibels versus distance R_k on the logarithmic scale over the range 0.1 to 10 km for a city using (2–9), and for a suburban area using (2–12).

(b) Assume that the standard deviation σ_{SF} of the shadow fading is 8 dB in a city and 6 dB in suburban areas. Draw bands extending $\pm\sigma_{SF}$ about the corresponding mean path loss in the plot of part (a).

(c) Assume that the base station receiver sensitivity is –113 dBm, the mobile transmits 5 W, and the combined gain of the transmitting and receiving antennas is 5 dB. Find the maximum path loss that the system will tolerate, and indicate this value by a horizontal line in the plot of part (b). Find the value of R_k where the line crosses the path loss curve $L + \sigma_{SF}$ for the two environments. This is the range out to which the system can be expected to operate (neglecting fast fading).

2.5 In the Har formulas (2–15 – 2–20) for LOS paths and for S-T and lateral paths in residential areas, assume that $f_G = 1.8$ GHz, $h_{BS} = 13$ m, $H_B = 10$ m, and R_k ranges between 0.1 and 5 km.

(a) On semilog paper plot the dependence of the mean path loss on R_k for the three types of paths.

(b) The power radiated by a hand-held portable is limited by government regulations to 800 mW. Assume that the base station receiver sensitivity is –113 dBm and the combined gain of the antennas is 3 dB. Find the maximum path loss that the system can tolerate, and plot this value as a horizontal line on the plot of part (a).

(c) From the crossing points with the ST and lateral path curves in part (b), determine the maximum range over which the system can operate on such paths. Also find the maximum range over which the system can operate on LOS paths.

2.6 Look up three or more of the references [17–22] for pulse propagation over outdoor links. For each paper find the following measurement parameters and observations, if listed, and prepare a table listing them: (1) center or carrier frequency; (2) transmitted pulse width; (3) bandwidth; (4) measurement environment; (5) path lengths; (6) typical received pulse widths; (7) range of τ_{RMS}; and (8) the mean and standard deviation of τ_{RMS}.

2.7 Repeat Problem 2–6 for indoor links using three or more of references [24], [25], [27], [29], and [35].

2.8 A base station transmits a pulsed signal at $f = 1.8$ GHz whose envelope is

$$w(t) = \begin{cases} \sin\left(\dfrac{\pi t}{0.3 \times 10^{-6}}\right) & \text{for } 0 \le t \le 0.3 \times 10^{-6} \\ 0 & \text{otherwise} \end{cases} \tag{2–33}$$

Versions of this signal arrive at a subscriber along 10 separate paths whose delay relative to the first arrival is τ_i, where for the first arrival $\tau_1 = 0$. The relative amplitudes of the arrivals are $\exp(-1.3 \times 10^6 \tau_i)$ and the values of τ_i are listed below.

i	1	2	3	4	5	6	7	8	9	10
$\tau_i \times 10^6$	0	0.228	0.541	0.636	1.032	1.112	1.148	1.316	1.602	1.692

(a) Compute the complex received voltage amplitude $V(t)$ and plot its magnitude, where

$$V(t) = \sum_{i=1}^{10} w(t - \tau_i)\exp(-1.3 \times 10^6 \tau_i)\exp(-j2\pi f \tau_i)$$

(b) Using the results of part (a), compute and plot the power delay profile $|V(t)|^2$.

(c) Compute the mean excess delay and the RMS delay spread of (2–27).

References

1. Y. Okumura, E. Ohmori, T. Kawano, and K. Fukuda, Field Strength and Its Variability in VHF and UHF Land-Mobile Radio Service, *Rec. Elec. Commun. Lab.*, vol. 16, pp. 825–873, 1968.
2. A. P. Barsis, Determination of Service Area for VHF/UHF Land Mobile and Broadcast Operation over Irregular Terrain, *IEEE Trans. Veh. Technol.*, vol. VT-22, pp. 21–29, 1973.
3. D. O. Reudink, Properties of Mobile Radio Propagation above 400 MHz, *IEEE Trans. Veh. Technol.*, vol. VT-23, pp. 143–159, 1974.
4. K. Allsebrook and J. D. Parson, Mobile Radio Propagation in British Cities at Frequencies in the VHF and UHF Bands, *IEEE Trans. Veh. Technol.*, vol. VT-26, pp. 313–322, 1977.
5. G. D. Ott and A. Plitkins, Urban Path-Loss Characteristics at 820 MHz, *IEEE Trans. Veh. Technol.*, vol. VT-27, pp. 189–197, 1978.
6. K. K. Kelly, Flat Suburban Area Propagation at 820 MHz, *IEEE Trans. Veh. Technol.*, vol. VT-27, pp. 198–204, 1978.
7. N. H. Shepherd et al., Coverage Prediction for Mobile Radio Systems Operating in the 800/900 MHz Frequency Range, *Special Issue IEEE Trans. Veh. Technol.*, vol. VT-37, pp. 3–72, 1988.
8. D. C. Cox, R. R. Murray, and A. W. Norris, 800 MHz Attenuation Measured in and around Suburban Houses, *AT&T Bell Lab. Tech. J.*, vol. 63, pp. 921–954, 1984.
9. S. T. S. Chia, R. Steele, E. Green, and A. Baran, Propagation and Bit Error Ratio Measurements for a Microcellular System, *J. IRE*, vol. 57, pp. S255–S266, 1987.
10. J. H. Whitteker, Measurements of Path Loss at 910 MHz for Proposed Microcell Urban Mobile Systems, *IEEE Trans. Veh. Technol.*, vol. VT-37, pp. 125–129, 1988.
11. P. Harley, Short Distance Attenuation Measurements at 900 MHz and 1.8 GHz Using Low Antenna Heights for Microcells, *IEEE J. Select. Areas Commun.*, vol. SAC-7, pp. 5–11, 1989.

12. A. J. Rustako, Jr., N. Amitay, G. J. Owens, and R. S. Roman, Radio Propagation at Microwave Frequencies for Line-of-Sight Microcellular Mobile and Personal Communications, *IEEE Trans. Veh. Technol.*, vol. 40, pp. 203–210, 1991.
13. P. E. Mogensen, P. Eggers, C. Jensen, and J. B. Andersen, Urban Area Radio Propagation Measurements at 955 and 1845 MHz for Small and Micro Cells, *Proc. GLOBECOM'91*, Phoenix, Ariz., pp. 1297–1302, 1991.
14. A. J. Goldsmith and L. J. Greenstein, A Measurement-Based Model for Predicting Coverage Areas of Urban Microcells, *IEEE J. Select. Areas Commun.*, vol. SAC-11, pp. 1013–1023, 1993.
15. H. H. Xia, H. L. Bertoni, L. R. Maciel, A. Lindsay-Stewart, and R. Rowe, Radio Propagation Characteristics for Line-of-Sight Microcellular and Personal Communications, *IEEE Trans. Antennas Propagat.*, vol. AP-41, no. 10, pp. 1439–1447, 1993.
16. H. H. Xia, H. L. Bertoni, L. R. Maciel, A. Lindsay-Stewart, and R. Rowe, Microcellular Propagation Characteristics for Personal Communication in Urban and Suburban Environments, *IEEE Trans. Veh. Technol.*, vol. VT-43, no. 3, pp. 743–752, 1994.
17. D. C. Cox and R. P. Leck, Correlation Bandwidth and Delay Spread in Multipath Propagation Statistics for 910 MHz Urban Radio Channels, *IEEE Trans. Commun.*, Vol. COM-23, pp. 1271–1280, 1975.
18. D. M. J. Devasirvatham, Time Delay Spread and Signal Level Measurements of 850 MHz Radio Waves in Building Environments, *IEEE Trans. Antennas Propagat.*, vol. AP-34, pp. 1300–1305, 1986.
19. S. Y. Seidel, T. S. Rappaport, S. Jain, M. L. Lord, and R. Singh, Path Loss, Scattering and Multipath Delay Statistics in Four European Cities for Digital Cellular and Microcellular Radiotelephone, *IEEE Trans. Veh. Technol.*, vol. VT-40, pp. 721–730, 1991.
20. S. Kozono and A. Taguchi, Mobile Propagation Loss and Delay Spread Characteristics with a Low Base Station Antenna on an Urban Road, *IEEE Trans. Veh. Technol.*, vol. VT-42, pp. 103-108, 1993.
21. E. S. Sousa, V. M. Jovanovic, and C. Daigneault, Delay Spread Measurements for the Digital Cellular Channel in Toronto, *IEEE Trans. Veh. Technol.*, vol. VT-43, pp. 837–847, 1994.
22. J. A. Wepman, J. R. Hoffman, and L. H. Loew, Analysis of Impulse Response Measurements for PCS Channel Modeling Applications, *IEEE Trans. Veh. Technol.*, vol. VT-44, pp. 613–618, 1995.
23. J. Fuhl, J.-P. Rossi, and E. Bonek, High-Resolution 3-D Direction-of-Arrival Determination for Urban Mobile Radio, *IEEE Trans. Antennas Propagat.*, vol. AP-45, pp. 672–682, 1997.
24. D. M. J. Devasirvatham, Time Delay Spread Measurements of Wideband Radio Signals within a Building, *Electron. Lett.*, vol. 20, pp. 950–951, 1984.
25. D. M. J. Devasirvatham, Multipath Time Delay Spread in the Digital Portable Radio Environment, *IEEE Commun.*, vol. 25, pp. 13–21, 1987.
26. D. M. J. Devasirvatham, Multipath Time Delay Jitter Measured at 850 MHz in the Portable Radio Environment, *IEEE J. Select. Areas Commun.*, vol. SAC-5, pp. 855–861, 1987.
27. A. A. M. Saleh and R. A. Valenzuela, A Statistical Model for Indoor Multipath Propagation, *IEEE J. Select. Areas in Commun.*, 5, vol. SAC-1987, pp. 128–137.
28. D. M. J. Devasirvatham, C. Banerjee, M. J. Krain, and D. A. Rappaport, Multi-frequency Radiowave Propagation Measurements in the Portable Radio Environment, *Proc. IEEE ICC'90*, pp. 1334–1340, 1990.
29. R. J. C. Bultitude, S. Mahmoud, and W. Sullivan, A Comparison of Indoor Radio Propagation Characteristics at 910 MHz and 1.75 GHz, *IEEE J. Select. Areas Commun.*, vol. SAC-7, pp. 20–30, 1989.
30. T. S. Rappaport and C. D. McGillen, UHF Fading in Factories, *IEEE J. on Sel. Areas in Commun.*, vol. SAC-7, pp. 40–48, 1989.

31. S. Y. Seidel and T. S. Rappaport, 914 MHz Path Loss Prediction Models for Indoor Wireless Communications in Multifloored Buildings, *IEEE Trans. on Antennas Propagat.*, vol. AP-40, pp. 207–217, 1992.
32. J. F. Lafortune and M. Lecours, Measurement and Modeling of Propagation Losses in a Building at 900 MHz, *IEEE Trans. Veh. Technol.*, vol. VT-39, pp. 101–108, 1990.
33. R. Bultitude, R. Melangon, H. Zaghloul, G. Morrison, and M. Prokki, The Dependence of Indoor Radio Channel Multipath Characteristics on Transmit/Receive Ranges, *IEEE J. Select. Areas Commun.*, vol. SAC-11, pp. 979–990, 1993.
34. H. Hashemi, The Indoor Radio Propagation Channel, *Proc. IEEE*, vol. 81, pp. 943–968, 1993.
35. H. Hashemi and D. Tholl, Statistical Modeling and Simulation of the RMS Delay Spread of Indoor Radio Propagation Channels, *IEEE Trans. Veh. Technol.*, vol. VT-43, pp. 110–120, 1994.
36. S. Kim, H. L. Bertoni, and M. Stern, Pulse Propagation Characteristics at 2.4 GHz Inside Buildings, *IEEE Trans. Veh. Technol.*, vol. 45, pp. 579–592, 1996.
37. Y. P. Zhang and Y. Hwang, Characterization of UHF Radio Propagation Channels in Tunnel Environments for Microcellular and Personal Communications, *IEEE Trans. Veh. Technol.*, vol. VT-47, pp. 283–296, 1998.
38. T. Taga, Analysis for Mean Effective Gain of Mobile Antennas in Land Mobile Radio Environments, *IEEE Trans. Veh. Technol.*, vol. VT-39, pp. 117–131, 1990.
39. M. Lecours, I.Y. Chouinard, G.Y. Delisle, and J. Roy, Statistical Modeling of the Received Signal Envelope in a Mobile Radio Channel, *IEEE Trans. Veh. Technol.*, vol. VT-37, pp. 204–212, 1988.
40. W. C. Y. Lee, Mobile Communications Engineering, McGraw-Hill, New York, Chap. 1 and 6, 1982.
41. H. W. Nylund, Characteristics of Small-Area Signal Fading on Mobile Circuits in the 150 MHz Band, *IEEE Trans. Veh. Technol.*, vol. VT-17, pp. 24–30, 1968.
42. G. L. Turin et al., A Statistical Model of Urban Multipath Propagation," *IEEE Trans. Veh. Technol.*, vol. VT-21, pp. 1–9, 1972.
43. D. Parsons, The Mobile Radio Propagation Channel, Wiley, Chichester, West Sussex, England, pp. 134–135, 1996.
44. M. Abramowitz and I. A. Stegun, eds., Handbook of Mathematical Functions, Dover Publications, New York, pp. 376, 487, 1965.
45. S. Kozono and K. Watanabe, Influence of Environmental Buildings on UHF Land Mobile Radio Propagation, *IEEE Trans. Commun.*, vol. COM-25, pp. 1133–1145, 1977.
46. A. Mawira, Models for the Spatial Correlation Functions of the (Log)-Normal Component of the Variability of VHF/UHF Field Strength in Urban Environments, IEEE Press, New York, pp. 436–440, 1992.
47. M. Gudmundson, Correlation Model for Shadow Fading in Mobile Radio Systems, *Electron. Lett.*, vol. 27, pp. 2145–2146, 1991.
48. M. Hata, Empirical Formula for Propagation Loss in Land Mobile Radio Service, *IEEE Trans. Veh. Technol.*, vol. VT-29, pp. 317–325, 1980.
49. D. Har, H. H Xia, and H. L. Bertoni, Path Loss Prediction Model for Microcells, *IEEE Trans. Veh. Technol.*, vol. 48, pp. 1453–1462, 1999.
50. N. H. Shepherd et al., Special Issue on Radio Propagation, *IEEE Trans. Veh. Technol.*, vol. VT-37, p. 45, 1988.
51. K. I. Pedersen, P. E. Mogensen, B. H. Fleury, F. Frederiksen, K. Olesen, and S. L. Larsen, Analysis of Time, Azimuth and Doppler Dispersion in Outdoor Radio Channels, *Proc. ACTS*, 1997.
52. W. C. Y. Lee and Y. S. Yeh, Polarization Diversity System for Mobile Radio, *IEEE Trans. on Commun.*, vol. COM-20, pp. 912–913, 1972,
53. S. A. Bergmann and H. W. Arnold, Polarization Diversity in Portable Communications Environments, *Electron. Lett.*, vol. 22, pp. 609–610, 1986.

54. R. G. Vaughn, Polarization Diversity in Mobile Communications, *IEEE Trans. Veh. Technol.*, vol. VT-39, pp. 177–186, 1990.
55. J. J. A. Lempiainen, J. K. Laiho-Steffens, and A. F. Wacker, Experimental Results of Cross Polarization Discrimination and Signal Correlation Values for a Polarization Diversity Scheme, *Proc. VTC'97*, pp. 1498–1502.
56. W. Honcharenko, H. L. Bertoni, and J. Dailing, Bi-lateral Averaging over Receiving and Transmitting Areas for Accurate Measurements of Sector Average Signal Strength inside Buildings, *IEEE Trans. on Antennas Propagat.*, AP-43, pp. 508–512, 1995.
57. R. A. Valenzuela, O. Landron, and D. L. Jacobs, Estimating Local Mean Signal Strength of Indoor Multipath Propagation, *IEEE Trans. Veh. Technol.*, vol. VT-46, pp. 203–212, 1997.
58. W. Honcharenko, H. L. Bertoni, and J. Dailing, Mechanism Governing Propagation between Different Floors in Buildings, *IEEE Trans. Veh. Technol.*, vol. VT-42, pp. 787–790, 1993.
59. A. Davidson and C. Hill, Measurement of Building Penetration into Medium Buildings at 900 and 1500 MHz, *IEEE Trans. Veh. Technol.*, vol. VT-40, pp. 161–168, 1997.
60. R. Gahleitner and E. Bonek, Radio Wave Penetration into Urban Buildings in Small Cells and Microcells, *Proc. IEEE VTC'94*, pp. 887–891, 1994.
61. D. M. J. Devasirvatham, Radio Propagation Studies in a Small City for Universal Portable Communications, *Proc. IEEE VTC'88*, pp. 100–104, 1988.
62. H. L. Bertoni, S. Kim, and W. Honcharenko, Review of In-Building Propagation Phenomena at UHF Frequencies, *Proc. IEEE ASILOMAR-29*, pp. 761–765, 1996.
63. A. Zogg, Multipath Delay Spread in a Hilly Region at 210 MHz, *IEEE Trans. Veh. Technol.*, vol. VT-36, pp. 184–187, 1987.
64. Li Yuanqing, A Theoretical Formulation for the Distribution Density of Multipath Delay Spread in a Land Mobile Radio Environment, *IEEE Trans. Veh. Technol.*, vol. VT-43, pp. 379–388, 1994.
65. J. G. Proakis, *Digital Communications*, McGraw-Hill, New York, pp. 702–719, 1989.
66. S. B. Rhee and G. I. Zysman, Results of Suburban Base Station Spatial Diversity Measurements in the UHF Band, *IEEE Trans. Commun.*, vol. COM-22, pp. 1630–1636, 1974.
67. F. Adachi, M. T. Feeney, A. G. Williamson, and J. D. Parsons, Crosscorrelation between the Envelopes of 900 MHz Signals Received at a Mobile Radio Base Station Site, *IEE Proc.*, vol. 133, Pt. F, pp. 506–512, 1980.

CHAPTER 3

Plane Wave Propagation, Reflection, and Transmission

Many excellent texts, such as references [1] to [3], discuss Maxwell's equations and their solutions under various conditions. The solutions give the space and time dependence of the electric and magnetic intensities. These two field quantities are vectors, each of whose components is therefore a function of the spatial coordinates x, y, z and time t. They are usually designated by the vectors **E** and **H** and have units of V/m and A/m, respectively. In simple media of interest for wireless propagation, such as air, earth, and masonry, the electric flux density **D** and magnetic flux density **B** are related to **E** and **H** through the linear relations [1,2]

$$\begin{aligned} \mathbf{D} &= \varepsilon_r \varepsilon_0 \mathbf{E} \\ \mathbf{B} &= \mu_r \mu_0 \mathbf{H} \end{aligned} \tag{3–1}$$

Here ε_r is the relative dielectric constant of the material in which the fields are being computed. At UHF frequencies ε_r takes on values ranging from very close to unity for air to around 80 for water. The relative permeability μ_r is very close to unity for materials of interest at UHF frequencies, and we will assume that $\mu_r = 1$ in the remainder of this work. The permeability of free space in MKS units is $\mu_0 = (4\pi)10^{-7}$ H/m. The value of the permeativity of free space ε_0 is found from the speed of light c using the relation

$$c = \frac{1}{\sqrt{\varepsilon_0 \mu_0}} \tag{3–2}$$

The measured value of c to three decimal places is 2.998×10^8 m/s [3], which for the purposes of studying radio propagation in cities may be approximated as $c \approx 3 \times 10^8$ m/s. Using the value of μ_0 and the approximate value for c, it is easily seen from (3–2) that $\varepsilon_0 \approx 10^{-9}/(36\pi)$ F/m.

When using phasor notation $e^{j\omega t}$ with $\omega = 2\pi f$, for harmonic time dependence the conductivity σ in S/m of the medium can be accounted for by treating ε_r as a complex quantity having both a positive real part and a negative imaginary part, so that $\varepsilon_r = \varepsilon_r' - j\varepsilon_r''$, where $\varepsilon_r'' = \sigma/\omega\varepsilon_0$. Physical process other than conductivity can lead to an imaginary part ε_r'', which in any case represents energy absorption by the dielectric. However, these other physical absorption processes will have a different dependence on frequency. Values of ε_r' and σ or ε_r'' that have been reported for various materials of interest [1,2,4–8] are listed in Table 3–1 but should be considered only nominal since many of the materials are not well defined. For example, bricks may be fabricated from different materials, window glass may be metallized and be very reflective, and so on. Second, because of the very high value of the dielectric constant of water, the water content of materials such as brick, concrete, and the ground has a major effect on the real and imaginary parts of their dielectric constant. Thus dried bricks measured in a laboratory will have a lower dielectric constant and lower conductivity than bricks in a wall exposed to rain. The dielectric constant of the earth, as well as its conductivity, increase significantly with water content. However, propagation over terrestrial path is not sensitive to the value of ε_r' for earth, and many studies assume a value of 15. At UHF frequencies the imaginary part ε_r'' is smaller than the real part ε_r' for most materials, as indicated in the table.

3.1 Plane waves in an unbounded region

Simple wave solutions to Maxwell's equations are in the form of a plane wave propagating along the x axes in a rectangular coordinate system. For this idealized case the vector fields have no variation with y or z and point in directions that are perpendicular to each other and to the direction of propagation. Figure 3–1 shows a snapshot of the variation of **E** and **H** along x for harmonic time dependence, assuming that **E** is polarized along the y axis and that ε_r is purely real. In Figure 3–1, **E** is perpendicular to the direction propagation, as called for by Maxwell's equations, and **H** is perpendicular to both **E** and the direction of propagation. The cross product between **E** and **H**, called the *Poynting vector* $\mathbf{p} = \mathbf{E} \times \mathbf{H}$, has units of W/m^2. The Poynting vector represents the power flowing through a surface, per unit area, and for a plane wave points in the direction of propagation, which is taken to be the positive x direction in Figure 3–1. Not only are **E** and **H** perpendicular to each other, but they must be such that **p** is in the direction of wave propagation.

The mathematical expressions for the dependence of fields on x and t corresponding to Figure 3–1 are

$$\mathbf{E}(x, t) = \mathbf{y}_0 A \cos(\omega t - k_d x + \phi) \tag{3–3}$$

$$\mathbf{H}(x, t) = \mathbf{z}_0 \frac{1}{\eta_d} A \cos(\omega t - k_d x + \phi)$$

In (3–3), $\mathbf{y}_0$ and $\mathbf{z}_0$ are unit vectors pointing in the y and z directions, respectively, and the symbols A and ϕ represent the amplitude and reference phase of the electric field. The magnetic field

Table 3-1 Dielectric constants and conductivity for some common materials

Material	ε_r'	σ S/m	ε_r'' @ Frequency	References
Glass	3.8–8		$< 3 \times 10^{-3}$ @ 3 GHz	[4, p. 310]
Wood	1.5–2.1		< 0.07 @ 3 GHz	[4, p. 359]
Gypsum board	2.8		0.046 @ 60 GHz	[5]
Chip board	2.9		0.16 @ 60 GHz	[5]
Dry brick	4		0.05–0.1 @ 4, 3 GHz	[6, 7]
Dry concrete	4–6		0.1–0.3 @ 3, 60 GHz	[5, 6]
Aerated concrete	2–3		0.1–0.5 @3, 60 GHz	[5, 8]
Limestone	7.5	0.03		[7]
Marble	11.6		0.078 @ 60 GHz	[5]
Ground	7–30	0.001–0.030		[2, pp. 633–638]
Fresh water	81	0.01		[1, pp. 50, 62]
Seawater	81	4		[1, pp. 50, 62]
Snow	1.2–1.5		$< 6 \times 10^{-3}$ @ 3 GHz	[4, p. 301]
Ice	3.2		2.9×10^{-3} @ 3 GHz	[4, p. 301]

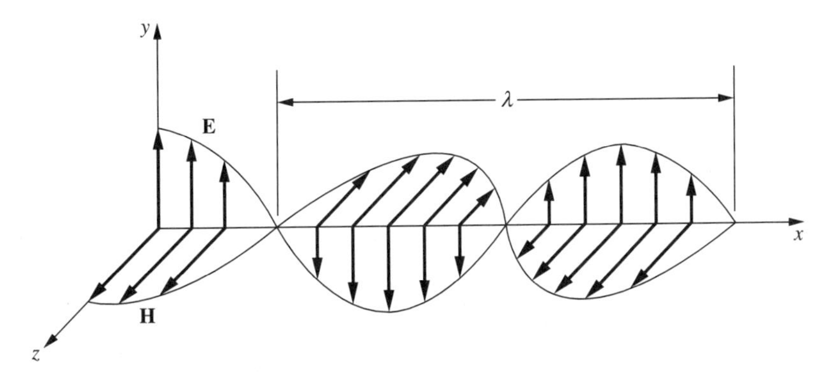

Figure 3-1 Harmonic plane wave propagating along x.

has the same phase as the electric field, but its amplitude differs by the factor $1/\eta_d$, where η_d is the wave impedance of the medium, with units of ohms, and is given by

$$\eta_d = \sqrt{\frac{\mu_0}{\varepsilon_r \varepsilon_0}} \approx \frac{377}{\sqrt{\varepsilon_r}} \tag{3–4}$$

When viewed as a function of time at fixed x, the expressions in (3–3) undergo harmonic time dependence at frequency $f = \omega/2\pi$. Alternatively, when viewed as a function of x at a given instant t, the cosine terms in (3–3) repeat their values when x changes by a distance equal to the wavelength $\lambda_d = 2\pi/k_d$. A point in the waveform is defined by the value of the total phase $(\omega t - k_{dt} x + \phi)$. For the total phase to remain constant as time increases, the value of x must also increase at a rate $dx/dt = \omega/k_d$, which is the velocity v of the wave in the medium. Thus the wavenumber k_d of the dielectric medium is

$$k_d = \frac{\omega}{v} = \omega\sqrt{\varepsilon_r \varepsilon_0 \mu_0} \tag{3–5}$$

where $v = c/\sqrt{\varepsilon_r}$ m/s is the velocity of propagation in the medium. For example, at 900 MHz, $\omega = 18\pi \times 10^8$, so that in air ($\varepsilon_r = 1$), $k = 6\pi$ m^{-1} and $\lambda = 1/3$ m. In a dry brick wall with $\varepsilon_r = 4$, for the same frequency $k_d = 12\pi$ m^{-1} and $\lambda_d = 1/6$ m.

When describing propagation in air, with ε_r very close to unity, we will omit the subscript d on the wavenumber and wave impedance. Because there is no variation of the plane wave fields with (y,z), the plane wave occupies all of space. Although this is an idealization that can never be realized in practice, over limited regions of space the fields radiated by actual sources are approximately those of a plane wave. For example, the fields of a laser beam have approximately the variation given by (3–3) with the amplitude A slowly varying to zero at points (y,z) outside the beam. When viewed over a limited region far from the antenna, even the fields radiated by a radio antenna have approximately the same variation as the plane wave fields.

3.1a Phasor notation

As in the case of harmonic voltages and currents, we can avoid writing the time dependence explicitly by using the phasor electric and magnetic fields, which depend on position only. Thus if $\mathbf{E}(x,y,z)$ and $\mathbf{H}(x,y,z)$ represent the phasor fields, the actual time-dependent fields $\mathbf{E}(x,y,z;t)$ and $\mathbf{H}(x,y,z;t)$ are found using the expressions

$$\mathbf{E}(x,y,z;t) = \mathrm{Re}\{\mathbf{E}(x,y,z)e^{j\omega t}\} \tag{3–6}$$

$$\mathbf{H}(x,y,z;t) = \mathrm{Re}\{\mathbf{H}(x,y,z)e^{j\omega t}\} \tag{3–7}$$

where $\mathrm{Re}\{\cdot\}$ implies taking the real part of the quantity in the braces. For the plane wave fields of (3–3), the phasor fields depend only on x and are seen to be

$$\mathbf{E}(x) = \mathbf{y}_0 A e^{j\phi} e^{-jk_d x} \tag{3–8}$$

$$\mathbf{H}(x) = \mathbf{z}_0 \frac{1}{\eta_d} A e^{j\phi} e^{-jk_d x}$$

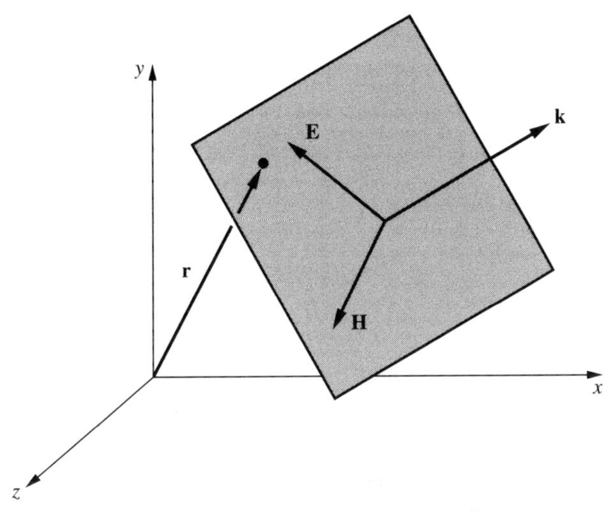

Figure 3-2 Plane wave propagation oblique to the coordinate axes.

Complex dielectric constants are easily accommodated by the phasor notation. In this case, k_d of (3–5) has a positive real part and a negative imaginary part, while η_d of (3–4) has positive real and imaginary parts. For example, 1 GHz waves in aerated concrete with $\varepsilon_r = 2 - j0.1$ have $k_d = (20\pi/3)\sqrt{2 - j0.1} \approx 29.62 - j0.74$ m^{-1} and $\eta_d = 377/\sqrt{2 - j0.1} \approx 266.6 + j6.7\Omega$. The wave propagates with phase variation exp($-j29.62x$) and attenuation exp($-0.74x$), so that the field amplitude decreases by a factor $1/e$ in a distance 1/0.74 = 1.35 m. Using the phasor notation, the period average power flow **P** W/m^2 can be found from the expression

$$\mathbf{P} = \frac{1}{2} Re\{\mathbf{E} \times \mathbf{H}^*\} \tag{3–9}$$

where $\mathbf{H}^*$ represents the complex conjugate of the magnetic field phasor. If the power flow in air is $|\mathbf{P}|$ = W/m^2, then in view of (3–7) and (3–8) the amplitude of the electric field is $A = \sqrt{2\eta}$ = 27.5 V/m.

3.1b Propagation oblique to the coordinate axes

Expression (3–7) can easily be generalized to represent propagation oblique to the coordinate axes. Let **k** be a vector that points in the direction of propagation and whose magnitude is the wavenumber $k_d = \omega/v$, We refer to **k** as the wavevector and can write it in terms of its components (k_1,k_2,k_3) and the unit vectors $(\mathbf{x}_0,\mathbf{y}_0,\mathbf{z}_0)$ as

$$\mathrm{k} = \mathbf{x}_0 k_1 + \mathbf{y}_0 k_2 + \mathbf{z}_0 k_3 \tag{3–10}$$

Referring to Figure 3–2, if **r** is the displacement vector from the origin to any point on a plane of constant phase perpendicular to **k**, then the dot product

$$\mathbf{k} \cdot \mathbf{r} = k_1 x + k_2 x + k_3 x \tag{3–11}$$

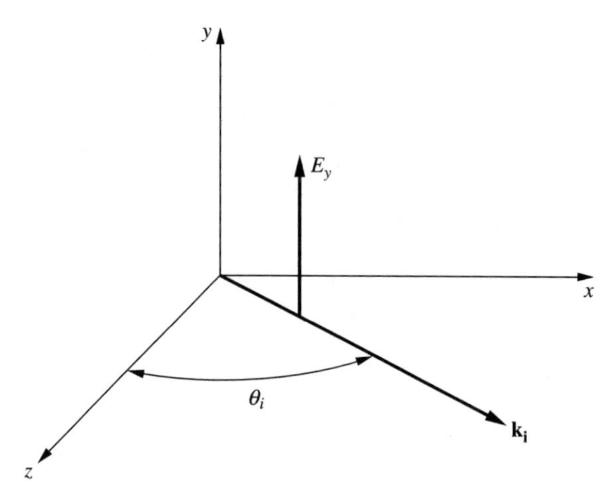

Figure 3-3 Plane waves propagating in the (x,y) plane are used to simulate fast fading.

represents the perpendicular distance from the origin to the plane, multiplied by the wavenumber k_d and is the generalization for oblique propagation of the phase term $k_d x$ in (3–7). If the polarization of the electric field is given by the unit vector $\mathbf{e}_0$ and that of the magnetic field by the unit vector $\mathbf{h}_0$, both must be perpendicular to each other and to $\mathbf{k}$ and must satisfy the relation

$$\mathbf{e}_0 \times \mathbf{h}_0 = \mathbf{k}/k_d \tag{3–12}$$

where we have used the fact that $|\mathbf{k}| = k_d$. The phasor electric and magnetic fields of the wave can thus be written as

$$\mathbf{E}(x,y,z) = \mathbf{e}_0 A e^{j\phi} e^{-j\mathbf{k}\cdot\mathbf{r}} \tag{3–13}$$

$$\mathbf{H}(x,y,z) = \mathbf{h}_0 \frac{1}{\eta_d} A e^{j\phi} e^{-j\mathbf{k}\cdot\mathbf{r}} \tag{3–14}$$

3.1c Fast fading due to several plane waves

We can use several oblique plane waves, of the form given by (3–12), to show how multipath leads to fast fading of the received signal. Consider plane waves propagating in air parallel to the (x,z) plane with wavevector $\mathbf{k}$ making an angle θ_i with the positive z axis, as shown in Figure 3–3. In this case the components of the wavevector are $k_1 = k\sin\theta_i$, while $k_2 = 0$ and $k_3 = k\cos\theta_i$ in (3–11). If we further assume that the electric fields of the plane waves are all polarized along y, and if we wish to plot the variation of the total field only along the x axes ($z = 0$), the y component of the electric field is found by summing the individual plane wave contributions to give

$$E_y(x,0) = \sum_i V_i e^{j\phi_i} e^{-jkx\sin\theta_i} \tag{3–15}$$

where V_i and ϕ_i are the amplitudes and phases of the plane waves.

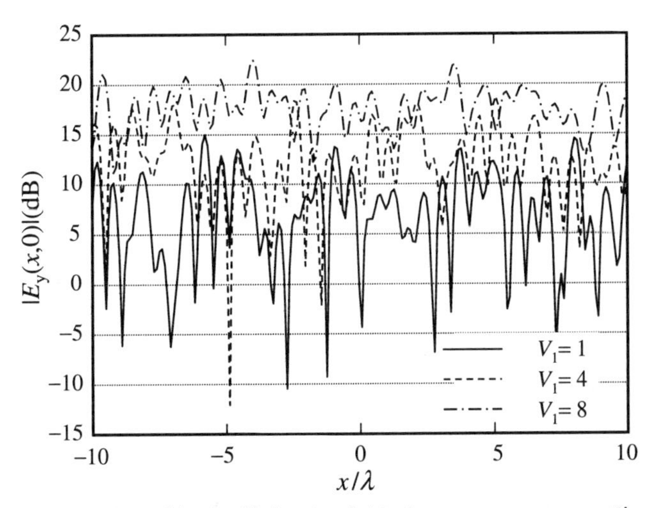

Figure 3-4 Variation of $|E_y(x,0)|$ due to eight plane waves propagating at randomly chosen angles θ_i and having randomly chosen phases ϕ_i for the cases when (a) $V_i = 1$ for $i = 1$ to 8; (b) $V_1 = 4$ and $V_i = 1$ for $i = 2$ to 8; (c) $V_1 = 8$ and $V_i = 1$ for $i = 2$ to 8.

Expression (3–13) has been used to simulate the variation of the electric field for different conditions. Figure 3–4 shows the variation in $|E_y(x,0)|$ over a distance of 20λ for eight plane waves whose propagation directions θ_i and phases ϕ_i have been chosen randomly over the range 0 to 2π. Three curves have been plotted: one for the case when all eight waves have the same amplitude $V_i = 1$ for $i = 1$ to 8; the second when $V_1 = 4$ and the remaining amplitudes $V_i = 1$ for $i = 2$ to 8; and the last case when $V_1 = 8$ and the remaining amplitudes $V_i = 1$ for $i = 2$ to 8. It is seen that when the amplitudes are equal, or nearly equal, the variation has the appearance of a Rayleigh fading signal with fades of 20 dB or more. Note that the use of a logarithmic scale for the field strength accentuates the depth of the fades when the signals nearly cancel, as compared to the places where they add.

When one plane wave is significantly stronger that the others ($V_1 = 8$), the variation of the total field about its average is seen from Figure 3–4 to be relatively smaller. The scale length of the fading is also seen to be on the order of $\lambda/2$, as found for measured signal variations. To demonstrate the statistical properties of the signal variation, each curve of Figure 3–4 was normalized to its average value and then sampled at 200 points, from which a cumulative distribution function was formed. The resulting distribution functions are plotted in Figure 3–5 along with that of the Rayleigh distribution. In confirmation of the previous observation, it is seen that the curves for equal, or nearly equal, values of V_i are close to the Rayleigh distribution, while the curve for $V_1 = 8$ is distinctly different.

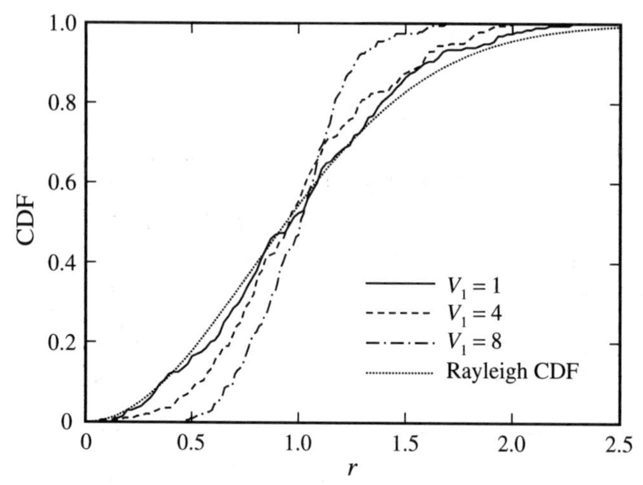

Figure 3-5 Cumulative distribution functions for the three fast-fading curves of Figure 3–4, each normalized to its average value. The Rayleigh distribution is shown for comparison.

3.1d Correlation function and Doppler spread

Expression (3–13) for the x dependence of the field can be used to find the autocorrelation function, and from it the correlation length of the fading and Doppler spread when the subscriber is moving. The autocorrelation $C(s)$ is defined by the integral

$$C(s) = \frac{1}{2W}\int_{-W}^{W} E_y(x, 0)E_y^*(x-s, 0)dx \tag{3–16}$$

where $2W >> s$ is the distance over which the field is known. Substituting (3–13) for E_y and E_y^*, and using a different index of summation in each expression for the field, the orders of summation and integration can be interchanged to give

$$C(s) = \sum_i \sum_j V_i V_j e^{j(\phi_i - \phi_j)} e^{-jks\sin\theta_j} \frac{1}{2W}\int_{-W}^{W} e^{-jkx(\sin\theta_i - \sin\theta_j)}dx \tag{3–17}$$

Carrying out the integration in (3–15) results in the expression

$$C(s) = \sum_i \sum_j V_i V_j e^{j(\phi_i - \phi_j)} e^{-jks\sin\theta_j} \frac{\sin[kW(\sin\theta_i - \sin\theta_j)]}{kW(\sin\theta_i - \sin\theta_j)} \tag{3–18}$$

If W is large compared to wavelength, then $kW >> 1$ and the sinc function will be very small except for the terms having $i = j$, so that (3–16) reduces to

$$C(s) = \sum_i V_i^2 e^{-jks\sin\theta_i} \tag{3–19}$$

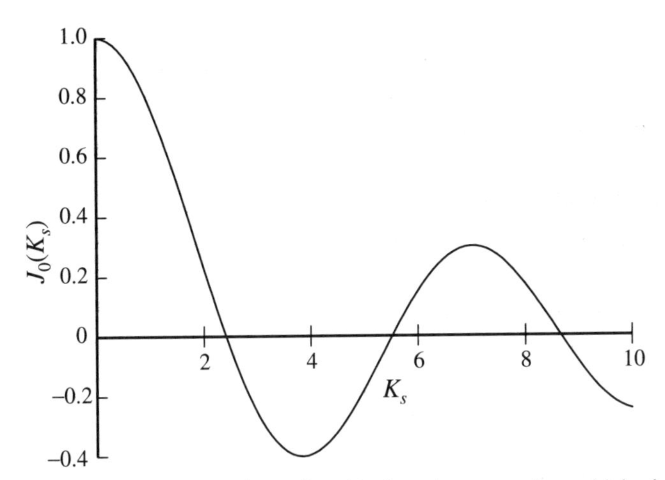

Figure 3-6 Bessel function of the first kind and zero order, which describes the spatial correlation function for a continuous distribution of arriving plane waves that are uniformly distributed in the horizontal plane.

The classic closed-form expression for the autocorrelation is obtained by passing from discrete arrival angles to a continuous distribution by replacing the summation in (3–17) with an integration over θ. Assuming that the amplitudes V_i are the same for all directions of propagation, taken here to be unity, the continuous limit is

$$C(s) = \frac{1}{2\pi}\int_{-\pi}^{\pi} e^{-jks\sin\theta}d\theta = \frac{1}{\pi}\int_{0}^{\pi}\cos(ks\sin\theta)d\theta = J_0(ks) \tag{3–20}$$

where J_0 is the zero-order Bessel function [9] plotted in Figure 3–6. The first zero of the Bessel function occurs when $k|s| = 2.4$, after which $|C(s)| < 0.5$, so that the field at x and at $x - s$ are decorrelated. Recalling that $k = 2\pi/\lambda$, the fast-fading pattern is seen to be decorrelated for $|s| > 0.4\lambda$, as has been observed in measurements [10].

For a subscriber traveling along the street with velocity v_s, the distance x in (3–13) can be replaced by $v_s t$, and the subscriber is seen to experience a time variation of the received signal that is in addition to the harmonic time dependence $e^{j\omega t}$ of the carrier. This time variation has a power spectral distribution about ω that can be found from its autocorrelation function [11]. The autocorrelation function in (3–18) can be viewed as resulting from the time delay τ between the point x as compared to the point $x - s$ by replacing s in (3–18) by $v_s\tau$. If δ is the deviation of the radian frequency from ω, the power spectral density $P(\delta)$ is then given by the Fourier transform of $C(v_s\tau)$, which can be evaluated in closed form [9, p. 486] as

$$P(\delta) = \int_{-\infty}^{\infty} J_0(kv_s\tau)e^{-j\delta\tau}d\tau = \frac{2/kv_s}{\sqrt{1-(\delta/kv_s)^2}}U(kv_s - |\delta|) \tag{3–21}$$

where $U(\cdot)$ is the unit step function. In (3–19), $kv_s = 2\pi v_s/\lambda$ is the radian frequency of the maximum Doppler shift found when the wave propagates in the same or opposite direction to the motion of the subscriber. The power spectrum of (3–19) is plotted as the smooth curve in Figure 2–13. If the propagation directions of the plane waves also have a vertical component of travel ($k_2 \neq 0$), the Doppler spectrum will have a difference in shape that is most pronounced near the maximum Doppler frequency $\pm kv_s$, where (3–21) diverges [12].

3.1e Fading at elevated base stations

The paths by which waves arrive at an elevated base station from a subscriber located among the buildings can be seen from Figure 2–19 by reversing the directions of the arrows. In typical suburban regions where there are no tall buildings or mountains to scatter the signals through large angles, the waves arriving at the base station come from a limited range of angles in the horizontal plane about the direction to the subscriber. Because of this limited angle range, the received signal at the base station location will exhibit spatial fading much like that shown in Figure 3–4, except that the scale length of the fading will be much larger than $\lambda/2$.

To estimate the scale of fading, assume that the waves arrive within a narrow wedge of angles $-\theta_{max} < \theta_i < \theta_{max}$ about the z axis in Figure 3–3. If the wave directions are uniformly distributed over this wedge, the same arguments as those used in the preceding section may be employed to find the autocorrelation function for the displacement s of the base station antenna. With the help of the approximation $\sin\theta \approx \theta$ for angles in radians, it is seen that

$$C(s) = \frac{1}{2\theta_{max}}\int_{-\theta_{max}}^{\theta_{max}} e^{-jks\sin\theta}\,d\theta \approx \frac{\sin(ks\theta_{max})}{\theta_{max}ks} \tag{3–22}$$

The correlation function has its first zero when $ks\theta_{max} = \pi$, or since $k = 2\pi/\lambda$, the zero occurs at $s = \lambda/2\theta_{max}$. For s greater than this value, $|C(s)| < 0.5$, so that the field at x and at $x - s$ are decorrelated. Other distributions of wave arrivals give slightly different results [13].

For a region of fixed size about the subscriber that gives rise to the scattered rays, the angle θ_{max} decreases with distance R to the subscriber, and hence the value of s for decorrelation increases. For example, if $R = 2$ km and the scattering region extends 100 m from the subscriber, then $\theta_{max} = 0.05$ rad or 1.4°, which gives the decorrelation distance of 10λ. At $R = 4$ km the decorrelation distance would be twice as large. Base stations are typically fitted with diversity receivers that use two or more receiving antennas separated by a distance greater than the decorrelation distance. The diversity receiver makes disproportionate use of the antenna with the greatest received power when detecting the subscriber's signal, thereby avoiding the possibility of being in a deep fade, which would cause loss of the received signal.

3.2 Reflection of plane waves at planar boundaries

When a plane wave propagating in one medium encounters a boundary with a second medium having different properties, it will be partially reflected back into the first medium and partially transmitted (refracted) into the second medium. The direction of propagation and amplitudes of the reflected and transmitted plane waves are determined by the boundary conditions at the interface [1,2], which require that the tangential components of the total electric field **E** and the total

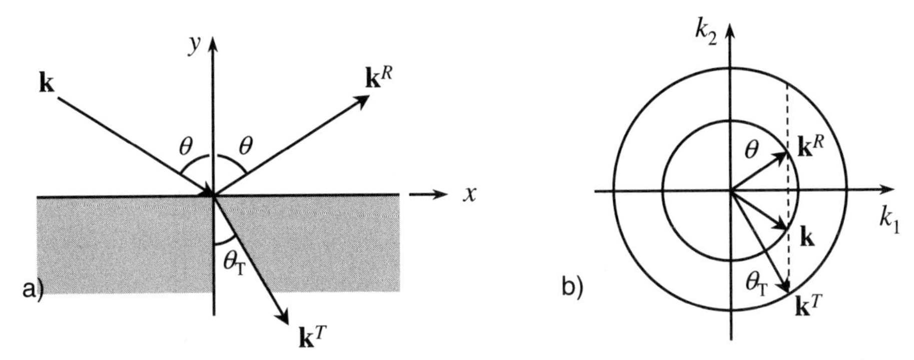

Figure 3-7 Graphical interpretation of Snell's law giving the direction of transmitted and reflected waves, as seen in (a) physical space; (b) the wavevector plane.

magnetic field **H** be continuous across the boundary. A diagram showing the directions of the incident, reflected and transmitted plane waves for a planar boundary between two media is shown in Figure 3–7a, which has been drawn assuming that the incident plane wave propagates parallel to the (x,y) plane, making an angle θ with the normal to the boundary. The reflected and transmitted waves in this case will also propagate parallel to the (x,y) plane, as will be argued subsequently.

3.2a Snell's law

For the electric and magnetic fields to satisfy the boundary conditions at all points along the planar boundary in Figure 3–7a, it is necessary that all three plane waves have the same variation with x. In this way satisfying the boundary conditions at $x = 0$ by appropriate choice of wave amplitudes will ensure that they are satisfied for all x. Thus using (3–11), (3–13) and (3–14) we require that

$$\exp(-jk_1x) = \exp(-jk_1{}^Rx) = \exp(-jk_1{}^Tx) \tag{3–23}$$

Because (3–23) can only be satisfied if the exponents themselves are equal,

$$k_1 = k_1{}^R = k_1{}^T \tag{3–24}$$

which is to say that the component of the wavevectors along the interface for all three plane waves must be equal. Similarly, when the incident plane wave propagates parallel to the (x,y) plane, $k_3 = 0$, so that the z components of the wavevectors for the reflected and transmitted waves must also vanish [i.e., these waves also propagate parallel to the (x,y) plane]. Note that (3–24) is valid even for dielectrics having complex dielectric constant ε_r.

A simple graphical interpretation of (3–24) is shown in Figure 3–7b for the case when the lower medium is a dielectric, with ε_r assumed to be real, and the upper medium is air. The wavevector of a plane wave propagating parallel to the (x,y) plane in air must lie on a circle in the (k_1,k_2) plane that has radius $k = \omega/c$, while that of the transmitted wave must lie on a larger circle of radius $k_d = \omega/v = (\omega/c)\sqrt{\varepsilon_r}$. Snell's law (3–24) requires that all three wavevectors have

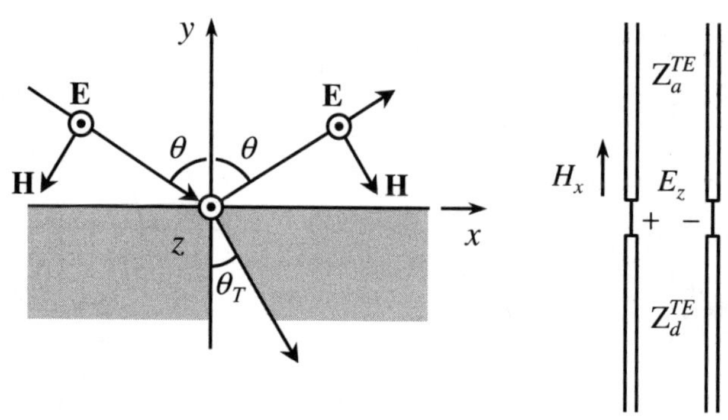

Figure 3-8 Reflection and transmission of a TE polarized plane wave at a dielectric surface showing the field components in space and the transmission line analog.

the same component along x, as indicated in Figure 3–7b. It is clear from the diagram that the angle of reflection θ_R is equal to the angle of incidence θ. Also, for $k_d > k$ it is seen that the angle θ_T must be smaller than θ (i.e., a wave entering a denser medium is bent away from the normal). Expressing the x components of the wavevectors in terms of the angles gives the familiar form of Snell's law

$$\sin\theta = \sin\theta_R = \sqrt{\varepsilon_r}\sin\theta_T \tag{3–25}$$

As an example of (3–25), if $\theta = 60°$ and $\varepsilon_r = 3$, then (3–25) becomes $\sqrt{3}/2 = \sqrt{3}\sin\theta_T$ and $\theta_T = 30°$.

With the help of Snell's law we can now find the amplitudes of the reflected and transmitted waves. There are two possible polarizations of the incident plane wave that must be considered separately. One polarization has the electric field transverse to the plane of incidence, which is the plane defined by the normal to the boundary and the incident wavevector (the plane of the paper in Figure 3–7). The other has the magnetic field transverse to the plane of incidence. These two polarizations are called TE and TM, respectively.

3.2b Reflection and transmission coefficients for TE polarization

Letting V^I and V^R represent the strength of the electric field for the incident and reflected waves in the air, then with the help of Figure 3–8 it is seen that the components of the electric and magnetic fields can be written in the form

$$E_z(x,y) = (V^R e^{-jky\cos\theta} + V^I e^{+jky\cos\theta})e^{-jkx\sin\theta} \tag{3–26}$$

$$H_x(x,y) = \frac{\cos\theta}{\eta}(V^R e^{-jky\cos\theta} - V^I e^{+jky\cos\theta})e^{-jkx\sin\theta} \tag{3–27}$$

$$H_y(x,y) = \frac{\sin\theta}{\eta}E_z(x,y) \tag{3–28}$$

The first term in parentheses on the right-hand side of (3–26) and (3–27) represents the reflected wave, which has a component of travel in the positive y direction in Figure 3–8 and therefore has a negative sign in the exponent, while the second term gives the incident wave, which has a component of travel in the negative y direction. The negative sign before V^I in the expression for H_x is a result of the fact that the wave with a component of travel in the negative y direction will have a y component of the Poynting vector (3–8) that is negative.

The y dependencies of E_z and H_x in (3–24) are analogous to those of voltage and current on a transmission [2,3] line with wavenumber β and impedance Z^{TE} given by

$$\beta = k\cos\theta \tag{3–29}$$

$$Z^{\mathrm{TE}} = \frac{\eta}{\cos\theta} = \frac{\eta k}{\beta} = \frac{\eta}{\sqrt{1-\sin^2\theta}} \tag{3–30}$$

Similar expressions can be written for the fields in the dielectric in terms of the amplitude V^T of the transmitted wave's electric field. The transmitted fields will have the same dependence on x as shown in (3–26) and (3–27). The y dependence of the fields in the dielectric is also analogous to voltage and current on a transmission line with wavenumber β_d and impedance Z_d^{TE}, which can be expressed via Snell's law and the definitions (3–4) and (3–5), in the various equivalent forms

$$\beta_d = k_d\cos\theta_T = \sqrt{k_d^2 - k^2\sin^2\theta} = k\sqrt{\varepsilon_r - \sin^2\theta} \tag{3–31}$$

$$Z_d^{\mathrm{TE}} = \frac{\eta_d}{\cos\theta_T} = \frac{\eta_d k_d}{\beta_d} = \frac{\eta}{\sqrt{\varepsilon_r - \sin^2\theta}} \tag{3–32}$$

The transmission line analog to the reflection problem is shown in Figure 3–8. The electromagnetic boundary conditions require that E_z and H_x be continuous at the boundary $y = 0$, and are analogous to continuity of voltage and current at a direct connection of transmission lines. Using the analogy, we may draw on the expressions derived for transmission lines to find the reflection and transmission coefficients [2,3]. Thus the reflection coefficient of the electric field for the TE polarization is

$$\Gamma_E \equiv \frac{V^R}{V^I} = \frac{Z_d^{\mathrm{TE}} - Z^{\mathrm{TE}}}{Z_d^{\mathrm{TE}} + Z^{\mathrm{TE}}} = \frac{\cos\theta - \sqrt{\varepsilon_r}\cos\theta_T}{\cos\theta + \sqrt{\varepsilon_r}\cos\theta_T} \tag{3–33}$$

and the transmission coefficient of the electric field is

$$T_E \equiv \frac{V^T}{V^I} = 1 + \Gamma_E = \frac{2\cos\theta}{\cos\theta + \sqrt{\varepsilon_r}\cos\theta_T} \tag{3–34}$$

When the dielectric has complex ε_r, then $\cos\theta_T$ in (3–31) and (3–32) is complex and can be found from the expression $\cos\theta_T = \beta_d/k_d = \sqrt{1-(\sin\theta^2/\varepsilon_r)}$, which is derived from (3–5) and (3–31).

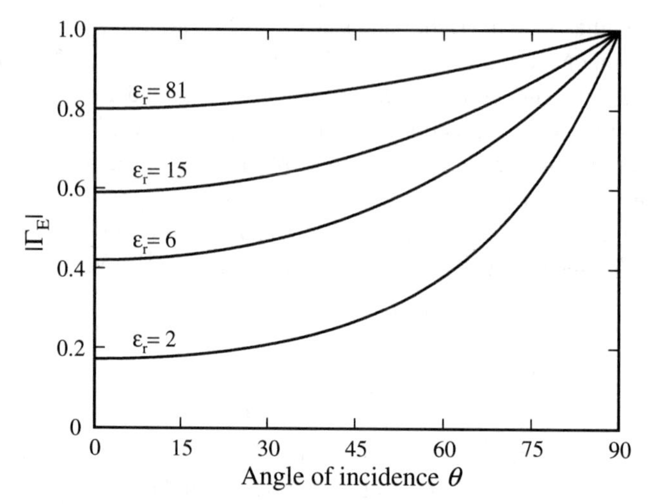

Figure 3-9 Variation of the magnitude of the TE reflection coefficient with angle of incidence for a plane wave reflected at a dielectric ε_r.

For ε_r real, the reflection coefficient in (3–33) is negative in going from a less dense medium, such as air, to a more dense medium, such as earth. When going from a more dense to a less dense medium, the reflection coefficient has the same magnitude but opposite sign. The magnitude of the reflection coefficient for waves incident from air onto a dielectric is plotted in Figure 3–9 for several values of ε_r and is seen to go from a finite value to unity as θ goes from 0 to 90°. By definition, the reflection coefficient in (3–33) is the ratio of the electric field component of the reflected wave to that of the incident wave. Using (3–8) it is seen that the ratio of the period average reflected power to the period average incident power is given by $|\Gamma_E|^2$. Arguing on the basis of power conservation, the fraction of the incident power P_y flowing in the y direction that is transmitted can be found from $1 - |\Gamma_E|^2$. Alternatively, since the transmission coefficient is the ratio of electric fields, the fraction of the incident power that is transmitted can be found from $(Z^{TE}/Z_d^{TE})|T_E|^2$, which can directly be shown to equal $1 - |\Gamma_E|^2$ using (3–33) and (3–34).

3.2c Reflection and transmission coefficients for TM polarization

Figure 3–10a shows the field components for the TM polarization. In this case it is most convenient to describe the fields in terms of the amplitudes of the magnetic fields of the incident, reflected, and transmitted waves, which we denote as I^I, I^R, and I^T. With the help of Figure 3–10a it is seen that

$$H_z(x,y) = (I^R e^{-jky\cos\theta} + I^I e^{+jky\cos\theta})e^{-jkx\sin\theta} \tag{3–35}$$

$$E_x(x,y) = -\eta \cos\theta (I^R e^{-jky\cos\theta} - I^I e^{+jky\cos\theta})e^{-jkx\sin\theta} \tag{3–36}$$

$$E_y(x,y) = \eta \sin\theta H_z(x,y) \tag{3–37}$$

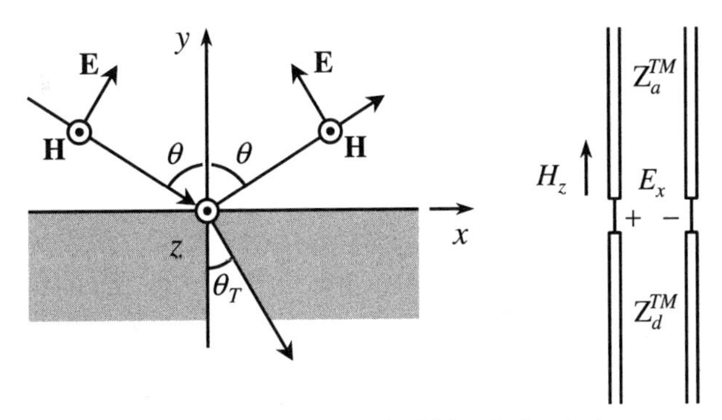

Figure 3-10 Reflection and transmission of a TM polarized plane wave at a dielectric surface showing the field components in space and the transmission line analog.

As for the TE polarization, the first term in the parentheses on the right represents a wave having a component of travel in the positive y direction and hence is the reflected wave in Figure 3–10a, while the second travels in the negative y direction and is therefore the incident wave. Again, the y dependence in (3–35) and (3–36) is analogous to that for the current and voltage on a transmission line of wavenumber β, which is the same as given in (3–29) for the TE polarization and impedance Z^{TM}. Similar expressions for the fields of the transmitted plane wave have y dependence that is analogous to that of the voltage and current on a transmission line having wavenumber β_d, which is the same as given in (3–31) for the TE polarization and impedance Z_d^{TM}. Using Snell's law and the definitions (3–4) and (3–5), expressions for the impedances can be written in the following equivalent forms:

$$Z^{TM} = \eta\cos\theta = \eta\frac{\beta}{k} = \eta\sqrt{1-\sin\theta^2} \tag{3–38}$$

$$Z_d^{TM} = \eta_d\cos\theta_T = \eta_d\frac{\beta_d}{k_d} = \eta\frac{1}{\varepsilon_r}\sqrt{\varepsilon_r - \sin\theta^2} \tag{3–39}$$

The equivalent transmission line problem is shown in Figure 3–10, from which the reflection and transmission coefficients can be determined [2,3].

The reflection coefficient of the magnetic field for the TM polarization is the ratio of the I^R to I^I and is given by

$$\Gamma_H \equiv \frac{I^R}{I^I} = \frac{Z^{TM} - Z_d^{TM}}{Z^{TM} + Z_d^{TM}} = \frac{\sqrt{\varepsilon_r}\cos\theta - \cos\theta_T}{\sqrt{\varepsilon_r}\cos\theta + \cos\theta_T} \tag{3–40}$$

while the transmission coefficient of the magnetic field is

$$T_H \equiv \frac{I^T}{I^I} = 1 + \Gamma_H = \frac{2\sqrt{\varepsilon_r}\cos\theta}{\sqrt{\varepsilon_r}\cos\theta + \cos\theta_T} \tag{3–41}$$

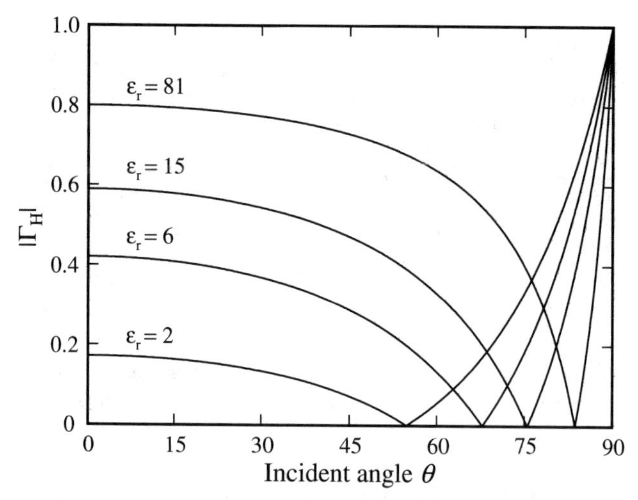

Figure 3-11 Variation of the magnitude of the TE reflection coefficient with angle of incidence for a plane wave reflected at a dielectric ε_r. The reflection coefficient vanishes at Brewster's angle θ_B.

For ε_r real, the reflection coefficient in (3–40) is positive for $\theta = 0$ and approaches –1 at $\theta = 90°$. In going from a positive to a negative value, Γ_H passes through zero at Brewster's angle θ_B, which can be found from the expression

$$\theta_B = \arctan\sqrt{\varepsilon_r} \tag{3–42}$$

The magnitude of Γ_H is plotted in Figure 3–11 as a function of the angle of incidence θ for several values of ε_r, and is seen to be zero at an angle given by (3–33). As for the TE polarization, the ratio of the period average reflected power to the period average incident power is given by $|\Gamma_H|^2$. The fraction of the incident power P_y flowing in the y direction that is transmitted can again be found from $1 - |\Gamma_H|^2$. However, since the transmission coefficient is a ratio of magnetic fields rather than electric fields, when expressed in terms of the transmission coefficient, the fraction of the power that is transmitted is given by $(Z_d^{\mathrm{TE}}/Z^{\mathrm{TE}})|\mathrm{T_H}|^2$.

3.2d Height gain for antennas above ground

In radio systems located on the earth's surface, the received signal is typically found to increase as the antenna is raised and is referred to as *height gain*. This effect can be modeled using plane waves when the transmitting antenna is far from the receiving antenna and the earth is flat. In this case the fields in the vicinity of the receiver are approximately a combination of an incident plane wave and a ground reflected plane wave, as illustrated in Figure 3–12. If the transmitter and receiver are polarized for the vertical component of electric field, then using the reflection coefficient in (3–40), the spatial dependence of the vertical field component E_y is given by

$$E_y(x,y) = \eta \sin\theta\, I^I (e^{jky\cos\theta} + \Gamma_H e^{-jky\cos\theta}) e^{-jkx\sin\theta} \tag{3–43}$$

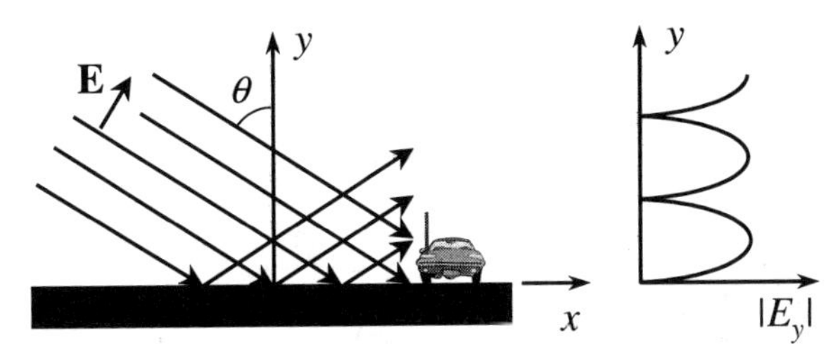

Figure 3-12 Incident fields from a distant transmitter are approximated by a plane wave, which together with the reflected plane wave, lead to the standing-wave variation of $|E_y|$ with vertical distance, as shown.

In typical radio links the distance R to the transmitter is much greater than its elevation h_{BS}, so that the angle θ in Figure 3–12 is close to 90°. It is therefore appropriate to take the limit as $\theta \to 90°$, in which case and $\sin\theta \to 1$, and $\Gamma_H \to -1$ from (3–40), so that

$$|E_y(x,y)| \to 2\eta I^I|\sin(ky\cos\theta)| \tag{3–44}$$

The variation in (3–35) is sketched in Figure 3–12, where it is seen that the electric field, and hence the received voltage, increases linearly with height for small y satisfying the condition $ky\cos\theta << \pi$. This linear height gain corresponds to 6 dB of gain per doubling of height. For example, if $R = 10$ km and $h_{BS} = 100$ m, the limitation on y for linear height gain is $y << 50\lambda$, or 15 m for 1-GHz signals. When the receiver is located on a city street and is not visible from the transmitter, we will see that the angle of incidence θ is determined by the surrounding buildings, so that the height gain will be different.

3.2e Reflection of circularly polarized waves

We have considered previously waves that are linearly polarized along some particular coordinate direction, as in (3–3) and (3–8). More generally, the wave can have two vector components that are perpendicular to each other and to the direction of propagation. For example, the generalization of (3–8) has components along both y and z and for waves in air can be written in the form

$$\mathbf{E}(x) = (\mathbf{y}_0 A + \mathbf{z}_0 B)\, e^{j\phi}\, e^{-jkx} \tag{3–45}$$

$$\mathbf{H}(x) = \frac{1}{\eta}(\mathbf{z}_0 A - \mathbf{y}_0 B)\, e^{j\phi}\, e^{-jkx} \tag{3–46}$$

where A and B are complex amplitudes. For both sets of field components the power (3–9) is carried in the positive x direction.

If A and B have the same phase, for example both are real, new coordinates (y',z') can be defined by a rotation through an angle $\arctan(B/A)$ such that the fields in (3–45) lie along the y' axis. In this case the wave is still said to be linearly polarized. Alternatively, if A and B have dif-

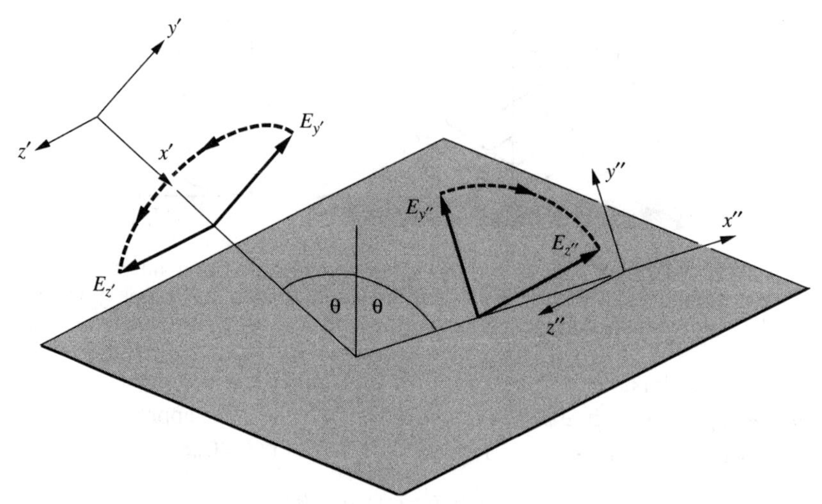

Figure 3-13 Reflection of a circularly polarized plane wave by a conductor, or a dielectric for $\theta < \theta_B$, reverses the sense of the rotation, shown dashed.

ferent phases, B/A is complex and no rotation angle exists that will result in **E** being polarized along y'. Of special interest for satellite and other applications is the case when $B = \pm jA$, so that

$$\mathbf{E}(x) = A(\mathbf{y}_0 \pm j\mathbf{z}_0)\, e^{-jkx} \tag{3–47}$$

Using (3–6), and assuming that A is real for simplicity, the instantaneous fields are given by

$$\mathbf{E}(x,t) = A[\mathbf{y}_0 \cos(\omega t - kx) \mp \mathbf{z}_0 \sin(\omega t - kx)] \tag{3–48}$$

At any given value of x, the field in (3–48) has a fixed amplitude and its direction rotates in time about the x axis. When viewed in the direction of propagation from behind, the rotation is counterclockwise for the upper sign in (3–48) and clockwise for the lower sign. These waves are said to be left- and right-hand circularly polarized. If A and B are not equal in magnitude, or have a phase difference other than 90°, the tip of the **E** vector will trace out an ellipse and the wave is said to be elliptically polarized.

To treat reflection at surfaces, it is convenient to decompose the incident wave (3–36) into TE and TM components whose reflection coefficients are then computed, as discussed in previous sections. The reflected field is the sum of the reflected TE and TM waves. An example of this is shown in Figure 3–13, where a right-hand circularly polarized wave is incident on a reflecting surface. If the surface is that of a metal, or if the angle of incidence θ is less than the Brewster angle for a dielectric, then for both the TE and TM reflected waves, the components of **E** tangent to the surface will be reversed in direction, as shown in Figure 3–13. Because the normal component of **E** does not reverse, **E** of the reflected wave will have positive y'' component. Thus the reflected wave exhibits left-hand rotation, which is of opposite sense to that of the incident wave. For a good conductor the two components of the field will be reflected with equal amplitude so that a circularly polarized incident wave will remain circularly polarized but have

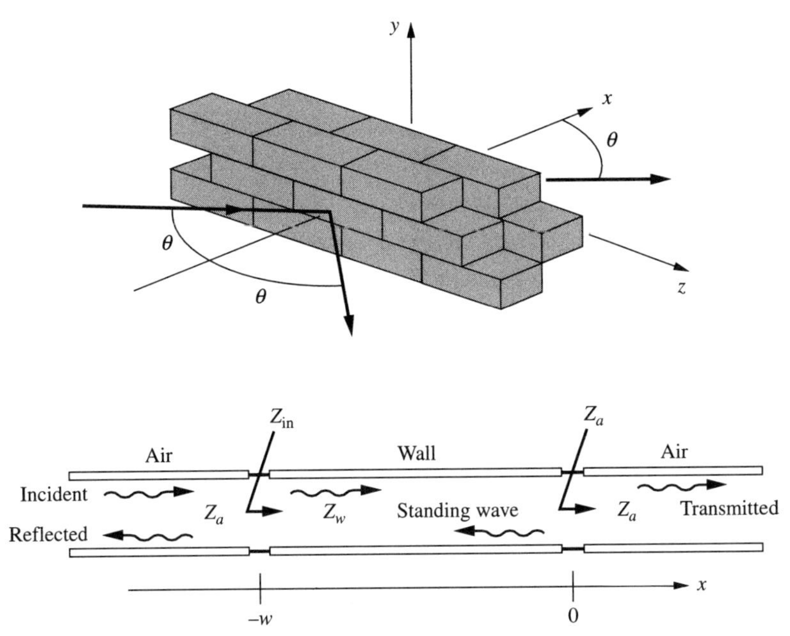

Figure 3-14 Brick wall treated by the transmission line approach to compute reflected and transmitted waves. The impedances are those for either the TE or TM case, depending on the polarization of the incident wave.

the opposite rotation. However, at a dielectric, the amplitudes will be different, except for normal incident, and the polarization of the reflected wave will be elliptic. For incidence on a dielectric with $\theta > \theta_B$ the sense of the rotation will be preserved. The reversal of rotation upon reflection is of interest in some applications [14], since using circularly polarized transmitting and receiving antennas can reduce interference caused by signals undergoing a single reflection.

3.3 Plane wave incidence on dielectric layers

Some walls and floors have the appearance of one or more dielectric layers, as in the case of the brick wall in Figure 3–14. Whenever the faces of the dielectric are parallel to each other, as in Figure 3–14, Snell's law applied successively to each boundary shows that the wavenumber parallel to the boundaries must be the same in each layer. Thus in Figure 3–14, if the incident wave travels parallel to the horizontal plane and makes an angle θ with the normal, the wave transmitted into the air on the other side of the wall will also propagate parallel to the horizontal plane at the angle θ to the normal. To find the fraction of the power that is reflected and transmitted at the wall, we can use the transmission line analog shown in Figure 3–14.

Because of the impedance mismatch at $x = 0$ and at $x = -w$, standing waves will be set up inside the wall. All the field quantities will have the same transverse variation $\exp(-jkz \sin\theta)$.

Neglecting this factor, for either polarization the transverse electric and magnetic field components in the wall will have x dependence, given by

$$V(x) = V^{+} e^{-j\beta_w x} + V^{-} e^{+j\beta_w x} \tag{3–49}$$

$$I(x) = \frac{1}{Z_w}\left(V^{+} e^{-j\beta_w x} - V^{-} e^{+j\beta_w x}\right) \tag{3–50}$$

respectively. Here V^{+} and V^{-} are the amplitudes of the transverse electric field component for the waves traveling in the positive and negative x directions, respectively, $\beta_w = k_w \cos\theta_w$ is the wavenumber along x in the wall, and Z_w is the wave impedance of the wall for either TE or TM polarization. The ratio of voltage to current gives the x dependent impedance $Z(x)$ seen along the segment of transmission line, or

$$Z(x) = \frac{V(x)}{I(x)} = Z_w \frac{V^{+} e^{-j\beta_w x} + V^{-} e^{+j\beta_w x}}{V^{+} e^{-j\beta_w x} - V^{-} e^{+j\beta_w x}} \tag{3–51}$$

At the junction $x = 0$ the impedance of (3–51) must be equal to the load impedance Z_L seen looking to the right. In Figure 3–14 this load impedance is simply the impedance of air Z. Thus, evaluating the impedance in (3–51) at $x = 0$ and setting it equal to Z_L, we can solve for V^{-} in terms of V^{+} to arrive at

$$V^{-} = V^{+} \frac{Z_L - Z_w}{Z_L + Z_w} \tag{3–52}$$

Substituting (3–52) back into (3–51) and evaluating (3–51) at $x = -w$, after some manipulation the input impedance seen at the wall is found to be

$$Z_{\text{in}} = Z(-w) = Z_w \frac{Z_L(e^{+j\beta_w w} + e^{-j\beta_w w}) + Z_w(e^{+j\beta_w w} - e^{-j\beta_w w})}{Z_w(e^{+j\beta_w w} + e^{-j\beta_w w}) + Z_L(e^{+j\beta_w w} - e^{-j\beta_w w})} \tag{3–53}$$

The formulation in (3–53) is useful when the wall material has loss, but for lossless dielectrics the input impedance can be expressed more conveniently in terms of the trigonometric functions as

$$Z_{\text{in}} = Z_w \frac{Z_L \cos\beta_w w + jZ_w \sin\beta_w w}{Z_w \cos\beta_w w + jZ_L \sin\beta_w w} \tag{3–54}$$

3.3a Reflection at a brick wall

With the foregoing expression for the input impedance, the reflection coefficient Γ, at the brick wall, which is the ratio of the transverse components of electric field of the reflected and incident waves, is then given by

$$\Gamma = \frac{Z_{\text{in}} - Z}{Z_{\text{in}} + Z} \tag{3–55}$$

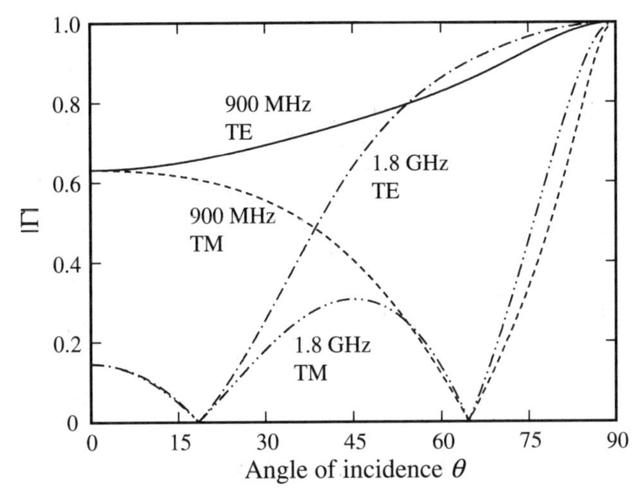

Figure 3-15 Magnitude of the reflection coefficient versus angle of incidence at a 20 cm thick brick wall ($\varepsilon_w = 4.44$) at 900 and 1800 MHz for TE and TM polarizations.

where Z is the TE or TM wave impedance for air. The fraction of the incident power that is reflected is equal to $|\Gamma|^2$, and if no loss is present in the wall, the fraction of the power that is transmitted into the air on the other side of the wall is $1 - |\Gamma|^2$. If loss is present, the attenuation of the fields in the wall must be accounted for in computing the transmitted power.

Two special cases for which there is no reflected wave from the wall ($\Gamma = 0$) can be seen from (3–54). Note that for air on the other side of the wall, $Z_L = Z$. Then for TM polarization and incidence at Brewster's angle θ_B, the impedance Z_w of the wall material will be equal to Z of the air. In this case (3–54) reduces to $Z_{\text{in}} = Z$, and from (3–55), $\Gamma = 0$. The second case applies to both polarizations and occurs when the frequency and angle of incidence are such that the phase shift $\beta_w w$ is a multiple of π. In this case $\sin \beta_w w = 0$, and again $Z_{\text{in}} = Z$ in (3–43), so that from (3–55), $\Gamma = 0$.

Figure 3–15 shows a plot of $|\Gamma|$ as a function of the angle of incidence θ for both TE and TM polarized waves reflected from a brick wall of thickness w = 20 cm, assuming a relative dielectric constant $\varepsilon_r = 4.44$ and frequencies of 900 and 1800 MHz. At $\theta = 0$ there is some finite reflection, which is the same for both polarizations, while at 90° total reflection is obtained in all cases. This behavior at the two limiting angles is the same for all types of wall construction. For the TM polarization, the Brewster angle condition at $\theta_B = 64.6°$ causes the reflection coefficient to vanish for both frequencies, as noted above. At 1800 MHz there is an additional impedance match for both polarizations at $\theta \approx 18°$ due to the fact that $\beta_w w = 5\pi$ there, and the reflection coefficient vanishes, as discussed above. Because the TM polarized wave at 1800 MHz has two zeros of Γ at $\theta \approx 18°$ and 65°, its magnitude is small over the range between these angles.

3.3b Reflection at walls with loss

In the case when the wall material is lossy, the input impedance must be computed accounting for ε_r''. To do so, recognize that the wavenumber of the medium, which is given by

$$k_W = \frac{\omega}{c}\sqrt{\varepsilon_r' - j\varepsilon_r''} \tag{3–56}$$

has a positive real part and a negative imaginary part. Using (3–31), the wavenumber in the direction perpendicular to the boundary (the transmission line direction) is then complex:

$$\beta_W = k\sqrt{\varepsilon_r' - j\varepsilon_r'' - \sin\theta^2} \equiv \kappa_w - j\alpha_w \tag{3–57}$$

with a positive real part and a negative imaginary part. The wave impedances defined in (3–32) and (3–39) are also found to have complex values

$$Z_w^{\text{TE}} = \frac{\eta}{\sqrt{\varepsilon_r' - j\varepsilon_r'' - \sin\theta^2}} \tag{3–58}$$

$$Z_w^{\text{TE}} = \frac{\eta}{\varepsilon_r' - j\varepsilon_r''}\sqrt{\varepsilon_r' - j\varepsilon_r'' - \sin\theta^2} \tag{3–59}$$

If (3–57) is used in expression (3–53), the impedance Z_{in} at a wall with loss is

$$Z_{\text{in}} = Z_W \frac{Z_L(e^{j\kappa_w w}e^{\alpha_w w} + e^{-j\kappa_w w}e^{-\alpha_w w}) + Z_w(e^{j\kappa_w w}e^{\alpha_w w} + e^{-j\kappa_w w}e^{-\alpha_w w})}{Z_w(e^{j\kappa_w w}e^{\alpha_w w} + e^{-j\kappa_w w}e^{-\alpha_w w}) + Z_L(e^{j\kappa_w w}e^{\alpha_w w} + e^{-j\kappa_w w}e^{-\alpha_w w})} \tag{3–60}$$

When the width of the wall is such that $\alpha_w w \geq 1$, the first term in each pair of parentheses in (3–60) is larger than the second term, so that (3–48) reduces to $Z_{\text{in}} \approx Z_W$. In other words, the loss dampens the multiple reflections within the wall so that the wall appears to the incident wave to be infinitely thick. To demonstrate the effect of loss in the wall, we have computed the reflection coefficient as a function of the angle of incidence at 4 GHz assuming that $\varepsilon_r = 4 - j0.1$. Expressions (3–33) and (3–40) were used for the case of a half-space ($w \to \infty$), and (3–60) was used for a wall of thickness w = 30 cm. The results of the computations for both polarizations are shown in Figure 3–16, from which it is seen that the standing waves inside the wall are strongly damped so that the value of $|\Gamma|$ for the finite thickness wall shows only a small oscillation about the curve for a dielectric half-space.

Few measurements of reflection coefficients for walls have been reported. The results of one set of 4-GHz measurements made on a large, windowless brick wall using directional antennas [7] have been replotted for both TE and TM polarizations in Figure 3–16. It is difficult to compare these results with theory since the thickness of the wall, the actual construction behind the first layer of bricks, and the interior finish were not described. Both the models for a finite-thickness wall and a dielectric half-space show about the same agreement with the measure-

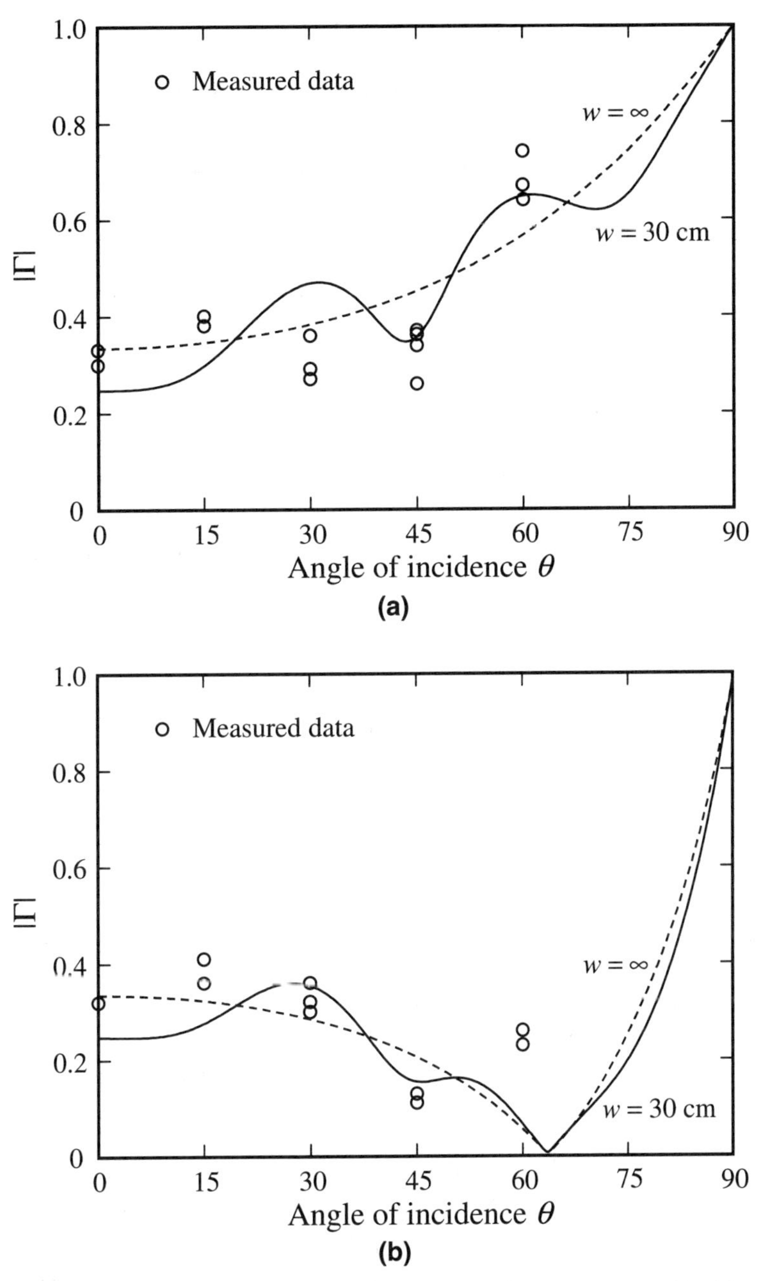

Figure 3-16 Measured and theoretical values of $|\Gamma|$ versus angle of incidence at a brick wall for 4 GHz and (a) TE polarization; (b) TM polarization. The measurements reported in [7] have been replotted for comparison with calculations assuming that $\varepsilon_w' = 4$, $\varepsilon_w'' = 0.1$ for a wall of thickness $w = 30$ cm, shown by the solid curves, and for a dielectric half-space, shown by the dashed curves.

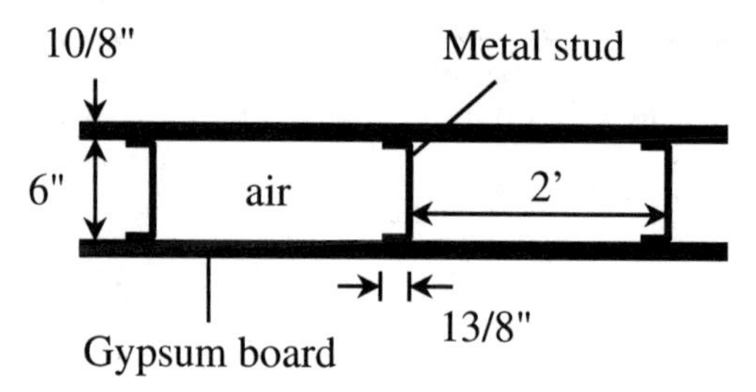

Figure 3-17 Dimensions for a gypsum board walls built on metal studs [16] (©1994 IEEE).

ments. The similarity of reflection coefficients for finite-thickness walls and a dielectric half-space is significant for site-specific propagation prediction models, as discussed in Chapter 6.

3.3c Transmission through walls of uniform construction

A few researchers have reported using directive antennas to measure transmission loss through uniform walls [5,15,16]. The most common type of interior wall construction consists of layers of gypsum board mounted on either side of studs, as shown in cross section in Figure 3–17 for a wall constructed with a double thickness of gypsum board. Neglecting the supporting studs, reflection from this wall can be analyzed by cascading the impedance transformation of (3–54) through one layer of gypsum board, then through the air gap between the layers, and finally through the second layer. The fraction $1-|\Gamma|^2$ of the incident power having TE polarization that is transmitted through the wall of Figure 3–17 is plotted as a function of the angle of incidence by the solid curve in Figure 3–18 for 2.4 GHz.

In modern construction the supporting studs are often made of steel. A more complicated analysis that includes the scattering from the steel studs has been carried out [16], and results for the transmitted power are also shown in Figure 3–18. The studs are seen to reduce the transmission by a factor of 0.7 to 0.6 (1.5 to 2.2 dB) at normal incidence and to greatly reduce the peaks at ±50° and ±70° that are caused by multiple reflections between the two layers of gypsum board. Because the studs create a periodic structure, like a grating in optics, there can be additional scattering into diffraction orders. The power transmitted into the first orders are also plotted in Figure 3–18 and are seen to be generally small. The curves lack symmetry about $\theta = 0$ because of the nonsymmetric shape of the studs. It is again seen from these curves that there is total reflection as θ approached 90°.

Measurement of transmission through walls can be made using directive antennas, as suggested in Figure 3–19. A horn antenna directs radiation at the wall along an axis making an angle θ_i with the normal and lying in the horizontal plane. To evaluate the scattering by metal studs, the receiving horn is aimed at the opposite side of the wall and its axis scanned through angles θ_r in the horizontal plane. The results of such measurements at 2.6 GHz for normal incidence on a wall having the dimensions listed in Figure 3–17 are shown in Figure 3–20 [16]. In this plot the received power is normalized to that found when the two horns face each other in air

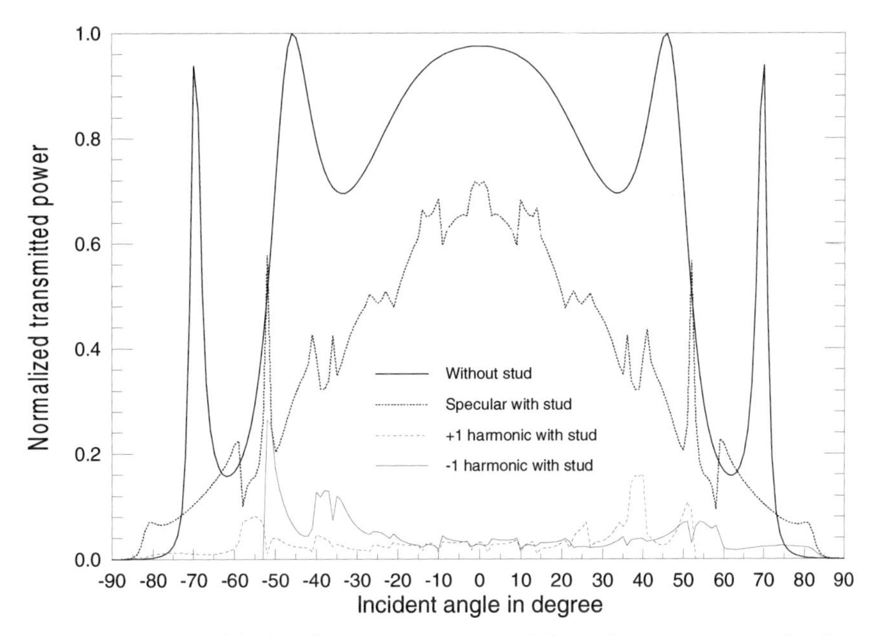

Figure 3-18 Fraction of the incident power transmitted through a gypsum board wall at 2.4 GHz computed with and without the metal studs. Also plotted are the fractions of the power transmitted into the first-order diffraction orders accounting for the studs. Calculations are for the dimensions in Figure 3–17 [16] (©1994 IEEE).

and are separated by the same total distance. Although the horn pattern functions must be taken into account for large angular displacements, for small angles the normalized received power approximates the plane wave transmission loss at normal incidence, which is seen to be a few decibels. Standard gain horns with beam widths of about 30° were located at a radial distance of 3.5 m from the aiming point on the wall, so that the illuminated spot was wider than the stud separation. It is seen from Figure 3–20 that no systematic difference occurs when the aiming point is located at a stud or between two studs. The corresponding theoretical results for both the case when the studs are included (asymmetric curve with fades) and that when they are omitted (smooth symmetric curve), and accounting for the pattern functions of the two horns, are shown for comparison in Figure 3–20. There is general agreement between the theory, either including or excluding the studs, and measurements for $|\theta_r| < 60°$, but greater scattering is measured than is predicted at larger angles. For oblique incidence with $\theta_i = 30°$, similar agreement was found between the measurements and theory within ±60° of the beam axis, but the measured scattering was stronger outside the main beam [16].

Some walls are constructed from concrete blocks, whose electromagnetic properties are made more complicated by the presence of the internal webs and voids, which are indicated in Figure 3–21, and causes the wall to act like a periodic structure [15,16] such as an optical grating. For common concrete blocks, the periodicity will generate diffracted waves (grating lobes)

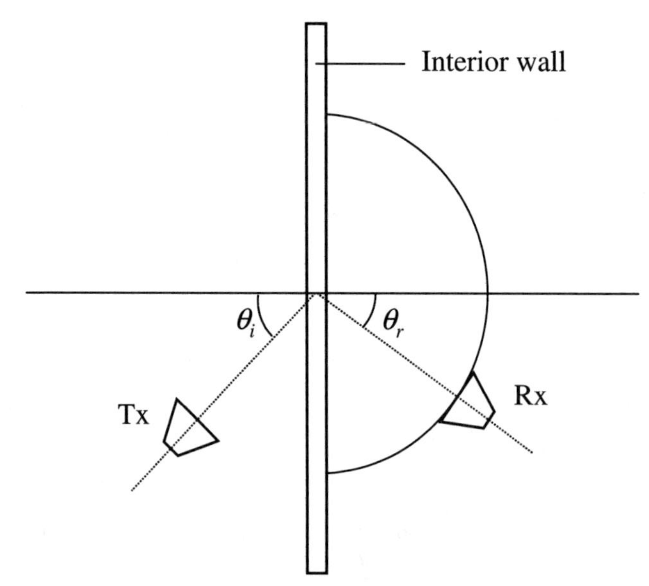

Figure 3-19 Using horn antennas to measure the transmission properties of interior walls.

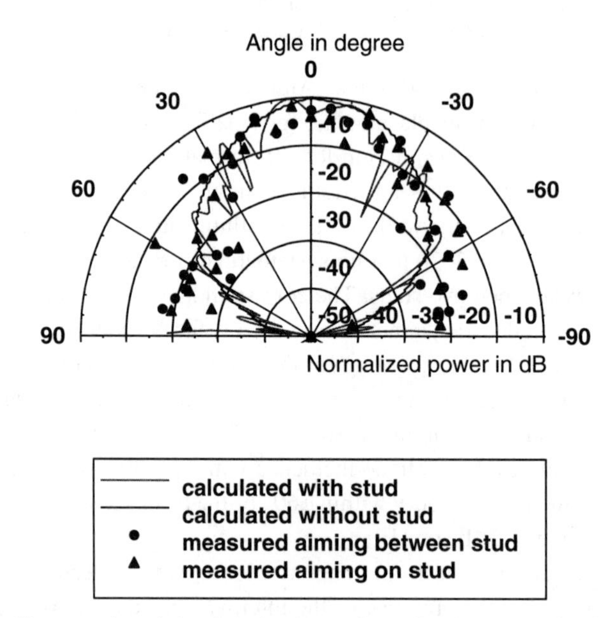

Figure 3-20 Measured and simulated values of received power using the approach in Figure 3–19 for normal incidence on a gypsum board wall at 2.6 GHz. The wall dimensions are the same as those for Figure 3–17 [16] (©1994 IEEE).

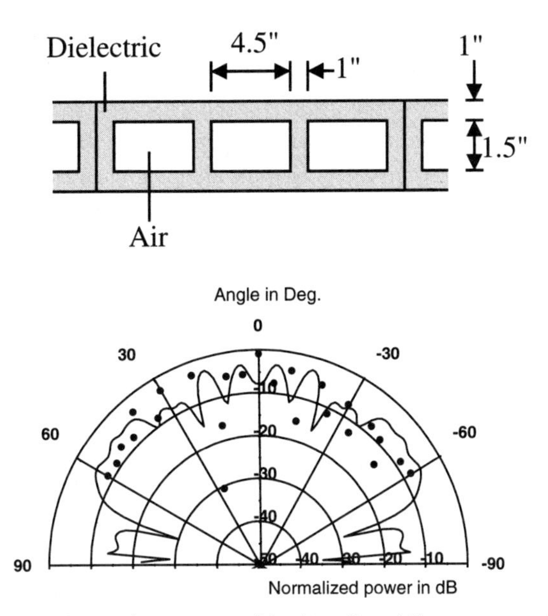

Figure 3-21 Dimensions of a concrete block wall and the measured and computed received power at 2.6 GHz using horn antennas as indicated in Figure 3–19 [16] (©1994 IEEE).

when the frequency is near or above 1800 MHz. At lower frequencies, only specularly reflected and transmitted waves exist, and their amplitudes can be found approximately using an equivalent homogeneous dielectric constant that is a weighted average of the dielectric constants in the actual wall [17]. Measurements of the type suggested in Figure 3–19 were made at 2.6 GHz on a dry interior wall constructed from the blocks of Figure 3–21 [16], and the results are shown in Figure 3–21. It is seen that the internal structure of the blocks scatters the transmitted power somewhat uniformly over the entire 120° measurement range. Theoretical analysis is possible if the block wall is approximated by a regular periodic structure [15,16], the results of which are also shown in Figure 3–21 and are seen to be in overall agreement with the measurements. Concrete block walls, wooden floors with joists, and precast concrete floors are examples of multi-layered periodic dielectric structures that can be analyzed using the rigorous procedures described in reference [18], which are beyond the scope of this book.

3.3d Transmission through in-situ walls and floors

Actual wall and floor construction makes use of various layers to provide insulation and act as a moisture barrier. In some buildings highly reflective layers, such as aluminum siding and/ or aluminized vapor barrier insulation are used in exterior walls. Walls are made even more complicated by the existence of windows, doors, supporting beams, and furnishings placed in the windows and on or against the wall. Floors also have supporting beams, as well as suspended air-handling ducts, lighting fixtures, and so on. In some cases these features may be more significant than the wall or floor construction itself and can be highly variable.

To account for these complex features in an average sense, measurements have been made in occupied buildings. Estimation of the transmission loss through exterior building walls is made by comparing the powers received by a dipole antenna when it is placed in front of an isolated building and when it is placed inside the building. Interior wall and floor losses are estimated from the received power when transmitting and receiving dipole antennas are located in different rooms or on different floors. Measurements of 800-MHz penetration loss into free-standing houses were in the range 4 to 7 dB [19], whereas metal buildings showed loss up to 24 dB [20]. Interior walls made of concrete blocks had a measured transmission loss of 1.5 to 2.4 dB at 900 MHz [21,22]. Similar measurements at 5.85 GHz found transmission losses of 14.5 and 8.8 dB at brick and wood siding exterior walls, respectively, and 4.7 dB at interior walls [23].

Penetration loss measurements in isolated suburban office buildings have been made at interior locations that the waves reach after transmission through an exterior wall and one or more interior walls [24]. The mean loss for 10 such buildings is reported to be 10.8 dB at 900 MHz and 10.2 dB at 1500 MHz [24]. The penetration loss is about the same on all floors above the first floor, which typically has an open lobby design. When penetration loss measurements are made inside office buildings located in a congested urban setting, shadowing by surrounding buildings results in higher penetration loss at the first floor, and significantly lower loss on higher floors [25]. When viewed over the frequency scale from under 100 MHz to about 2 GHz, the penetration loss into office buildings is found to decrease with frequency [24,25]. This frequency dependence may be due to the fact that the wave propagates through openings, such as windows, rather than through the wall, so that fields with a smaller wavelength do better. It is not known if this trend will continue for frequencies above 2 GHz.

Floors of modern office and commercial buildings are constructed of poured concrete and may contain steel reinforcing rods or may be cast into corrugated steel pans that are left in place. In the first case the reinforcing rods may be separated by 20 cm or more and will not strongly scatter the fields. However, corrugated steel pans that are left in place will be highly reflective. Precast concrete floors contain both steel reinforcing rods and hollow spaces. Transmission loss through concrete floors without steel pans has been found to be 13 dB [22]. In tall buildings it is found that the transmission loss does not increase linearly with the number of floors between transmitter and receiver [22]. This behavior is due to the presence of other paths by which the signal can travel from the transmitter to the receiver, as discussed in Chapter 8.

3.4 Summary

The electric and magnetic fields throughout small regions of space surrounding individual ray paths are approximately those of a plane wave. Therefore, studying the propagation of plane waves can give insight into the behavior of more complex waves. For example, interference patterns and related phenomena produced by multipath arrivals can be modeled using the superposition of plane waves. In later chapters Snell's law and the reflection and transmission coefficients found for plane waves are also used for spherical waves radiated by antennas to account for interactions with the ground and walls. In modeling propagation in cities, we will find it particularly important that the reflection coefficient is finite at normal incidence, and approaches unity at glancing incidence, no matter what the wall construction or polarization.

Problems

3.1 Assume that $f = 2.4$ GHz.

(a) Find the wavenumber k_d, wavelength λ_d, and impedance η_d for glass if $\varepsilon_r = 3.8$.

(b) For dry concrete with $\varepsilon_r = 5 - j0.3$, the wavenumber k_d will be complex. Find $\mathrm{Re}\{k_d\}$ and $\mathrm{Im}\{k_d\}$, as well as $\lambda_d = 2\pi/\mathrm{Re}\{k_d\}$. In the concrete, the amplitude of a radio wave attenuates with distance d as $\exp[-d\,\mathrm{Im}\{k_d\}]$. At what distances does the amplitude decay by a factor $1/e$ and a factor $1/e^3$?

3.2 Assume that there are $N = 6$ plane waves propagating parallel to the (x,z) plane at randomly distributed angles θ_i to the x axis and polarized along y. Then the variation of the electric field seen along the x axis is given by

$$E_y(x, 0) = \sum_{i=1}^{N} V_i e^{j\phi_i} e^{-jkx\sin\theta_i} \tag{3–61}$$

where V_i and ϕ_i are the amplitudes and phases of the plane waves.

(a) Using a random number generator or table of random numbers, select values for θ_i in the interval from 0 to 2π. Repeat this process to obtain an independent set of values for ϕ_i. List the values of θ_i and ϕ_i.

(b) For a frequency of $f = 1$ GHz, and assuming that all $V_i = 1$, compute and plot $|E(x,0)|$ at 400 points over the range $-3 \le x < 3$ m.

(c) For the 400 values u_n (n = 1, 2,...,400) computed in part (b), find the averages $\langle u_n \rangle$ and $\langle (u_n)^2 \rangle$, and compare the latter to $(\langle u_n \rangle)^2$.

(d) Normalize the values u_n in part (c) to $\langle u_n \rangle$. From the resulting set of numbers $r_n = u_n/\langle u_n \rangle$, compute and plot the cumulative distribution function. On the same graph, plot the Rayleigh distribution $P(r) = 1 - \exp(-\pi r^2/4)$.

(e) Using the values of r_n, compute and plot the correlation function $C(s)$ defined in (14) for $0 \le s < 0.3$ m.

3.3 Repeat Problem (3–2) for the cases when $V_1 = 3$ and all other $V_i = 1$, and when $V_1 = 6$ and all other $V_i = 1$.

3.4 Assuming that ε_r is real, show from (3–27) and (3–28) that for TE polarization $1 - \Gamma_E^2 = (Z^{TE}/Z_d^{TE})T_E^2$, so that either expression gives the same fraction of the power transmitted into the dielectric, as discussed after (3–28).

3.5 A wave propagating in air is incident on a frozen lake ($\varepsilon_r = 3.2$) at an angle θ to the normal. Find Brewster's angle θ_B. If $\theta = 45°$, find θ_T and Γ_H for TM polarization. Repeat for $\theta = 75°$.

3.6 A plane wave of frequency $f = 1$ GHz and TE polarization is reflected from a brick wall of thickness $w = 16$ cm. Neglecting loss, $\varepsilon_r = 4$. Find the reflection coefficient Γ_E for normal incidence ($\theta = 0$). What is the limiting value of Γ_E as θ approaches 90°? Find any angles of incidence θ for which $\Gamma_E = 0$.

3.7 Consider a 900-MHz plane wave that is incident with TE polarization onto a concrete wall having complex dielectric constant $\varepsilon_r = 4 - j0.2$. Compute and plot $|\Gamma_E|^2$ as a function of the angle of incidence θ for the case when the wall is very thick, and for the case of finite thickness $w = 25$ cm.

3.8 A common interior wall construction consists of two layers of gypsum board ($\varepsilon_r = 2.8$) of thickness 5/8 in. separated by 3.5 in. Neglecting the effect of the studs, compute and plot $|\Gamma_E|^2$ as a function of the angle of incidence θ for TE polarized plane waves incident on such a wall.

References

1. C. A. Balanis, Advanced Engineering Electromagnetics, Wiley, New York, 1989.
2. E. C. Jordan and K. G. Balmain, Electromagnetic Waves and Radiating Systems, 2nd ed., Prentice Hall, Upper Saddle River, N.J., 1968.
3. S. Ramo, J. R. Whinnery, and T. Van Duzer, Fields and Waves in Communication Electronics, 3rd ed., Wiley, New York, 1994.
4. A. von Hippel, ed., Dielectric Materials and Applications, Artech House, Norwood, Mass., 1995.
5. L. M. Correia, Transmission and Isolation of Signals in Buildings at 60 GHz, *Proc. IEEE International Symposium PIMRC'95*, pp. 510–513, 1995.
6. J. B. Hasted and M. A. Shah, Microwave Absorption by Water in Building Material, *Br. J. Appl. Phys.*, vol. 15, pp. 825–836, 1964.
7. O. Landron, M. J. Feuerstein, and T. S. Rappaport, A Comparison of Theoretical and Empirical Reflection Coefficients for Typical Exterior Wall Surfaces in a Mobile Radio Environment, *IEEE Trans. Antennas Propagat.*, vol. 44, pp. 341–351, 1996.
8. M. A. Shah, J. B. Hasted, and L. Moore, Microwave Absorption by Water in Building Materials: Aerated Concrete, *Br. J. Appl. Phys.*, vol. 16, pp. 1747–1754, 1965.
9. M. Abramowitz and I. A. Stegun, eds., Handbook of Mathematical Functions, Dover Publications, New York, Chap. 9, 1965.
10. W. C. Y. Lee, Mobile Communications Engineering, McGraw-Hill, New York, p. 276, 1982.
11. A. Papoulis, Probability, Random Variables, and Stochastic Processes, 3rd ed., McGraw-Hill, New York, pp. 319–332, 1991.
12. T. Aulin, A Modified Model for the Fading Signal at a Mobile Radio Channel, *IEEE Trans. Veh. Technol.*, vol. VT-28, pp. 182–203, 1979.
13. F. Adachi, M. T. Feeney, A. G. Williamson, and J. D. Parsons, Crosscorrelation between the Envelopes of 900 MHz Signals Received at a Mobile Radio Base Station Site, *IEE Proc.*, vol. 133, Pt. F, pp. 506–512, 1986.
14. A. Kajiwara, Line-of-Sight Indoor Radio Communication Using Circular Polarized Waves, *IEEE Trans. Veh. Technol.*, vol. 44, pp. 487–493, 1995.
15. W. Honcharenko and H. L. Bertoni, Transmission and Reflection Characteristics at Concrete Block Walls in the UHF Bands Proposed for Future PCS, *IEEE Trans. Antennas Propagat.*, vol. 42, pp. 232–239, 1994.
16. S. Kim, B. Bougerolles, and H. L. Bertoni, Transmission and Reflection Properties of Interior Walls, *Proc. IEEE ICUPC'94*, pp. 124–128, 1994.
17. C. L. Holloway, P. L. Perini, R. R. DeLyser, and K.C. Allen, Analysis of Composite Walls and Their Effects on Short-Path Propagation Modeling, *IEEE Trans. Veh. Technol.*, vol. 46, pp. 730–738, 1992
18. T. Tamir and S. Zhang, Modal Transmission-Line Theory of Multilayered Grating Structures, *IEEE J. Lightwave Technol.*, vol. 14, pp. 914-927, 1996.
19. D. C. Cox, R. R. Murray, and A. W. Norris, 800-MHz Attenuation Measured in and around Suburban Houses, *AT&T Bell Lab. Tech. J.*, vol. 63, pp. 921–953, 1984.

20. D. C. Cox, R. R. Murray, and A. W. Norris, Measurements of 800-MHz Radio Transmission into Buildings with Metallic Walls, *AT&T Bell Lab. Tech. J.*, vol. 62, pp. 2695–2717, 1983.
21. J. F. Lafortune and M. Lecours, Measurement and Modeling of Propagation Losses in a Building at 900 MHz, *IEEE Trans. Veh. Technol.*, vol. 39, pp. 101–108, 1990.
22. S. Y. Seidel and T. S. Rappaport, 914 MHz Path Loss Prediction Models for Indoor Wireless Communications in Multifloored Buildings, *IEEE Trans. Antennas Propagat.*, vol. 40, pp. 207–217, 1992.
23. G. Durgin, T. S. Rappaport, and H. Xu, 5.85-GHz Radio Path Loss and Penetration Loss Measurements in and around Homes and Trees, *IEEE Commun. Let*, vol. 2, pp. 70–72, 1998.
24. A. Davidson and C. Hill, Measurement of Building Penetration into Medium Buildings at 900 and 1500 MHz, *IEEE Trans. Veh. Technol.*, vol. 46, pp. 161–167, 1997.
25. A. M. D. Turkmani and A. F. de Toledo, Modelling of Radio Transmissions into and within Multistory Buildings at 900, 1800 and 2300 MHz, *IEE Proc.-1*, vol. 140, pp. 462–470, 1993.

CHAPTER 4

Antennas and Radiation

Antennas are used to convert the currents and voltages generated by the transmitter circuit into electromagnetic fields propagating through space. In the absence of material boundaries that reflect or scatter the waves, the fields propagate in the form of spherical waves, whose amplitudes vary inversely with the distance from the antenna. The field strength and polarization can also depend on the direction of propagation. This variation is described by the antenna pattern function and is one of the characteristics that distinguishes various types of antennas. The other distinguishing characteristic is the terminal impedance. These quantities, and their dependence on frequency, are determined by the shape of the conductors and dielectrics used to make the antenna. At the receiver, a second antenna is used to convert the electromagnetic fields back into voltages and currents that are detected by the electronic circuitry. The same properties of polarization, antenna pattern, and terminal impedance that characterize an antenna acting as a transmitter will be shown to characterize it when acting as a receiver.

Although just about any piece of wire can be used for an antenna, to have a relatively large and primarily resistive terminal impedance, which is important for matching to the transmitter or receiver circuits, the antenna must have critical dimensions that are about one-half wavelength or greater. In general, increasing the antenna size makes the radiation pattern more directive. In this chapter we discuss some of the properties of antennas and of the electromagnetic waves they radiate that are directly applicable to understanding the characteristics of the wireless radio channel. In particular, we are concerned with antennas as receivers as well as radiators, the concepts of path gain and path loss, the spherical nature of the radiated waves, and how these waves are influenced by the presence of ground. Much fuller discussions of antennas than presented here can be found in books on antennas [1,2] and electromagnetics [3,4].

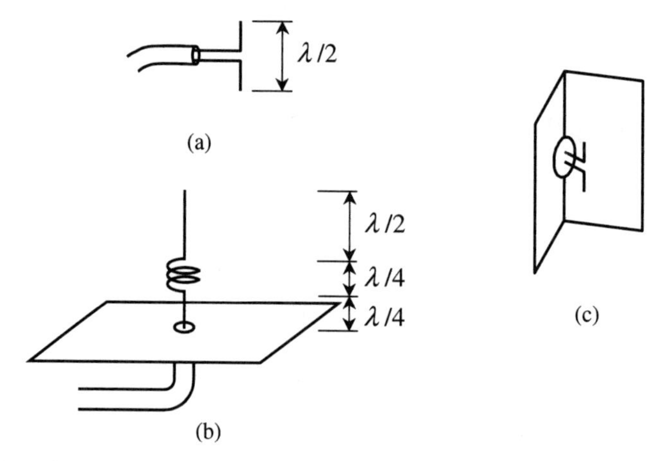

Figure 4-1 Simple antennas used in wireless communications systems: (a) half-wave dipole; (b) full-wave monopole above a conducting plane; and (c) dipole with corner reflector.

4.1 Radiation of spherical waves

Some examples of the antennas used in wireless are illustrated in Figure 4–1, and include a half-wave dipole, a full-wave monopole with isolation coil above a ground plane, and a dipole in front of a corner reflector. In the far-field zone of any type of antenna, the radiate waves are spherical with electric and magnetic fields given by

$$\mathbf{E} = \mathbf{a}_E Z I \frac{e^{-jkr}}{r} f(\theta, \phi) \qquad (4\text{–}1)$$

$$\mathbf{H} = \frac{1}{\eta}\mathbf{a}_r \times \mathbf{E}$$

Referring to Figure 4–2, r is the radial distance from the antenna, and $k = \omega/c = 2\pi/\lambda$ is the wavenumber, so that the phase kr in (4–1) is constant over spheres centered at the antenna. The unit vector $\mathbf{a}_E$ describing the polarization of the electric field must be perpendicular to the unit vector $\mathbf{a}_r$ pointing in the radial direction away from the antenna, so that $\mathbf{E}$ is tangent to the spheres of constant phase. The magnetic field $\mathbf{H}$ is perpendicular to $\mathbf{E}$ and to $\mathbf{a}_r$, and in a direction such that the Poynting vector $(1/2)\mathrm{Re}\{\mathbf{E} \times \mathbf{H}^*\}$ is in the outward radial direction. The amplitude of $\mathbf{H}$ differs from that of $\mathbf{E}$ by the wave impedance $\eta = \sqrt{\mu_0/\varepsilon_o} \approx 377$. These properties of $\mathbf{E}$ and $\mathbf{H}$ for a spherical wave are locally like those of a plane wave propagating in the radial direction. In (4–1) I is the terminal current at the antenna, Z is a constant with units of ohms, and $f(\theta,\phi)$ describes the variation of the fields with the direction of propagation (θ,ϕ) shown in Figure 4–2. For antennas that are about a wavelength in size, the far-field zone in which (4–1) is valid starts a few wavelengths from the antenna. For larger antennas of size L, the far-field zone starts at about $2L^2/\lambda$ [1, p. 60], which in wireless is very small compared to the typical separation between the transmitter and receiver.

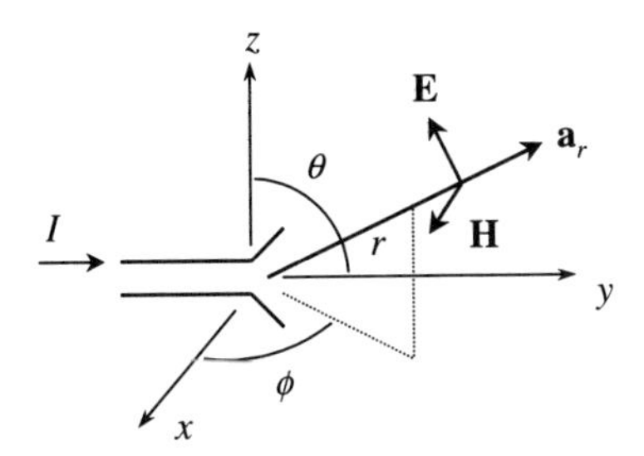

Figure 4-2 Spherical wave radiation into space by an antenna.

The power density in W/m^2 is given by the Poynting vector:

$$\mathbf{P} = \frac{1}{2}Re\{\mathbf{E} \times \mathbf{H}^*\} = \mathbf{a}_r \frac{1}{2\eta} \frac{|ZI|^2}{r^2} |f(\theta, \phi)|^2 \tag{4–2}$$

The power flows radically outward from the antenna and its density decreases as $1/r^2$ with distance. The direction dependence $|f(\theta,\phi)|^2$ represents the radiation pattern of the antenna. Let P_T be the total power in watts passing through a sphere of radius r centered at the antenna. Recognizing that the area element on the sphere is $dA = r^2 \sin\theta \, d\theta \, d\phi$, the total power is given by

$$P_T = \oiint_{\text{sphere}} \mathbf{P} \cdot \mathbf{a}_r dA = \int_{-\pi}^{\pi} \int_0^{\pi} \frac{|ZI|^2}{2\eta r^2} |f(\theta, \phi)|^2 r^2 \sin\theta d\theta d\phi \tag{4–3}$$

It is seen in (4–3) that the dependence on r in the numerator and in the denominator cancel, so that the total power passing through the sphere is independent of its size, as is to be expected from conservation of energy. Viewed another way, the electric and magnetic fields must have a $1/r$ dependence, shown in (4–1), to conserve energy.

Assume that the radiation pattern satisfies the normalization

$$\int_{-\pi}^{\pi} \int_0^{\pi} |f(\theta, \phi)|^2 \sin\theta d\theta d\phi = 4\pi \tag{4–4}$$

Then $|f(\theta,\phi)|^2$ is known as the directive gain $g(\theta,\phi)$. With this normalization it is seen that (4–3) can be written as

$$P_T = \frac{1}{2\eta} |ZI|^2 (4\pi) \tag{4–5}$$

Solving (4–5) for $|ZI|^2$ and substituting into (4–2) gives

$$\mathbf{P} = \mathbf{a}_r \frac{P_T}{4\pi r^2} |f(\theta, \phi)|^2 \equiv \mathbf{a}_r \frac{P_T}{4\pi r^2} g(\theta, \phi) \tag{4–6}$$

Equation (4–6) can be interpreted as saying that the power density at a sphere of radius r is the total radiated power divided by the area of the sphere, with the directive gain giving the devia-

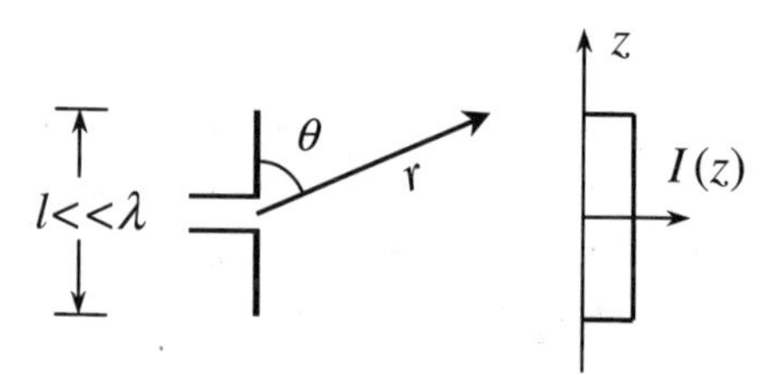

Figure 4-3 Radiation by a Hertzian dipole (a) antenna; (b) current distribution.

tion of the power from the average. The maximum value of $|f(\theta,\phi)|^2$ or $g(\theta,\phi)$ is called the antenna gain G. Because the electric and magnetic fields are transverse to the direction of propagation, symmetry precludes their being an antenna that is an isotropic radiator. However, if such an antenna could exist, $g(\theta,\phi)$ would be unity, the power would be uniformly distributed over the sphere, and G = 1. Although no real antenna can be isotropic, we will make use of this convenient concept. For example, if an isotropic antenna radiates the maximum power $P_T = 5$ W permitted for vehicle-mounted cellular telephones into free space, the power density at a distance of 1 km is $|\mathbf{P}| = 5/(4\pi \times 10^6) = 0.4\ \mu\text{W/m}^2$, and the magnitude of the electric field $|\mathbf{E}| = \sqrt{2\eta|\mathbf{P}|}$ is 17.4 mV/m.

Because the radiated fields are proportional to the terminal current (Maxwell's equations are linear), a transmission line connected to the terminals of an antenna sees a load impedance. Neglecting resistance in the antenna conductors, the power dissipated in the real part of the impedance R_r represents the power radiated by the antenna. This radiation resistance satisfies

$$P_T = \frac{1}{2}|I|^2 R_r \tag{4–7}$$

Comparing (4–7) and (4–5) it is seen that for the normalization (4–4), the factor Z satisfies

$$|Z| = \sqrt{\frac{R_r \eta}{4\pi}} \tag{4–8}$$

As an example, consider the Hertzian dipole of length $l \ll \lambda$, shown in Figure 4–3, whose current is assumed to be uniform along its length. In the far-field region [1, Chap. 2]

$$\mathbf{E} = \mathbf{a}_\theta j\eta \frac{l}{2\lambda} I \frac{e^{-jkr}}{r} \sin\theta \tag{4–9}$$

Because the antenna is symmetric about the z axis, the field **E** is independent of the direction ϕ and depends only on the polar angle θ. In (4–9) the unit vector $\mathbf{a}_\theta$ is tangent to the sphere of radius r and points in the direction of increasing θ. To put this expression in the form of (4–1), we define

$$f(\theta) = \sqrt{\frac{3}{2}} \sin\theta$$

$$Z = j\eta \sqrt{\frac{2}{3}} \frac{l}{2\lambda} \tag{4–10}$$

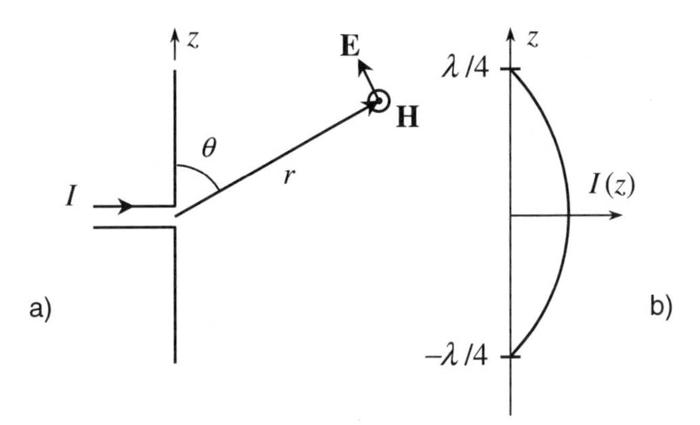

Figure 4-4 Radiation by a half-wave dipole: (a) antenna; (b) current distribution.

If (4–10) is substituted into (4–1), it is seen that (4–1) reduces to (4–9). It is easily shown that $f(\theta)$ satisfies the normalization condition (4–4). Solving (4–8) for R_r and using (4–10) gives

$$R_r = \eta\frac{2\pi}{3}\left(\frac{l}{\lambda}\right)^2 \tag{4–11}$$

Because $(l/\lambda)^2$ is small, the radiation resistance is only a small fraction of the wave impedance η. For this antenna $g(\theta) = (3/2)\sin^2\theta$, which is maximum at $\theta = 90°$, with gain $G = 3/2$.

A second example is offered by the half-wave dipole shown in Figure 4–4. The current in the antenna has a cosine variation with z, the electric field radiated by the dipole is given by [1, Chap. 2]

$$\mathbf{E} = \mathbf{a}_\theta\frac{j\eta I}{2\pi}\frac{e^{-jkr}}{r}\frac{\cos[(\pi/2)\cos\theta]}{\sin\theta} \tag{4–12}$$

It is not possible to find a closed-form expression for the integrated angular dependence over the sphere as called for in (4–4). However, the radiation resistance of a half-wave dipole is $R_r = 73\,\Omega = 0.194\eta$ [1]. Substituting this expression for R_r into (4–8) and incorporating the phasor j appearing in (4–12) into Z, it is seen after minor manipulation that

$$Z = j\frac{0.781}{2\pi}\eta \tag{4–13}$$

If the numerator and denominator of (4–12) are multiplied by 0.781, then using the foregoing expression for Z and comparing (4–12) with (4–1), the normalized pattern function is seen to be

$$f(\theta) = \frac{\cos[(\pi/2)\cos\theta]}{0.781\sin\theta} \tag{4–14}$$

The maximum value of $f(\theta)$ occurs at 90°, so that the dipole gain is $G = |f(90°)|^2 = 1.64$, which is equivalent to 2.2 dB. While the gain of the half-wave dipole is not much different than that of the Hertzian dipole, its radiation resistance is much higher and the reactive part of the terminal impedance is small.

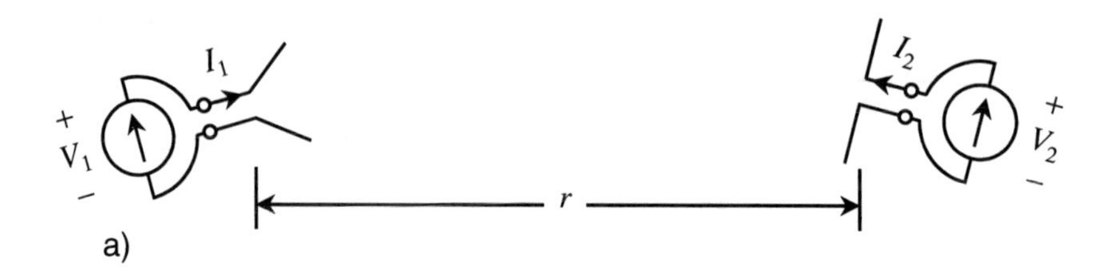

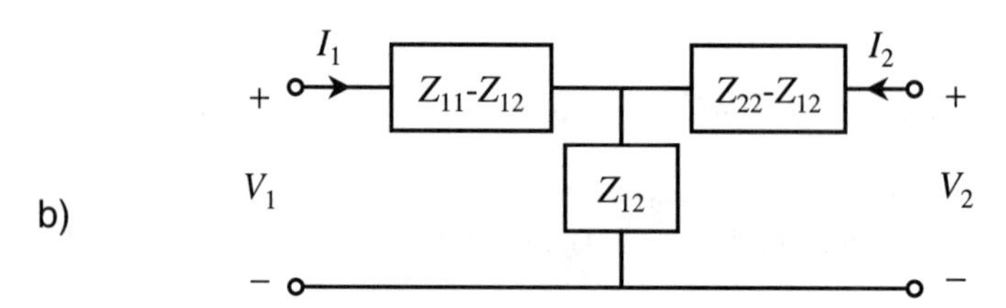

Figure 4-5 Pair of antennas for transmission and reception (a) with currents and voltages at the physical antenna terminals; (b) equivalent circuit.

4.2 Receiving antennas, reciprocity, and path gain or loss

Antennas are often studied in terms of their radiation characteristics rather than as detectors of electromagnetic radiation. When used to receive signals, antennas have the same characteristics as when used for transmission. This behavior is embodied in the reciprocity relation [4, pp. 479–482], which we will make use of but not prove. Consider the two antennas shown in Figure 4–5a, each with its terminal current and terminal voltage, as indicated. The antennas may be in free space, or may be in the presence of other objects, such as the ground and buildings. Because Maxwell's equations are linear, the terminal voltages and current satisfy the linear relations

$$V_1 = Z_{11}I_1 + Z_{12}I_2$$
$$V_2 = Z_{21}I_1 + Z_{22}I_2 \tag{4–15}$$

When the materials surrounding the antennas are ordinary dielectrics and conductors, the reciprocity condition implies that $Z_{12} = Z_{21}$, in which case (4–15) is equivalent to the circuit shown in Figure 4–5b.

Now suppose that antenna 1 is driven by a current source I_1 attached to its terminals, and the terminals of antennas 2 are left open. In that case the impedance seen at the terminals of antenna 1 is Z_{11}, so that the power delivered to the antenna terminals P_T is

$$P_T = \frac{1}{2}\mathrm{Re}\{V_1 I_1{}^*\} = \frac{1}{2}\mathrm{Re}\{Z_{11}|I_1|^2\} \tag{4–16}$$

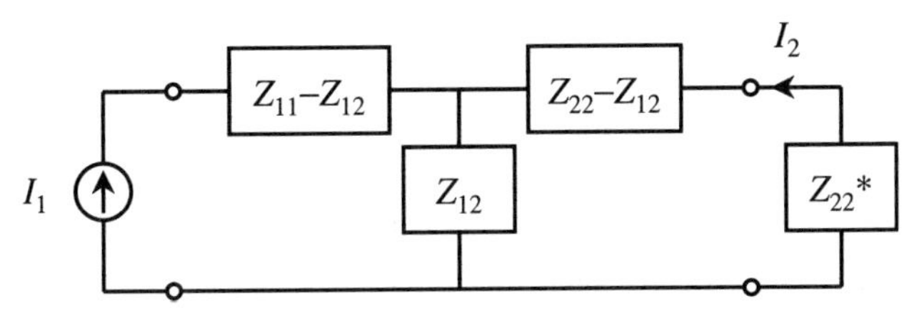

Figure 4-6 Equivalent circuit for a pair of antennas, with a current source applied to antenna 1 and conjugate matched load connected to antenna 2.

In the absence of loss in the antenna, this power is radiated into space, so that $\mathrm{Re}\{Z_{11}\}$ must be the radiation resistance R_{r1} of antenna 1, while $\mathrm{Im}\, Z_{11}$ is the terminal reactance X_1 of the antenna. The same reasoning can be applied to antenna 2, and thus

$$Z_{11} = R_{r1} + jX_1$$
$$Z_{22} = R_{r2} + jX_2 \tag{4–17}$$

The radiation and terminal reactances of an antenna for some frequency are fixed quantities independent of radial separation r between the antennas, provided that $r >> \lambda$. If the terminals of antenna 2 are left open, the voltage V_2 will be proportional to the electric field radiated by antenna 1 at the separation distance r. Thus the ratio $Z_{12} = V_2/I_1$ will decrease with separation, and for well-separated antennas, $|Z_{12}|$ will be much smaller than $|Z_{11}|$ or $|Z_{22}|$.

4.2a Path gain or loss

Now suppose that antenna 1 is driven by a current source I_1 and the terminals of antenna 2 are connected to a conjugate matched load $Z_{22}{}^*$, as suggested in Figure 4–6. It is easily shown that I_2 is given by

$$I_2 = -I_1 \frac{Z_{12}}{(Z_{22} - Z_{12} + Z_{22}{}^*) + Z_{12}} = -I_1 \frac{Z_{12}}{2R_{r2}} \tag{4–18}$$

Thus the power P_R received by the load resistance R_{r2} is given by

$$P_R = \frac{1}{2} R_{r2} |I_2|^2 = \frac{1}{8} \frac{|Z_{12}|^2}{R_{r2}} |I_1|^2 \tag{4–19}$$

For the conjugate matched load conditions in Figure 4–6, simple circuit considerations will show that the power delivered by current I_1 is

$$P_T = \frac{1}{2}|I_1|^2 \mathrm{Re}\left\{(Z_{11} - Z_{12}) + \frac{(2R_{r2} - Z_{12})Z_{12}}{(2R_{r2} - Z_{12}) + Z_{12}}\right\} = \frac{1}{2}|I_1|^2 \frac{2R_{r1}R_{r2} - \mathrm{Re}\{Z_{12}^2\}}{2R_{r2}} \tag{4–20}$$

This expression for the radiated power differs from that given by (4–16) and (4–17) as a result of the coupling to antenna 2. However, for antennas that are separated by a few wavelengths in free space, or on terrestrial links, the coupling impedance $|Z_{12}|$ is small compared to R_{r1} and R_{r2}, so that (4–20) reduces to (4–16).

It is conventional to define the *path gain* PG as the ratio of the received power P_R to the transmitted power P_T, and its inverse, the *path loss* PL, as the ratio of P_T to P_R. These quantities are characteristic of the propagation path and independent of the radiated power. Since $P_R < P_T$, the path gain $PG = P_R/P_T$ is less than unity, and while independent of the radiated power, has the spatial variation of the received power. On the other hand, the path loss $PL = 1/PG = P_T/P_R$ is greater than unity and increases with the separation between the antennas.

From (4–19) and (4–20) it is seen that the path gain can be written in terms of the equivalent circuit impedances as

$$PG = \frac{P_R}{P_T} = \frac{|Z_{12}|^2}{4R_{r2}R_{r1} - 2\mathrm{Re}\{Z_{12}^2\}} \tag{4–21}$$

For antennas that are separated by a few wavelengths, the coupling impedance Z_{12} is very small and can be neglected in the denominator of (4–21). Because (4–21) depends symmetrically on R_{r1} and R_{r2}, the path gain or path loss is the same no matter which antenna is the transmitter and which is the receiver. In other words, if either antenna is the transmitter and the other is terminated in its conjugate matched impedance, the path gain will be the same. This result is a further consequence of the reciprocity condition. We have derived the reciprocity statement (4–21) under very general conditions, not just in free space, so that in finding the path loss or path gain for wireless systems, we may take either end of the link as the transmitter, knowing that the same results will hold in the reverse direction.

4.2b Effective area of a receiving antenna

When acting as a receiver to a matched load, the antenna can be thought of as having an effective area A_e with units m^2 such that the received power is the product of the A_e and the incident power density, which has units W/m^2. If antenna 1 is the transmitter, as in Figure 4–7a, then with the help of (4–6) for antennas having the same polarization of the electric and magnetic fields,

$$P_R = |\mathbf{P}|A_{e2} = P_T \frac{g_1(\theta, \phi)}{4\pi r^2} A_{e2} \tag{4–22}$$

where $\mathbf{P}$ is the power density, A_{e2} the effective area of antenna 2, and g_1 the gain of antenna 1. Alternatively, if antenna 2 is the transmitter, as in Figure 4–7b, then

$$P_R = |\mathbf{P}|A_{e1} = P_T \frac{g_2(\theta, \phi)}{4\pi r^2} A_{e1} \tag{4–23}$$

where A_{e1} is the effective area of antenna 1 and g_2 is the gain of antenna 2. However, as shown previously, the path gain is the same no matter which antenna is the transmitter, so that P_R is the same in (4–22) and (4–23). Hence it is seen from these two equations that $g_1 A_{e2} = g_2 A_{e1}$, or

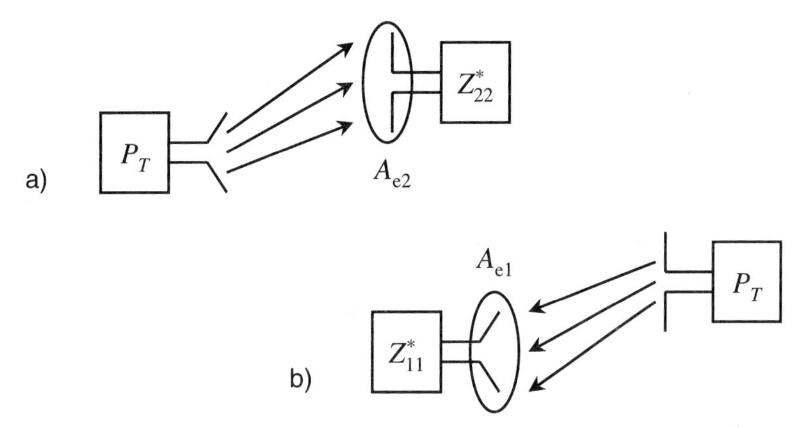

Figure 4-7 Effective area A_e of the receiving antenna when (a) antenna 1 transmits power P_T; (b) antenna 2 transmits power P_T.

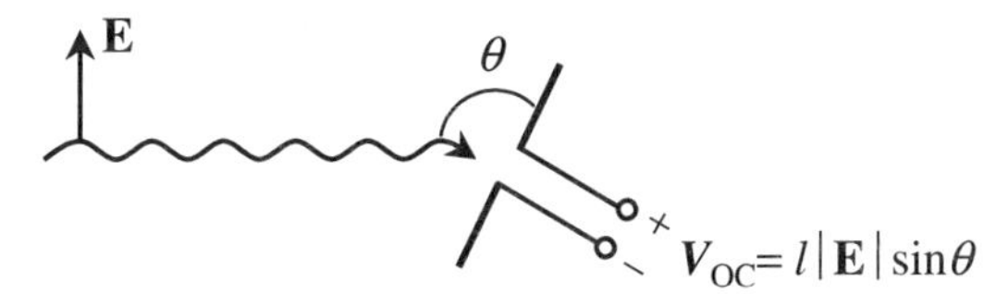

Figure 4-8 Hertzian dipole acting as a receiver of an obliquely arriving wave.

$$\frac{A_{e1}}{g_1} = \frac{A_{e2}}{g_2} \tag{4–24}$$

Since Figure 4–7 applies for any antennas, (4–24) implies that the ratio of the effective area to the gain is the same for any antenna.

To evaluate the ratio in (4–24), consider the Hertzian dipole in Figure 4–3 when acting as a receiver, as shown in Figure 4–8. In this case the antenna integrates the component of electric field along its axis. Thus the open-circuit voltage V_{oc} is given by

$$V_{oc} = lE\sin\theta \tag{4–25}$$

where E is the component of the incident field in the plane of incidence. In Figure 4–6, V_{oc} is the voltage at the receiving antenna when the matched termination is removed, and is also the voltage across the coupling impedance Z_{12}. Because the coupling impedance Z_{12} is very small, the voltage across it will be nearly the same when matched load impedance Z_{22}* is connected. Hence the power received by the matched load is

$$p_R = \frac{|V_{oc}|^2}{8Rr} = \frac{|E|^2(l\sin\theta)^2}{8\eta(l/\lambda)^2(2\pi/3)} = \frac{|E|^2}{2\eta}\frac{3(\lambda\sin\theta)^2}{8\pi} \tag{4–26}$$

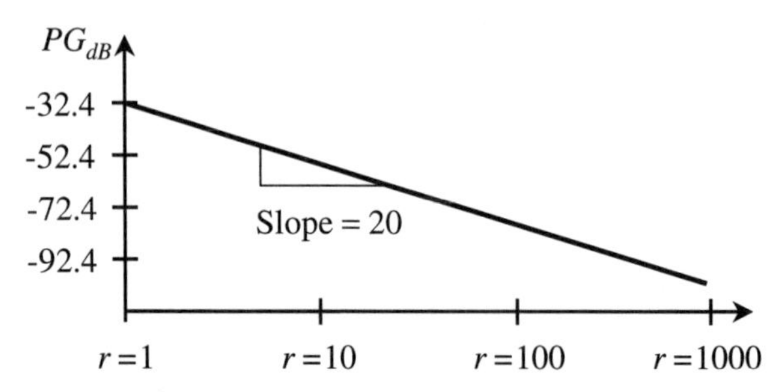

Figure 4-9 Variation of the path gain expressed in decibels for two isotropic antennas (g = 1) in free space with separation r on a log scale for f = 1 GHz.

where we have used (4–11) for R_r. Alternatively, the received power can be expressed in terms of the effective area and incident power density. Recalling relation (4–1) between **E** and **H**, the received power can be expressed as

$$P_R = A_e \frac{1}{2}|\mathrm{Re}\{\mathbf{E} \times \mathbf{H}^*\}| = A_e \frac{|E|^2}{2\eta} \tag{4–27}$$

Equating (4–26) and (4–27), we can solve for A_e, and making use of $g(\theta)$ for a Hertzian dipole, it is found that

$$\frac{A_e}{g} = \frac{(3/8\pi)(\lambda \sin\theta)^2}{(3/2)\sin\theta^2} = \frac{\lambda^2}{4\pi} \tag{4–28}$$

Although (4–28) was derived for a Hertzian dipole, expression (4–24) implies that for any antenna the effective area $A_e = (\lambda^2/4\pi)g$. Using this expression in (4–22) or (4–23), it is seen that the path gain for antennas in free space is given by

$$\mathrm{PG} = g_1 g_2 \left(\frac{\lambda}{4\pi r}\right)^2 \tag{4–29}$$

Isotropic antennas, for which g = 1, have an effective area A_e that is only $0.08\lambda^2$. If two such antennas are separated by a distance of 1 km in free space, the path gain is PG = 0.57×10^{-9}. For dipoles with g = 1.64, the path gain at 1 km is PG = 1.53×10^{-9}. Using this value in (4–21) and the radiation resistance R_r = 73 Ω for a dipole, it is seen that Z_{12} has a magnitude of only 5.7 mΩ.

Because the path gain is proportional to λ^2, it decreased with frequency as $1/f^2$. If f_G is the frequency in GHz, the path gain in free space can be expressed in decibels, and is given by

$$\mathrm{PG_{dB}} = 10 \log \mathrm{PG} = 10 \log g_1 g_2 - 32.4 - 20 \log f_G - 20 \log r \tag{4–30}$$

If $\mathrm{PG_{dB}}$ is plotted versus distance r on a log scale, it is seen from (4–30) that a straight line is obtained with a slope of –20. Such a plot is shown in Figure 4–9 for the case of isotropic antennas, for which $g_1 = g_2 = 1$ and for a frequency of 1 GHz. This plot is similar to that seen in Figure 2–6 for the range dependence of the overall average signal on terrestrial links, except that the coefficient of log r was around 36 rather than 20. The path loss in decibels $L = 10\log \mathrm{PL}$ is the negative of the path gain, so that when plotted versus log r for free space, an ascending straight line is obtained.

4.2c Received power in the presence of a multipath

The relations (4–24) and (4–28) between effective area and gain did make use of the free-space propagation conditions, in which there is a single direct propagation direction between the two antennas. In the presence of scatterers there may be several propagation paths that leave the transmitting antenna in different directions and arrive at the receiving antenna from different directions. In this case (4–24) would hold for each individual path, but in computing the total received power it is necessary to add the fields coherently due to the individual paths accounting for the antenna gains in the different directions. Moreover, in the presence of scatterers, the propagation will have a different dependence on antenna separation and position than for free space, so that (4–29) is no longer valid.

Much of the remainder of this book deals with finding the variation of the path gain, or its inverse the path loss, when the antennas are in the presence of objects such as the ground and buildings. To assist in this process we use (4–5) to express the electric field of the radiated wave given in (4–1) in terms of the total radiated power P_T, or

$$\mathbf{E} = \mathbf{a}_E\sqrt{\frac{\eta P_T}{2\pi}}e^{j\psi}\frac{e^{-jkr}}{r}f_R(\theta,\phi) \tag{4–31}$$

where $f_T(\theta,\phi)$ is the pattern function of the transmitting antenna and ψ is the phase of the product ZI, which is independent of direction.

We also need an expression for the power received by an antenna having conjugate match termination when two waves are incident from different directions (θ_1,ϕ_1) and (θ_2,ϕ_2). When receiving waves from the direction (θ,ϕ),

$$A_e = \frac{\lambda^2}{4\pi}g(\theta,\phi) = \frac{\lambda^2}{4\pi}|f_R(\theta,\phi)|^2 \tag{4–32}$$

where $f_R(\theta,\phi)$ is the pattern function of the receiving antenna. Assume that the incident electric fields E_1 and E_2 of the two waves have the same polarization as the fields radiated in those directions. Then the generalization of expression (4–27) for the received power due to two incident waves is

$$P_R = \frac{1}{2\eta}|E_1 f_R(\theta_1,\phi_1) + E_2 f_R(\theta_2,\phi_2)|^2\frac{\lambda^2}{4\pi} \tag{4–33}$$

Expressions (4–31) and (4–33) are used in the next section to find the effect of reflection from the ground.

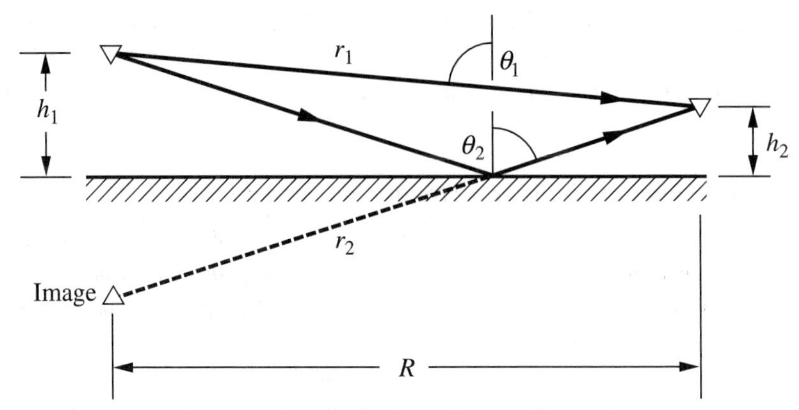

Figure 4-10 Two-ray model for propagation above a flat earth.

4.3 Two-ray model for propagation above a flat earth

For terrestrial links, where the antennas are placed above the earth, the earth itself acts as a scatterer of the fields. For short links we may neglect the earth's curvature, so that the geometry is as shown in Figure 4–10 for antennas separated by a horizontal distance R. If the antenna on the left is taken to be the transmitter, it sends out a spherical wave that is reflected from the ground, as well as propagating directly to the receiving antenna. We can think of the spherical wave as traveling along a family of radial lines, called *rays*, emanating from the transmitter. The power density (4–2) of the spherical wave implies that energy is conserved in any cone (tube) of rays. Moreover, in a small volume about a ray the electric and magnetic fields of the spherical wave (4–1) are like those of a plane wave. As a result, the ray fields that are incident on the earth will be reflected according to Snell's law, and their amplitude will be reduced by the same reflection coefficient found for plane waves [3, Chap. 7], [4, Chap. 16]. Because the angle of reflection of the ray will be the same as the angle of incidence, the rays reflected from a flat surface will appear to come from the image of the transmitter in the ground plane.

Making use of (4–31) and (4–33), the total received power due to the fields arriving along the direct and reflected ray paths in Figure 4–10 will be

$$P_R = P_T\left(\frac{\lambda}{4\pi}\right)^2 \left| f_1(\theta_1)f_2(\theta_1)\frac{e^{-jkr_1}}{r_1} + \Gamma(\theta_2)f_1(\theta_2)f_2(\theta_2)\frac{e^{-jkr_2}}{r_2}\right|^2 \qquad (4\text{–}34)$$

Here f_1 and f_2 are the normalized field pattern functions of the two antennas, which are both assumed to have the same vertical or horizontal polarization, and Γ is the plane wave reflection coefficient from the ground for the same vertical (TM) or horizontal (TE) polarization. The reflection coefficient is given by equation (3–33) for the horizontal (TE) polarization, and by (3–40) for the vertical (TM) polarization. The path lengths r_1 and r_2 are given in terms of the horizontal separation R by

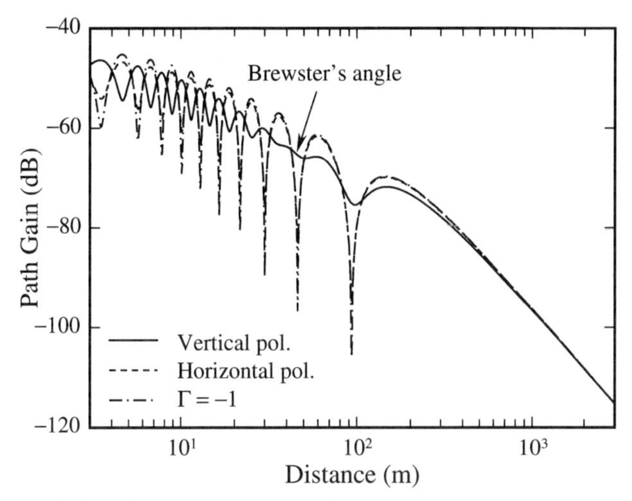

Figure 4-11 Path gain computed from the two-ray model using plane wave reflection coefficients for horizontal (TE) and vertical (TM) reflection coefficients, and using $\Gamma = -1$ ($h_1 = 8.7$ m, $h_2 = 1.8$ m, $f = 900$ MHz).

$$r_{1,2} = \sqrt{R^2 + (h_1 \mp h_2)^2} \tag{4–35}$$

The two-ray model (4–34) for radiation in the presence of the ground is valid when the antennas are at least several wavelengths above the ground, as is typical for UHF and microwave frequencies. At much lower frequencies, it may be necessary to account for an additional ground wave contribution [4, Chap. 16].

For typical wireless applications, the base station antenna height h_1 is a few tens of meters, the mobile antenna height h_2 is about 2 m, and the horizontal separation is 100 m or more. Hence the angles θ_1 and θ_2 in Figure 4–10 are both near 90°, and the differences due to the antenna patterns can be neglected when adding the fields of the two rays in (4–34). If we assume isotropic antennas for simplicity, then $f_1 = f_2 = 1$ in (4–34), and dividing both sides by P_T gives the path gain PG of the two-ray model as

$$\mathrm{PG} = \left(\frac{\lambda}{4\pi}\right)^2 \left| \frac{e^{-jkr_1}}{r_1} + \Gamma(\theta_2)\frac{e^{-jkr_2}}{r_2} \right|^2 \tag{4–36}$$

The path gain obtained from (4–36) is plotted on a decibel scale in Figure 4–11 using the reflection coefficients Γ_E for horizontal (TE) polarization, Γ_H vertical (TM) polarization, and $\Gamma = -1$ for a perfect conductor. The plots are for base station height $h_1 = 8.7$ m, mobile height $h_2 = 1.8$ m, and frequency $f = 900$ MHz.

As R goes from 2 m to about 150 m in Figure 4–11, it is seen that the path gain goes through variations that are due to the interference between the two terms in (4–36) as they go in and out of phase. As R approaches 200 m, the path distances r_1 and r_2 approach each other and the interference minima get deeper for $\Gamma = -1$. For horizontal (TE) polarization, Γ_E is negative for all angles of incidence, and its magnitude increases monotonically to unity with θ_2, as seen in Figure 3–9. As a result, the path gain in Figure 4–11 for horizontal polarization is similar to that for $\Gamma = -1$. For vertical (TM) polarization, Γ_H has a zero at the Brewster angle θ_B, which occurs at the point shown in Figure 4–11. For smaller values of R, $\theta_2 < \theta_B$ and Γ_H is positive, with the consequence that the minima for vertical polarization occur at the maxima for the case $\Gamma = -1$. Beyond the Brewster angle point, Γ_H is negative and the interference maxima and minima for vertical polarization occur at the same locations as for the horizontal polarization, although they are not as deep since $|\Gamma_H|$ is smaller than $|\Gamma_E|$.

The predictions described above are validated by comparison with measurements made at 800 MHz along a straight road on Sherman Island in the San Francisco area [5]. Sherman Island is flat with only low scrub growth and no buildings. The only obstacles were wooden telephone poles beside the road. The transmitting antenna was located on a mast atop a van parked to one side of the road, and the power was measured at a roof-mounted antenna on a very slowly moving station wagon, first using a pair of vertically polarized bicone antennas and then a pair of dipoles. The results are shown in Figure 4–12, along with predictions made using (4–36) for vertical polarization. For the measurements, the distance scale in Figure 4–12 is the separation between the bumpers of the van and station wagon, which is equivalent to $R - 3.2$ m. As applied to the predictions, the distance scale in Figure 4–12 represents R itself, which leads to a horizontal offset between the measurements and predictions that is significant for small distances but is not important at larger values of R.

The bicone and dipole antennas have different radiation patterns $f(\theta)$ and different gains G. For distances of less than 20 m, the differences in antenna patterns cause differences in the received power. For distances greater than 20 m, the primary difference between the two sets of measurements is the vertical offset of about 5 dB that is due to the different gains of the antennas pairs. The irregular signal variations of a few decibels are seen to be consistent in the two sets of measurements, and may be due to scattering from the telephone poles. The measured variation beyond 20 m is close to that predicted by (4–36) for isotropic antennas. In particular, the maximum at about 40 m and the rapid decrease for larger distances are in good agreement. The location of the minimum at 30 m is also in good agreement after accounting for the different definitions of distance used for the measurements and predictions.

4.3a Breakpoint distance

Beyond 150 m in Figure 4–11, $|\Gamma| \approx 1$ for both polarizations, and the path gain appears to decrease rapidly with distance. This behavior can be examined by further approximating (4–36) assuming that $R >> h_1, h_2$. For large R, the square root in (4–35) can be approximated as

$$r_{1,2} \approx R + \frac{(h_1 \mp h_2)^2}{2R} = R + \frac{h_1^2 + h_2^2}{2R} \mp \frac{h_1 h_2}{R} \tag{4–37}$$

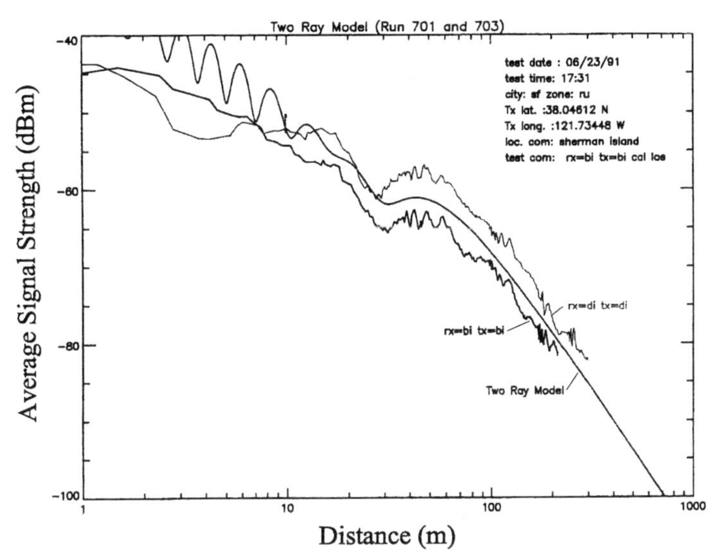

Figure 4-12 Variation of the received signal measured over flat terrain on Sherman Island, California [5](©1993 IEEE). The upper curve, for distances greater than 10 m, is for dipoles, and the lower is for bicone antennas. The smooth curve is calculated for isotropic antennas (h_1 = 3.2 m, h_2 = 1.6 m, and f = 800 MHz).

In the denominators of (4–36) we use the first term of (4–37) but retain all of (4–37) for the phase terms. With these approximations and assuming that $\Gamma \approx -1$, the path gain in (4–36) becomes

$$\mathrm{PG} \approx \left(\frac{\lambda}{4\pi R}\right)^2 \left|\exp\left(jk\frac{h_1h_2}{R}\right) - \exp\left(-jk\frac{h_1h_2}{R}\right)\right|^2 = 4\left(\frac{\lambda}{4\pi R}\right)^2 \left|\sin\left(2\pi\frac{h_1h_2}{\lambda R}\right)\right|^2 \tag{4–38}$$

As R increases, the argument of the sine function in (4–38) decreases, causing the maxima and minima observed in Figure 4–11, until the argument reaches $\pi/2$ and the last maximum is obtained. This last maximum occurs at the breakpoint R_b given by

$$R_b = \frac{4h_1h_2}{\lambda} \tag{4–39}$$

As an example, if h_1 = 10 m, h_2 = 1.5 m, and λ = 1/6 m, corresponding to f = 1.8 GHz, then R_b = 360 m.

Beyond the breakpoint, the argument of the sine function in (4–38) becomes small and the sine can be approximated by its argument. Thus for $R >> R_b$, the path gain reduces to

$$\mathrm{PG} \approx \frac{(h_1h_2)^2}{R^4} \tag{4–40}$$

Thus beyond the breakpoint the presence of a flat earth causes the received signal to vary as $1/R^4$ instead of the $1/R^2$ variation found in free space. In later chapters we will see that other mecha-

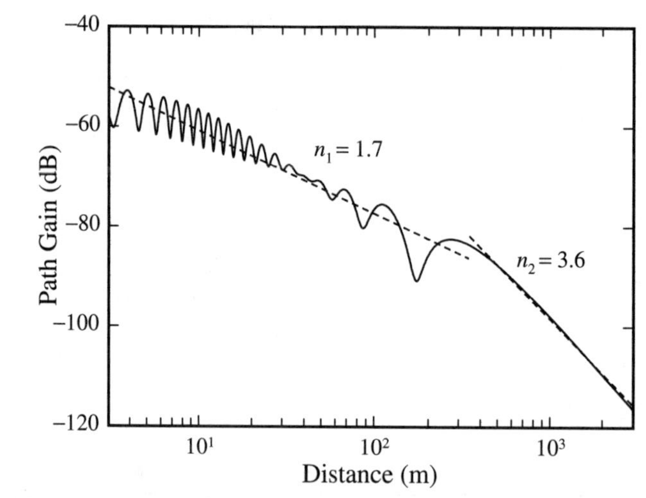

Figure 4-13 Regression fit lines in the regions $R < R_b$ and $R > R_b$ made to the predictions of the two-ray model for flat earth (h_1 = 8.7 m, h_2 = 1.6 m, f = 1850 MHz).

nisms besides reflection can lead to a $1/R^4$ variation. It is seen from (4–40) that raising either antenna increases the received signal. If the height of either antenna is doubled, the received signal will increase by a factor of 4, or 6 dB. If the antennas are raised very high for a fixed R, then at some height, R_b of (4–39) will increase above R and (4–40) will no longer be valid. In this case (4–38) shows that the height variation will be sinusoidal. The height gain obtained here is similar to that found in Chapter 3 for plane wave incidence. Provided that the frequency is low enough so that R_b of (4–39) is small compared to R, it is seen that the path gain in (4–40) is independent of frequency. This observation is supported by the measurements used to construct the LOS path loss model for microcells, as noted in the discussion following expression (2–17).

4.3b Two-slope regression fit

The break distance R_b for a flat earth given in (4–39) divides the signal variation into two regions whose average can be modeled separately by a least-square-error regression fit line. In the region before the breakpoint, the systematic variation of the signal with R is seen from Figures 4–11, 4–12, and 2–9 to have a lower slope than the variation after the breakpoint. This difference is indicated in Figure 4–13 for the received signal predicted by (4–36) at f = 1850 MHz for vertical polarization. The regression fits to the actual variation before and after the breakpoint are also shown. Before the breakpoint the slope index is n = 1.7, after the breakpoint it is n = 3.6. Even though the value of slope index before the breakpoint is less than 2, it does not imply that the signal is greater than in free space. Instead, it is a result of plotting the path gain versus the horizontal separation R, rather than the actual antenna separation, when the antennas are at different heights. Unless the two fit lines are constrained to join at the breakpoint during the least-squares fit process, there can be a jump at R_b, as seen in Figure 4–13. Use of such a

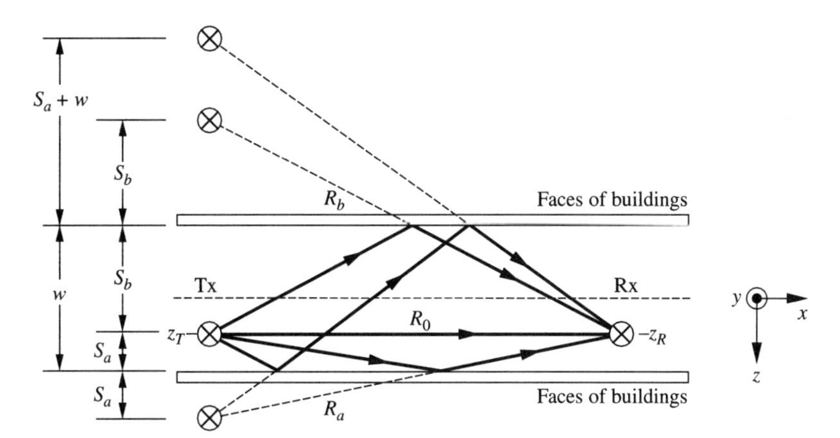

Figure 4-14 Top view of a street canyon showing a few of the rays that contribute to the received signal. Rays that are reflected from the faces of the buildings appear to come from the multiple images of the transmitter. Each ray shown in this top view represents two rays when viewed from the side, one of which is reflected by the ground.

two-slope fit was employed in analyzing the LOS data used for the statistical microcell model discussed in Chapter 2. In Chapter 5 we interpret the break distance in terms of Fresnel zones.

4.4 LOS Propagation in an urban canyon

Within a city, LOS propagation paths commonly lie down streets having buildings on either side of the street that can serve to channel the radiation. This channeling may be especially important in dense urban environments where microcells employing base station antennas below the roof tops are used to increase capacity. For such low antennas, rays reflected from the faces of the buildings contribute to the received signal as well as the direct and ground reflected rays. A simplified top view of the urban canyon is shown in Figure 4–14, where the faces of the buildings have been assumed to form a continuous smooth wall on either side of the street. The rays arriving at the subscriber antenna Rx can be found by first constructing the images of the base station antenna Tx in the building faces, a few of which are shown in Figure 4–14. The rays undergoing a single reflection in a building face appear to come from the first images on either side of the canyon, which are at a distance $s_{a,b} = (w/2) \mp z_T$ from the corresponding faces, where w is the street width and z_T is the distance from the center of the street to Tx. Note that the distances $s_{a,b}$ are equal to those from Tx to the building faces. Rays that are twice reflected from the building faces appear to come from images in each face of the first images created in the opposite face, as shown in Figure 4 –14 for one such double reflected ray. Repeating this process leads to an infinite series of images on both sides of the street that represent the multiply reflected rays.

The presence of the ground must also be taken into account by taking the image in the ground of each image generated by the building faces. Thus each ray shown in the top view in Figure 4–14 represents two rays when viewed from the side, one that goes directly from Tx to

Rx, and one that is reflected from the ground, as shown in Figure 4–10. Because these two rays partially cancel at large distances, they should be taken in pairs. Thus the lowest-order approximation to the received signal is found by accounting for the direct and ground reflected rays, as in the preceding section, while the next-order approximation accounts for six rays, which include the previous two and the four additional rays that undergo a single reflection at the building faces. Making higher-order approximations therefore requires adding rays in groups of four. Because they involve higher powers of the wall reflection coefficient and have longer path lengths, the additional groups of rays will give successively smaller contributions to the total signal.

As an example, consider the six-ray approximation for isotropic antennas and vertical polarization of the electric field. If x is the displacement of Tx and Rx along the center of the street and z_T and z_R are the displacements from the centerline, as shown in Figure 4–14, the horizontal displacements between Rx and the source and images are given by

$$R_0 = \sqrt{x^2 + (z_T - z_R)^2}$$

$$R_a = \sqrt{x^2 + (w - z_T - z_R)^2}$$

$$R_b = \sqrt{x^2 + (w + z_T + z_R)^2} \qquad (4\text{–}41)$$

where w is the street width from building face to building face. For each pair of rays $n = 0, a, b$, the lengths of the direct (1) and ground-reflected (2) rays are given in terms of R_n by

$$r_{n1,2} = \sqrt{(R_n)^2 + (h_1 \mp h_2)^2} \qquad (4\text{–}42)$$

where h_1 and h_2 are the antenna heights. The angle of reflection in the ground θ_n for the three ground reflected rays is

$$\theta_n = \arctan\frac{R_n}{h_1 + h_2} \qquad (4\text{–}43)$$

Because the building reflected rays travel obliquely to the face in both the vertical and horizontal planes, even for a vertically or horizontally polarized transmitter, the fields incident on the building face will not be pure TE or TM, so that polarization coupling will occur. However, if the horizontal displacement between Tx and Rx is large compared to their heights h_1 and h_2, the polarization coupling will be small and we may simply use the TM or TE reflection coefficient given in (3–27) or (3–31). Under these conditions, the angle of incidence for both rays of a pair will be approximately the angles ψ_a and ψ_b given by

$$\psi_{a,b} = \arctan\frac{x}{w \mp (z_T + z_R)} \qquad (4\text{–}44)$$

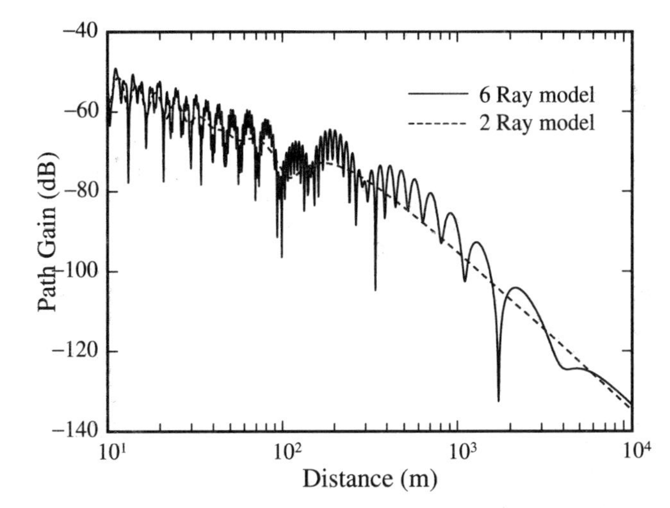

Figure 4-15 Path gain computed for the two- and six-ray models for vertical polarization of isotropic antennas located at the curb ($z_T = z_R = 8$ m) on the same side of a street of width $w = 30$ m for $h_1 = 10$ m, $h_2 = 1.8$ m, and $f = 900$ MHz.

Assuming that each pair of rays has the same reflection coefficient at the building faces, the interference between each pair of rays $n = 0, a, b$ is then embodied in the factor V_n, which for vertical polarization is defined by

$$V_n = \frac{e^{-jkr_{n1}}}{r_{n1}} + \Gamma_H(\theta_n)\frac{e^{-jkr_{n2}}}{r_{n2}} \tag{4–45}$$

The path gain for isotropic antennas, as found by considering the six rays, is then

$$\mathrm{PG} = \left(\frac{\lambda}{4\pi}\right)^2 |V_0 + \Gamma_E(\psi_a)V_a + \Gamma_E(\psi_b)V_b|^2 \tag{4–46}$$

where V_0, V_a, and V_b are found from (4–45). Expression (4–46) is plotted in Figure 4–15 assuming that Tx and Rx are located at the curb ($z_T = z_R = 8$ m) on the same side of a street of width $w = 30$ m for $h_1 = 10$ m, $h_2 = 1.8$ m, and $f = 900$ MHz. For comparison, the path gain computed for the two-ray model is also shown in Figure 4–15. It is seen that including the four building reflected rays causes rapid variations of the path gain about the value predicted by the two ray model. These variations bear a resemblance to those of the measured path gain shown in Figure 2–9, although in a real environment there are many additional scattered rays that can contribute to the fast fading. At large distances both models exhibit an inverse fourth-power dependence on distance, as demonstrated in some measurements [5,6], although other measurements exhibit a stronger decrease with distance. Attempts have also been made to account for the gaps between the buildings along the street [7,8].

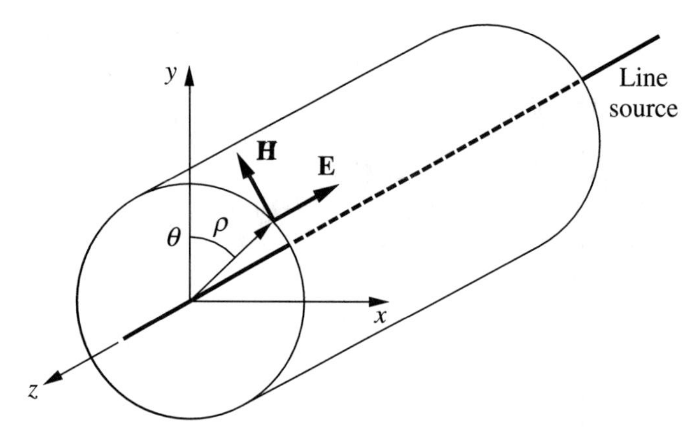

Figure 4-16 Cylindrical wave radiated by a line source along the z axis.

4.5 Cylindrical waves

In discussing diffraction by edges we will make use of the concept of a cylindrical wave. Cylindrical waves can be viewed as being generated by sources located continuously along an a straight line of infinite extent. If the sources are all in phase, they will radiate a wave whose phase fronts are cylinders, as suggested in Figure 4–16. Since the area of a cylinder increases as its radius ρ, in order to conserve power the fields must vary as $1/\sqrt{\rho}$ rather than the $1/r$ variation found for a spherical wave. Far enough from the line source, the radiated fields are given by [4, Chap. 11]

$$\mathbf{E} = \mathbf{a}_E ZI \frac{e^{-jk\rho}}{\sqrt{\rho}} f(\theta)$$

$$\mathbf{H} = \frac{1}{\eta}\mathbf{a}_\rho \times \mathbf{E} \qquad (4\text{–}47)$$

Here $\mathbf{a}_\rho$ is a unit vector in the radial direction and $\mathbf{a}_E$ is a unit vector perpendicular to the radial direction that gives the polarization of the electric field **E**. The magnetic field **H** is also perpendicular to the radial direction and such that $\mathrm{Re}\{\mathbf{E} \times \mathbf{H}^*\}$ is in the outward radial direction. In (4–47), $f(\theta)$ gives the direction dependence of the fields, I is the terminal current, and Z is a constant dependent on the source. In subsequent studies we will be concerned with the polarization and dependence on ρ and θ but will not need to define I or Z explicitly.

4.6 Summary

Far from an antenna the fields propagate in the radial direction with phase variation and polarization like that of a plane wave. Because the radiated power is spread over ever larger spheres, the amplitude of the fields must decrease as the inverse of the distance. The dependence of the fields on the direction of propagation cause radiated power to be greater in some directions, which is described by the antenna gain. Because of reciprocity, when an antenna is used as a receiver, it has the same directive properties as it does when transmitting. Path gain, defined as the ratio of the received to the transmitted power, and its inverse, path loss, are determined by the antenna gains, the frequency, and effects associated with the propagation path. In free space, the path gain will decrease as the inverse square of the distance. For antennas above a plane earth, the interference between direct and reflected waves for distances beyond the breakpoint causes the path gain to decrease as the inverse fourth power of distance.

Problems

4.1 For the coordinates of Figure 4–1, a directive antenna radiates into the hemisphere $y > 0$ a wave with electric field E_θ given by

$$E_\theta = j2\eta I \frac{e^{-jkr}}{r} \frac{\sin(2\pi\cos\phi)}{2\pi\cos\phi} \frac{\sin(4\pi\cos\theta)}{4\pi\cos\theta} \sin\theta \qquad (4\text{–}48)$$

If there is no field radiated into the hemisphere $y < 0$, find the pattern function $f(\theta,\phi)$ that satisfies the normalization (4–4). Note that this will require numerical integration. What are the radiation resistance R_r and gain G of the antenna?

4.2 Two half-wave dipoles are used to communicate in free space. If the receiver sensitivity is –113 dBm and the transmitter power is $P_T = 1$ W, what is the greatest separation R between the antennas for which communication is possible? At this separation, what are the power density $|\mathbf{P}|$ and the voltage across the resistance of the matched load?

4.3 For large aperture antennas such as a parabolic dish, the effective area in the direction of the main beam is close to the aperture area. For a 2-m-diameter dish operating at 9 GHz, compute the antenna gain in decibels.

4.4 Older cellular systems used 900 MHz antennas located on taller buildings ($h_1 = 30$ m), while more recent PCS systems use 1800 MHz antennas located at lamp post height ($h_1 = 10$ m). Assume that both systems are located above a flat earth ($\varepsilon_r = 15$), as in Figure 4–10, and both make use of isotropic antennas with vertical polarization (TM) and that the subscriber antenna height is $h_2 = 1.5$ m. For each system **(a)** find the breakpoint R_b, and **(b)** compute the path gain in decibels (PG_{dB}) and plot versus $\log R$ over the range 100 m $\le R \le$ 10 km.

4.5 For a narrow street canyon as shown in Figure 4–14, the width is $w = 20$. Assume that the base station is located on a lamppost at the curb ($z_T = 5$ m, $h_1 = 9$ m) and the subscriber is in the middle of the street ($z_R = 0$, $h_2 = 1.6$ m). Compute PG_{dB} for a 1800 MHz PCS system and plot over the range 10 m $\le x \le$ 3 km using a logarithmic scale for x. Assume that ε_r is 15 for the ground and 6 for the walls. On the same drawing plot the negative of the path loss L given by the Har model for LOS paths [equations (2–16) and (2–17)].

4.6 Walkie-talkies operating at 300 MHz have antennas at height $h_1 = h_2 = 2$ m. Two users are standing alongside a straight road, both at a distance $s = 4$ m from a high retaining wall with a vertical face. The geometry is like that of Figure 4–14, except that there is a wall on only one side of the street. Assuming that the antennas are isotropic, write down expressions from which the path gain PG can be computed as a function of the distance x between the users. Plot PG in decibels as a function of $\log x$ over the range 10 m $\le x \le$ 1 km.

References

1. J. D. Kraus, Antennas, 2nd ed., McGraw-Hill, New York, 1988.
2. K. Siwiak, Radiowave Propagation and Antennas for Personal Communications, 2nd ed., Artech House, Norwell, Mass., 1998.
3. E. C. Jordan and K. G. Balmain, Electromagnetic Waves and Radiating Systems, 2nd ed., Prentice Hall, Upper Saddle River, N.J., 1968.
4. C. A. Balanis, Advanced Engineering Electromagnetics, Wiley, New York, 1989.
5. H. H. Xia, H. L. Bertoni, L. R. Maciel, A. Lindsay-Stewart, and R. Rowe, Radio Propagation Characteristics for Line-of-Sight Microcellular and Personal Communications, *IEEE Trans. Antennas Propagat.*, vol. 41, pp. 1439–1447, 1993.
6. A. J. Rustako, Jr., N. Amitay, G. J. Owens, and R.S. Roman, Radio Propagation at Microwave Frequencies for Line-of-Sight Microcellular Mobile and Personal Communications, *IEEE Trans. Veh. Technol.*, vol. 40, pp. 203–210, 1991.
7. N. Blaunstein, R. Giladi, and M. Levin, Characteristics Prediction in Urban and Suburban Environments, *IEEE Trans. Veh. Technol.*, vol. 47, pp. 225–234, 1998.
8. R. Mazar, A. Bronshtein, and I.-T. Lu, Theoretical Analysis of UHF Propagation o a City Street Modeled as a Random Multislit Waveguide, *IEEE Trans. Antennas Propagat.*, vol. 46, pp. 864–871, 1998.

CHAPTER 5

Diffraction by Edges and Corners

In Chapters 3 and 4 we discussed, among other topics, the reflection and transmission of waves at material surfaces that were of large lateral extent. In this chapter we begin to examine the interaction of waves with material surfaces that are of finite lateral extent, to uncover the role that diffraction plays in directing signals to regions that would otherwise receive little or no signal. To describe the diffraction effects caused by the edges of surfaces using relatively simple mathematics, we employ the Kirchhoff–Huygens approximation [1]. Huygens recognized that the fields reaching any mathematical surface between a source and receiver can be thought of as producing secondary point sources on the surface that in turn generate the received fields. This concept is depicted in Figure 5–1, where the plane $x = 0$ is taken to be the location of the secondary sources. Later, Kirchhoff gave the mathematical formulation of Huygens' principle that is used here. Since then there have been many mathematical studies of diffraction for geometries that are amenable to rigorous analysis, and out of these studies have come ray methods for approximating the diffracted fields when rigorous solutions are not possible (see for example [2–4]).

Rigorous analysis of diffraction by an absorbing screen [2] indicates that the diffraction coefficient is the same for both possible polarizations of the electric field. As a consequence, orthogonal polarizations will not couple, and we can use the scalar Kirchhoff approximation to find the complex amplitude of the diffracted field [1,5]. The assumption of an absorbing screen, which is realistic in many wireless problems, thus avoids the added complexity required for a careful accounting of polarization [4, Chap. 6]. In this chapter we first use the Kirchhoff–Huygens approximation to find the physical significance of the Fresnel zone for propagation from source to receiver. We then study diffraction of plane waves by the edge of an absorbing half-screen and discuss the results for diffraction by obstacles with other boundary conditions. Polarization effects are discussed briefly in connection with diffraction by screens and corners

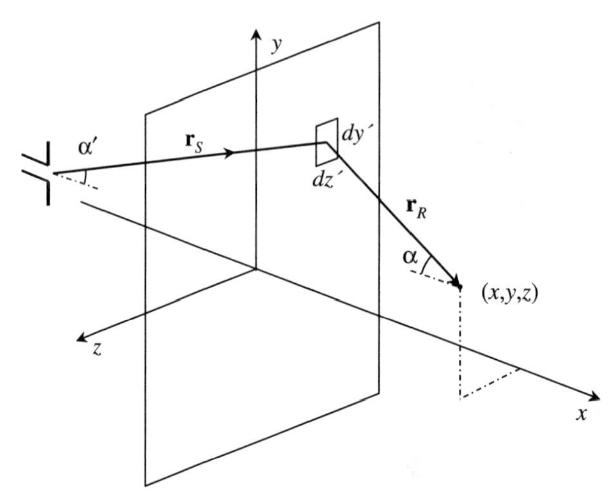

Figure 5-1 The field arriving at a receiver can be found by summing the contributions from equivalent sources generated in an intermediate plane by the field incident from the transmitter.

with conducting boundary conditions. Finally, we generalize the results to the diffraction of spherical waves by one or more edges.

The scalar Kirchhoff approximation for the complex amplitude E of the electric field, or H of the magnetic field, at a receiver point (x,y,z) with $x > 0$ due to sources in the region $x < 0$, as in Figure 5–1, can be written as an integral over the plane $x = 0$, which is given by

$$\left.\begin{matrix} E(x,y,z) \\ H(x,y,z) \end{matrix}\right\} \approx \int_{-\infty}^{\infty}\int_{-\infty}^{\infty} \Lambda(\alpha,\alpha')\left\{\begin{matrix} E^{\text{inc}}(0,y',z') \\ H^{\text{inc}}(0,y',z') \end{matrix}\right\}\frac{jke^{-jkr_R}}{4\pi r_R}dy'dz' \tag{5–1}$$

Here r_R is the distance from the secondary source point (y', z') in the plane $x = 0$ to the receiver point and

$$\Lambda(\alpha, \alpha') = \cos\alpha + \cos\alpha' \tag{5–2}$$

In (5–1) E^{inc} and H^{inc} are the fields due to the actual sources, which give the amplitude of the secondary sources in the plane $x = 0$. The integrand in (5–1) is in the form of the product of the equivalent source, the spherical wave dependence on r, and a pattern function $\Lambda(\alpha, \alpha')$. In more rigorous vector formulations, $\Lambda(\alpha, \alpha')$ is a differential operator that also describes the polarization of the diffracted field [6]. Integrating the spherical wave contributions over the distribution of secondary sources gives the total field at the receiver point.

5.1 Local nature of propagation

Consider a spherical wave propagating from a transmitter to a specific receiving point. It is useful to determine the local region of space that the fields propagate through in going from the transmitter to the receiving point. One way to do this is to make use of the shadowing concept by

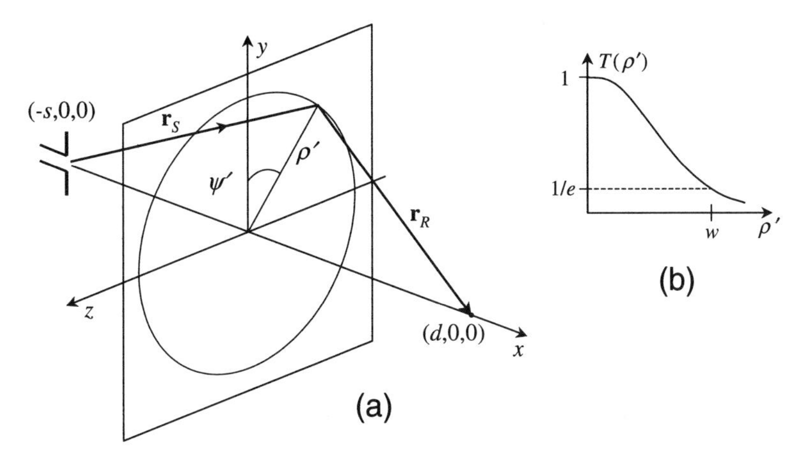

Figure 5-2 The circular window in (a) is centered on the line between the transmitter and receiver and is used to find the size of the region through which the wave propagates, assuming that its transmission coefficient (b) has continuous variation with radial distance.

placing an absorbing screen with a hole in it in a plane between the transmitter and receiving point. By adjusting the size, shape, and location of the hole to minimize the distortion of the field at the receiving point, we can find the cross section of the region through which the fields propagate. To simplify the mathematics, we use our everyday experience with the shadows cast by light, which indicates that light waves travel along the straight line connecting the transmitter and the receiving point. For waves having greater wavelength, we can also expect the region through which the waves propagate to be centered on the connecting line, except that the width of the region will be larger. For a screen placed perpendicular to the direction of propagation, symmetry suggests that the region will be circular. Our original task then becomes one of finding the size of the region by determining the relation between the size of the hole and the distortion of the field.

For simplicity, let the source be located at $(-s,0,0)$ along the negative x axis, and let the receiving point be located at $(d,0,0)$ along the positive x axis. The absorbing screen is placed in the plane $x = 0$ with a circular hole centered on the x axis, as shown in Figure 5–2a. A screen with a circular hole is mathematically equivalent to assigning a transmission coefficient $T(\rho')$ to the plane $x = 0$, where ρ' is the radial distance

$$\rho' = \sqrt{(y')^2 + (z')^2} \tag{5–3}$$

In the hole $T(\rho') = 1$, while $T(\rho')$ vanishes for ρ' large. If $T(\rho')$ switches abruptly from 1 inside the hole to 0 outside, diffraction effects will occur that can confuse the interpretation of the region of propagation. These diffraction effects can be avoided by using a continuous variation of $T(\rho')$, such as the Gaussian dependence

$$T(\rho') = \exp\left[-\frac{(\rho')^2}{w^2}\right] \tag{5–4}$$

which leads to simple mathematical expressions. The Gaussian variation is shown in Figure 5–2b, where w is seen to be the $1/e$ radius of the hole.

5.1a Evaluation of the field distortion

The electric field incident on the screen in the plane $x = 0$ is

$$E(0, y', z') = ZIf(\theta, \phi)\frac{e^{-jkr_s}}{r_s} \tag{5–5}$$

where $r_s = \sqrt{s^2 + (\rho')^2}$ is the distance from the transmitter to points in the plane x = 0. The secondary source field E^{inc} just to the right of the screen is then given by the product of (5–5) and the transmission function (5–4). The field $E(d,0,0)$ is found from the Kirchhoff–Huygens integral (5–1) with E^{inc} set equal to the product of (5–4) and (5–5). Because the hole is centered on the x axis, α and α' will be small, and from (5–2), $\Lambda(\alpha, \alpha') \approx 2$. Thus using the cylindrical variables (ρ', ψ') shown in Figure 5–2a for integration over the plane,

$$E(d,0,0) = \int_0^{\infty}\int_0^{2\pi} ZIf(\theta,\phi)\frac{e^{-jkr_s}}{r_s}\exp\left[-\frac{(\rho')^2}{w^2}\right]\frac{jke^{-jkr_R}}{2\pi r_R}d\psi'\rho' d\rho' \tag{5–6}$$

where $r_R = \sqrt{d^2 + (\rho')^2}$ is the distance from the integration point to the receiver point.

To evaluate the integral in (5–6), recognize that the Gaussian function limits the contribution to values of ρ' near 0. We may therefore approximate r_R and r_s as

$$r_R = \sqrt{d^2 + (\rho')^2} \approx d + \frac{(\rho')^2}{2d} \tag{5–7}$$

$$r_s = \sqrt{s^2 + (\rho')^2} \approx s + \frac{(\rho')^2}{2s}$$

In the denominator of (5–6), the second terms in (5–7) will cause little variation of the integrand, and we may therefore set $r_R = d$ and $r_s = s$ there. Because the variation of $f(\theta,\phi)$ over the hole will be small, it can be replaced by its value at the center of the hole. In this case the integrand is independent of ψ' so that the integration over ψ' gives 2π. With the foregoing approximations, (5–6) becomes

$$E(d,0,0) = ZIf(\theta,\phi)\frac{jke^{-jk(s+d)}}{sd}\int_0^{\infty}\exp\left\{-\left[\frac{1}{w^2} + j\frac{k}{2}\left(\frac{1}{s} + \frac{1}{d}\right)\right](\rho')^2\right\}\rho' d\rho' \tag{5–8}$$

The integration in (5–8) can be carried out in closed form, and after some manipulation yields the expression

$$E(d,0,0) = \left[ZIf(\theta,\phi)\frac{e^{-jk(s+d)}}{s+d}\right]\frac{1}{1-j\dfrac{2sd}{kw^2(s+d)}} \tag{5–9}$$

The first term in brackets in (5–9) gives the field that would have arrived at the receiver point if the transmitting screen were not placed in the plane $x = 0$. The term following the brackets gives the distortion of the fields due to the presence of the screen, which is small if the radius w of the hole is large. If we define the error ε by the expression

$$\varepsilon \equiv \frac{2sd}{kw^2(s+d)} = \frac{\lambda sd}{\pi w^2(s+d)} \tag{5–10}$$

then for ε small the distortion term in (5–9) is approximately $1 + j\varepsilon$. Thus the phase of the field is distorted by $\arctan\varepsilon$, and the amplitude by $\sqrt{1+\varepsilon^2} \approx 1 + \varepsilon^2/2$. For example, if $\varepsilon = 0.1$, the phase is distorted by about 18° and the amplitude by 0.5%.

For fixed error ε, the radius w of the hole gives a cross section of the local region in space through which the fields propagate from the source to the receiver point. For a given ε the radius can be found from (5–10) as

$$w = \sqrt{\frac{\lambda sd}{\pi\varepsilon(s+d)}} \tag{5–11}$$

If the total distance $s + d = r$ between source and receiver is held fixed as s and d are varied, (5–11) describes a cigar-shaped region about the direct line between the source and receiver. This region is widest at its midpoint when $s = d = r/2$, where its radius is $\sqrt{\lambda r/4\pi\varepsilon}$. Note that for $r \gg \lambda$, as is typically the case, the width of the region is small compared to s and d, which justifies the approximations used in evaluation the integrals in (5–6). We will explore the implications of (5–11) for wireless applications in the next section.

5.1b Interpretation of the local region in terms of Fresnel zones

The Fresnel zones are ellipsoids of revolution about the direct line from a transmitter to a receiving point, with the transmitter and receiving points serving as foci of the ellipse, as shown in Figure 5–3. For the nth Fresnel zone, the distance $r_s + r_R$ from the transmitter to any point on the ellipsoid and then to the receiving point is $n\lambda/2$ greater than the direct line distance $s + d$, or

$$(r_s + r_R) - (s + d) = n\frac{\lambda}{2} \tag{5–12}$$

Let w_{Fn} be the radius of the nth Fresnel zone at a given distance s from the transmitter. If $s + d \gg \lambda$, r_s and r_R may be approximated by (5–7) with ρ' replaced by w_{Fn}. Substituting these expressions into (5–12) and solving for w_{Fn} gives the expression

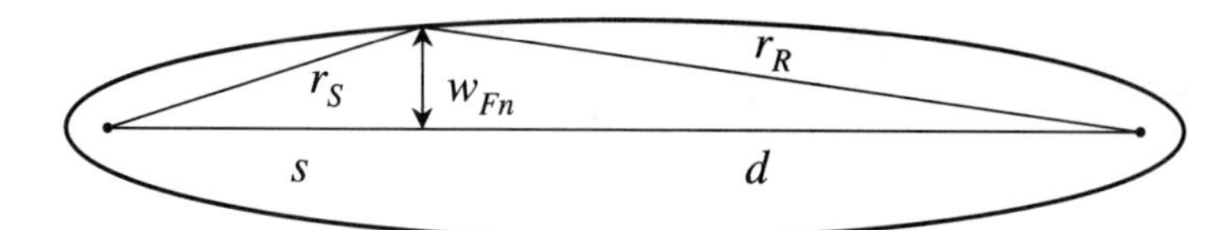

Figure 5-3 The Fresnel zone is a ellipsoid of revolution with the transmitter and receiver as foci.

$$w_{Fn} = \sqrt{n\frac{\lambda sd}{s+d}} \tag{5–13}$$

Comparing (5–11) and (5–13) it is seen that if w_{Fn} is taken to be the radius of the hole in the transmission screen, the distortion of the field will be $\varepsilon = 1/n\pi$. For $n = 1$, the phase error is about 60° but the amplitude error is only about 10%.

The foregoing result indicates that wave propagation from a source to a receiver point is a localized phenomena, with the local region being given by the Fresnel zones. Objects, such as the ground or buildings, that are located outside the Fresnel zones may introduce additional reflected or scattered contributions to the total field at the receiving point, but cause only a small distortion of the original wave. However, objects located inside the Fresnel zones, especially inside the $n = 1$ zone, will result in a significant perturbation of the direct wave. We have already seen an example of this in regard to the breakpoint R_b for the two-ray model of propagation over a flat earth (see Section 5–3a). It is easily shown that for the separation R_b between source and receiver, the $n = 1$ Fresnel zone just touches the ground at the point of reflection of the second ray. For greater separations, the Fresnel zone is wider and is partially blocked by the ground. As a result of Fresnel zone blockage, the received power decreases more rapidly with distance. A design criterion for point-to-point microwave links requires that the Fresnel zone be clear of the ground. If an object completely blocks the $n = 1$ Fresnel zone, it will prevent the direct wave from reaching the receiver point, as will be seen later.

The Fresnel zone is widest at the midpoint between the source and receiver. If the separation between source and receiver is r, at the midpoint $s = d = r/2$ the diameter of the $n = 1$ Fresnel zone is

$$2w_{F1} = \sqrt{\lambda r} \tag{5–14}$$

It is seen that the diameter increases with separation r and wavelength. If $r = 1$ km, at 900 MHz the diameter is $2w_{F1} = 18.3$ m, while at 1800 MHz it is 12.9 m. Thus for propagation over kilometer paths, the diameter of the first Fresnel zone is about the width of a medium-sized building, or a few houses. At distances $d \ll s$ near one end of a long path, the diameter of the first Fresnel zone is seen from (5–13) to be $2w_{F1} \approx 2\sqrt{\lambda d}$. For example, on a 1.625-GHz satellite link, if $d = 50$ m, $2w_{F1} \approx 6.1$ m. Thus even on satellite links where r is large, the Fresnel zone near the terrestrial end of the link, where it interacts with buildings, has a diameter that is about the width

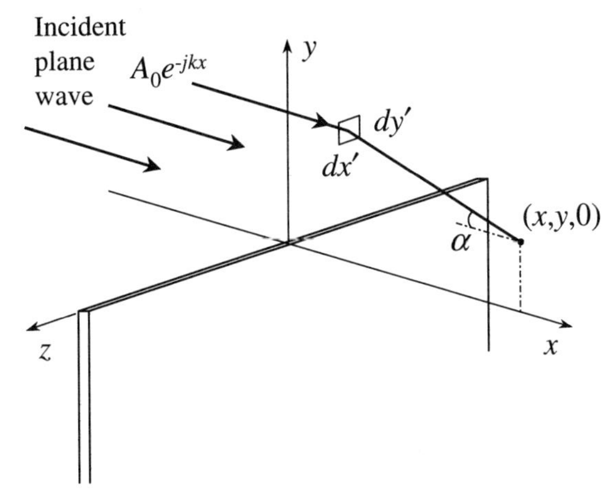

Figure 5-4 The field transmitted past an absorbing half-plane is found by summing the contributions from the equivalent sources due to an incident plane wave in the area above the half-plane.

of a house. The Fresnel zone concept is a convenient way to understand the behavior of waves and to set limits on various approximations.

5.2 Plane wave diffraction by an absorbing half-screen

The problem of diffraction of a plane wave by an edge serves as a prototype for many other diffraction problems. As shown in Figure 5–4, a plane wave generated by a distant source is incident from the left onto the $x = 0$ plane, the lower half of which contains an absorbing screen. To allow for diffraction through significant angles, consider the z component of either the **E** or **H** field, depending on the polarization of the plane wave. If the plane wave propagates along the x axis, $\alpha' = 0$ and the z component of the appropriate incident field will have spatial dependence given by

$$\left.\begin{matrix} E_z^{\text{inc}} \\ H_z^{\text{inc}} \end{matrix}\right\} = A_0 e^{-jkx} \tag{5–15}$$

The fields diffracted past the screen can be found from (5–1) by assuming the field just to the right of the screen to be that of the incident plane wave for $y > 0$ and to vanish for $y < 0$. Because of the translation symmetry along z, there is no loss in generality if the receiver points are assumed to lie in the plane $z = 0$. With the foregoing assumptions, the field diffracted past the plane is given by

$$\left.\begin{matrix} E_z(x, y, 0) \\ H_z(x, y, 0) \end{matrix}\right\} = A_0\frac{jk}{4\pi}\int_0^{\infty}\int_{-\infty}^{\infty}(1 + \cos\alpha)\frac{e^{-jkr_R}}{r_R}dz'dy' \tag{5–16}$$

where the distance from the integration point to the receiver point is

$$r_R = \sqrt{x^2 + (y - y')^2 + (z')^2} \tag{5–17}$$

To carry out the z' integration in (5–16), recognize from Section 5–1 that the primary contribution to the integral comes from a region of z' that is given by the Fresnel zone, which is small compared to x for $x \gg \lambda$. The center of this region is the stationary-phase point where the derivative of the exponent kr_R with respect to z' vanishes [2, p. 386; 6]. Within the region giving the primary contribution to the z' integral in (5–16), r_R in the denominator of (5–17) and $\cos\alpha$ will vary by a small fractional amount and can be treated as constants. However, r_R may vary by several wavelengths, so that in the exponent it is necessary to expand r_R to second order as

$$r_R \approx \rho_R + \frac{(z')^2}{2\rho_R} \tag{5–18}$$

where

$$\rho_R = \sqrt{x^2 + (y - y')^2} \tag{5–19}$$

Using the foregoing approximations, the integration over z' in (5–17) reduces to

$$\int_{-\infty}^{\infty} (1 + \cos\alpha)\frac{e^{-jkr_R}}{r_R}dz' \approx (1 + \cos\alpha)\frac{e^{-jk\rho_R}}{\rho_R}\int_{-\infty}^{\infty} \exp\left[-jk\frac{(z')^2}{2\rho_R}\right]dz' \tag{5–20}$$

The integration on the right-hand side of (5–20) can be evaluated in closed form using the substitution $u = z'e^{j\pi/4}\sqrt{k/2\rho_R}$, with the result that

$$\int_{-\infty}^{\infty} (1 + \cos\alpha)\frac{e^{-jkr_R}}{r_R}dz' \approx (1 + \cos\alpha)\frac{e^{-jk\rho_R}}{\rho_R}e^{-j\pi/4}\sqrt{\frac{2\pi\rho_R}{k}} \tag{5–21}$$

Substituting (5–21) back into (5–17) gives the field at the receiver point:

$$\left.\begin{matrix} E_z(x, y, 0) \\ H_z(x, y, 0) \end{matrix}\right\} = A_0\frac{e^{j\pi/4}}{2}\sqrt{\frac{k}{2\pi}}\int_0^{\infty} (1 + \cos\alpha)\frac{e^{-jk\rho_R}}{\sqrt{\rho_R}}dy' \tag{5–22}$$

Integration over y' will be discussed separately for receiver points (1) well inside the region $y > 0$ that is illuminated by the plane wave, (2) well inside the shadow region $y < 0$, and (3) in a transition region about the shadow boundary $y = 0$.

5.2a Field in the illuminated region $y > 0$

The stationary-phase method can also be used to evaluate the y' integration of (5–22). In the case of (5–22), the stationary point is the value $y' = y$ where the derivative $d\rho_R/dy'$ vanishes. Expanding ρ_R about this point to second order gives

$$\rho_R \approx x + \frac{1}{2x}(y' - y)^2 \tag{5–23}$$

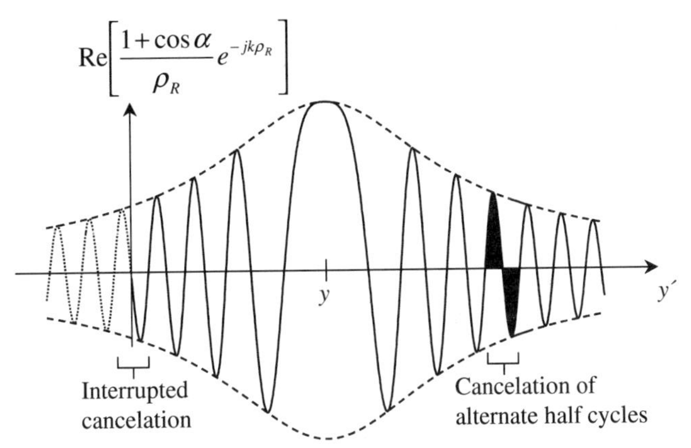

Figure 5-5 Variation of the real part of the integrand with y' about a stationary point $y' = y$ located in the illuminated region of the absorbing half-plane.

Because of the quadratic dependence (5–23) of the exponent, the exponential in (5–22) will oscillate with increasing spatial frequency as $|y' - y|$ increases. As a result, the real part of the integrand will have the dependence on y' shown in Figure 5–5, which has been drawn assuming that $x \gg \lambda$ so that the amplitude terms in (5–22) vary slowly as compared to the oscillations. The imaginary part has a similar dependence. The primary contribution to the integral will come from a limited interval about y, where the first few oscillations take place. This interval corresponds to the region through which the plane wave fields propagate to the receiver point (x,y) as discussed in Section 5–1b. Away from y, successive half-cycles of the oscillation cancel each other. However, the endpoint of the integration at $y = 0$ interrupts the cancellation, which results in a secondary contribution to the integral.

To separate the stationary point contribution from the endpoint contribution, (5–22) is rewritten as

$$\left.\begin{matrix} E_z(x,y,0) \\ H_z(x,y,0) \end{matrix}\right\} = A_0 \frac{e^{j\pi/4}}{2}\sqrt{\frac{k}{2\pi}}\left[\int_{-\infty}^{\infty}(1+\cos\alpha)\frac{e^{-jk\rho_R}}{\sqrt{\rho_R}}dy' - \int_{-\infty}^{0}(1+\cos\alpha)\frac{e^{-jk\rho_R}}{\sqrt{\rho_R}}dy'\right] \qquad (5\text{–}24)$$

The first integral in (5–24) can be approximated using the same approach as that taken for the z' integration. Thus the amplitude terms are evaluated at the stationary point $y' = y$, where $\alpha = 0$ and $\rho_R = x$, and taken outside the integral. In the exponent ρ_R is approximated by (5–23) to give

$$\int_{-\infty}^{\infty}(1+\cos\alpha)\frac{e^{-jk\rho_R}}{\sqrt{\rho_R}}dy' \approx \frac{2e^{-jkx}}{\sqrt{x}}\int_{-\infty}^{\infty}\exp\left[-j\frac{k}{2x}(y'-y)^2\right]dy' = 2e^{-j\pi/4}\sqrt{\frac{2\pi}{k}}e^{-jkx} \qquad (5\text{–}25)$$

The primary contribution to the second integral in (5–24) comes from the vicinity of the endpoint $y' = 0$. If y is sufficiently large so that the stationary point is well separated from the endpoint, the endpoint contribution can be approximated by expanding ρ_R in the exponent about the point $y' = 0$ and keeping the first two terms in the expansion. From (5–19) the expansion of ρ_R is seen to be

$$\rho_R \approx \rho - y'\frac{y}{\rho} \tag{5–26}$$

where

$$\rho = \sqrt{x^2 + y^2} \tag{5–27}$$

is the distance from the edge to the receiver point. With the foregoing approximations, the second integral in (5–24) becomes

$$\int_{-\infty}^{0} (1 + \cos\alpha)\frac{e^{-jk\rho_R}}{\sqrt{\rho_R}}dy' \approx (1 + \cos\theta)\frac{e^{-jk\rho}}{\sqrt{\rho}}\int_{-\infty}^{0} \exp\left(jky'\frac{y}{\rho}\right)dy' = \frac{1 + \cos\theta}{jk(y/\rho)}\frac{e^{-jk\rho}}{\sqrt{\rho}} \tag{5–28}$$

Here θ is the angle between the x axis and the line from the edge to the receiver point, so that $\sin\theta = y/\rho$. In carrying out the integration in (5–28), k is given a vanishingly small negative imaginary part, as appropriate for atmospheric absorption, to ensure convergence at the lower limit, but after the integration, k can be taken to be real.

Substituting (5–25) and (5–28) into (5–24), it is seen that well into the illuminated region $y > 0$ the field is given by

$$\left.\begin{matrix} E_z(x, y, 0) \\ H_z(x, y, 0) \end{matrix}\right\} = A_0 e^{-jkx} + A_0 e^{-j\pi/4}\frac{e^{-jk\rho}}{\sqrt{\rho}}D(\theta) \tag{5–29}$$

where $D(\theta)$ is the diffraction coefficient:

$$D(\theta) = -\frac{1}{\sqrt{2\pi k}}\frac{1 + \cos\theta}{2\sin\theta} \tag{5–30}$$

Equation (5–29) gives the field as the sum of the incident plane wave and a cylindrical wave propagating away from the edge and having direction dependence $D(\theta)$, as indicated in Figure 5–6.

On the shadow boundary $y = 0$ between the illuminated region and the shadow region, $D(\theta)$ is singular. This results from the fact that the endpoint and stationary points in (5–24) are not well separated and the approximation (5–28) is no longer valid. In fact, substituting (5–23) for $y = 0$ into (5–22), it is easily seen that the stationary point approximations lead to a value of $E_z(x,0^+,0)$ or $H_z(x,0^+,0)$ that is one-half of the incident plane wave field. In a subsequent section we derive a correction to the diffraction coefficient that makes the field finite and continuous at the shadow boundary.

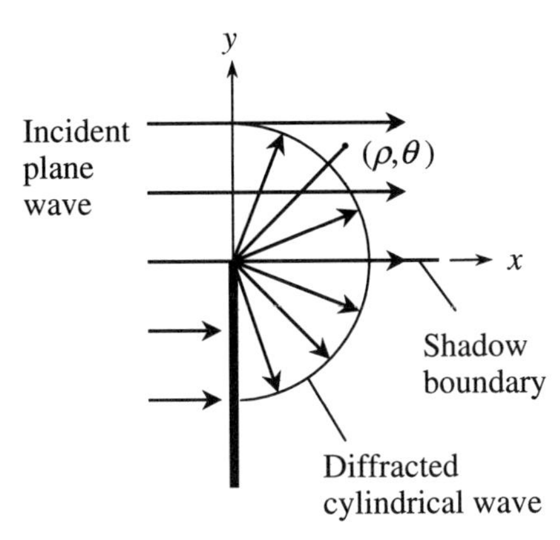

Figure 5-6 Plane wave transmitted past an absorbing half-plane in the illuminated region and the cylindrical wave generated by diffraction at the edge.

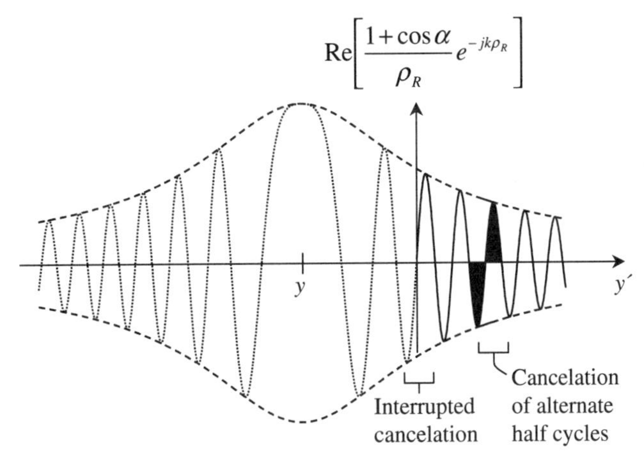

Figure 5-7 Variation of the real part of the integrand with y' about a stationary point $y' = y$ located in the shadowed region of the absorbing half-plane.

5.2b Field in the shadow region $y < 0$

When $y < 0$, the stationary point of the integrand in (5–22) lies outside the range of integration, as suggested in Figure 5–7 for the real part of the integrand. For $|y|$ large enough, the stationary point will be well separated from the endpoint, and we may use the endpoint approximation for the integral. Thus with the help of (5–26), (5–22) becomes

$$\left.\begin{matrix} E_z(x, y, 0) \\ H_z(x, y, 0) \end{matrix}\right\} = A_0 \frac{e^{j\pi/4}}{2}\sqrt{\frac{k}{2\pi}}(1 + \cos\theta)\frac{e^{-jk\rho}}{\sqrt{\rho}}\int_0^\infty \exp\left(jky'\frac{y}{\rho}\right)dy' \tag{5–31}$$

Carrying out the integration, with k given a vanishingly small negative imaginary part to ensure convergence at the upper limit, it is seen that

$$\left.\begin{matrix} E_z(x, y, 0) \\ H_z(x, y, 0) \end{matrix}\right\} = A_0 e^{-j\pi/4} \frac{e^{-jk\rho}}{\sqrt{\rho}} D(\theta) \tag{5–32}$$

where the diffraction coefficient $D(\theta)$ is the same as that in (5–30). In the shadow region, the field consists only of the cylindrical wave propagating away from the edge, as indicated in Figure 5–6.

5.2c Geometrical theory of diffraction

The results given in (5–29) and (5–32) are summarized in the following expression

$$\left.\begin{matrix} E_z(x, y, 0) \\ H_z(x, y, 0) \end{matrix}\right\} = A_0 e^{-jkx} U(\theta) + A_0 e^{-j\pi/4} \frac{e^{-jk\rho}}{\sqrt{\rho}} D(\theta) \tag{5–33}$$

Here the unit step function $U(\theta)$ in the first term indicates that the incident field is present in the illuminated region but not in the shadow region. The second term represents the cylindrical wave generated by the diffracting edge. As it now stands, (5–33) is not valid near the shadow boundary because of the singular nature of $D(\theta)$. In reality, the field must be bounded and continuous there and as discussed before, is equal to one-half the incident field. In the next section we derive a function that multiplies the diffraction coefficient so as to remove the singularity and results in the total field being continuous. Away from the shadow boundary, the diffraction term is small compared to the incident plane wave field. For example, if $\rho = 50\lambda$ and $\theta = \pm 10°$, then $|D(\pm 10°)|/\sqrt{50\lambda} = 0.129$, and the diffracted field is 17.8 dB smaller than the plane wave field. Because $D(\theta)$ depends inversely as the square root of $k = \omega/c$, at a given angle θ and distance ρ, the field in the shadow region decreases with increasing frequency.

Expression (5–33) represents the geometrical theory of diffraction (GTD) representation for the fields due to a plane wave incident on an absorbing screen. For wedges that have a finite angle, and for other boundary conditions, the GTD solution will be similar to that in (5–33). However, for reflecting boundary conditions, the geometrical optics field consists of the incident and reflected plane waves that illuminate portions of space, which are limited by appropriate shadow boundaries. The wave incident on the edge generates diffracted fields that are in the form of a cylindrical wave, as in (5–33). The diffraction coefficient $D(\theta)$ will depend on the wedge shape and boundary conditions, as we will see later in the chapter, but is singular at shadow boundaries.

5.2d Evaluating the Fresnel integral for *y* near the shadow boundary

When the receiver point is near the shadow boundary, the stationary point is near the endpoint of the integration, and the evaluation of the endpoint contributions in (5–28) and (5–31) is no longer valid. To find a more accurate expression for the endpoint contribution that is valid for all values of y, it is necessary to use a more accurate approximation for ρ_R in the exponent of (5–28). Thus we approximate ρ_R by a Taylor series retaining terms up to second order in y'. This expansion takes the form

$$\rho_R = \rho - \frac{y}{\rho}y' + \frac{x^2}{2\rho^3}(y')^2 \tag{5–34}$$

Using this expression, the end point integral in (5–28) for $y > 0$ can be approximated by

$$\int_{-\infty}^{0}(1+\cos\alpha)\frac{e^{-jk\rho_R}}{\sqrt{\rho_R}}dy' \approx (1+\cos\theta)\frac{e^{-jk\rho}}{\sqrt{\rho}}\int_{-\infty}^{0}\exp\left\{-jk\left[\frac{x^2}{2\rho^3}(y')^2 - \frac{y}{\rho}y'\right]\right\}dy' \tag{5–35}$$

To express (5–35) in terms of known functions, we complete the square in the exponent by adding and subtracting to the exponent the quantity

$$jS = j\frac{k\rho}{2}\frac{y^2}{x^2} = j\frac{k\rho}{2}\tan^2\theta \tag{5–36}$$

This approach leads to the following expression for the endpoint integral

$$\int_{-\infty}^{0}(1+\cos\alpha)\frac{e^{-jk\rho_R}}{\sqrt{\rho_R}}dy' \approx (1+\cos\theta)\frac{e^{-jk\rho}}{\sqrt{\rho}}e^{jS}\int_{-\infty}^{0}\exp\left[-j\left(\frac{x}{\rho}\sqrt{\frac{k}{2\rho}}y' - \sqrt{S}\right)^2\right]dy' \tag{5–37}$$

The variable of integration is now changed to

$$u = -\left(\frac{x}{\rho}\sqrt{\frac{k}{2\rho}}y' - \sqrt{S}\right) \tag{5–38}$$

for which

$$dy' = -\frac{\rho}{x}\sqrt{\frac{2\rho}{k}}du = -\frac{2\rho}{ky}\sqrt{S}du = -\frac{2\sqrt{S}}{k\sin\theta}du \tag{5–39}$$

Substituting (5–38) and (5–39) into (5–37), the endpoint integral becomes

$$\int_{-\infty}^{0}(1+\cos\alpha)\frac{e^{-jk\rho_R}}{\sqrt{\rho_R}}dy' \approx \frac{1+\cos\theta}{jk\sin\theta}\frac{e^{-jk\rho}}{\sqrt{\rho}}F(S) \tag{5–40}$$

The transition function $F(S)$ is related to the Fresnel integrals and is given by

$$F(S) = 2j\sqrt{S}e^{jS}\int_{\sqrt{S}}^{\infty}\exp(-ju^2)du \tag{5–41}$$

Expressions that can be used to compute $F(S)$ are given below.

The approach described above can also be used for $y < 0$. Approximation (5–34) is again used in (5–22), and after some manipulation it is found that

$$\int_0^\infty (1+\cos\alpha)\frac{e^{-jk\rho_R}}{\sqrt{\rho_R}}dy' \approx -\frac{1+\cos\theta}{jk\sin\theta}\frac{e^{-jk\rho}}{\sqrt{\rho}}F(S) \tag{5–42}$$

In (5–42), $\sin\theta < 0$ for $y < 0$, but in evaluating $F(S)$, we must still take $\sqrt{S} > 0$ in (5–41).

5.2e Uniform theory of diffraction

Substituting (5–40) into (5–24) for $y > 0$, and substituting (5–42) into (5–22) for $y < 0$, the total field in the transition region near the shadow boundary is found to be

$$\left.\begin{matrix} E_z(x,y,0) \\ H_z(x,y,0) \end{matrix}\right\} = A_0 e^{-jkx}U(y) + A_0 e^{-j\pi/4}\frac{e^{-jk\rho}}{\sqrt{\rho}}D_T(\theta) \tag{5–43}$$

In (5–43), the diffraction (UTD) coefficient of the uniform theory of

$$D_T(\theta) = D(\theta)F(S) \tag{5–44}$$

where $D(\theta)$ is the GTD diffraction coefficient defined in (5–30). The variation with y of the magnitude of the total field (5–43), expressed in decibels, is shown in Figure 5–8 for $A_0 = 1$, $x = 30$ m, and $f = 900$ MHz. The field is seen to be continuous at the shadow boundary $y = 0$, where its value is 1/2, or –6 dB. The oscillations about incident plane wave amplitude in the illuminated region are due to the interference between the direct and diffracted contributions. In the shadow region the field shows a strong monotonic decrease in amplitude. At $y = -10$ m, the diffraction angle is only $|\theta| = 18°$, and already the signal amplitude is more than 26 dB below that of the incident field. Thus diffraction around corners significantly reduces the received signal.

To make calculations, such as leading to Figure 5–8, the transition function $F(S)$ may be computed using the expression

$$F(S) = \sqrt{2\pi S}\left[f(\sqrt{2S/\pi}) + jg(\sqrt{2S/\pi})\right] \tag{5–45}$$

The functions $f(\xi)$ and $g(\xi)$ are associated with the Fresnel integrals, and for all positive values of ξ can be computed from the rational approximations [7]

$$f(\xi) = \frac{1+0.926\xi}{2+1.792\xi+3.104\xi^2}$$

$$g(\xi) = \frac{1}{2+4.142\xi+3.492\xi^2+6.670\xi^2} \tag{5–46}$$

Very near to the shadow boundary, where $S \ll 1$, the functions $f(\xi)$ and $g(\xi)$ approach 1/2, and $F(S) \approx \sqrt{\pi S}e^{j\pi/4}$. Substituting this expression into (5–41), it can be shown that the total field (5–43) is continuous across the shadow boundary and equal to one half the incident field there, as shown in Figure 5–8.

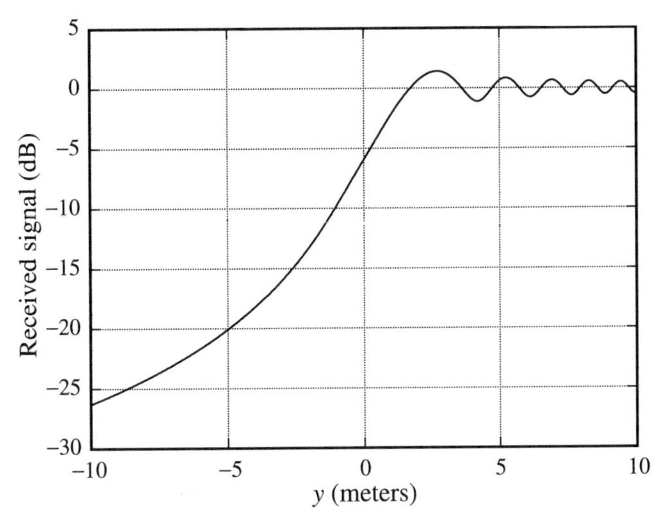

Figure 5-8 Variation of the total field with y at a distance x = 30 m beyond an absorbing half-plane due to a 900-MHz plane wave of unit amplitude. The total field is 1/2 of the incident field (–6 dB) on the shadow boundary.

When S is large, the transition function can be approximated by a series in negative powers of S, the first two terms of which are [4,8]

$$F(S) \approx 1 + j\frac{1}{2S} \tag{5–47}$$

Thus outside a transition region about the shadow boundary, the transition function approaches unity and $D_T(\theta)$ approaches the simple diffraction coefficient $D(\theta)$, so that the total field in (5–43) approaches that in (5–33). Because (5–43) is valid near the shadow boundary as well as in the illuminated and shadow regions, it represents the uniform theory of diffraction solution, whereas (5–33) represents the geometrical theory of diffraction solution. From (5–46) or (5–47), $F(S)$ will be near unity when $S > \pi$. If we take the condition $S \le \pi$ as defining the transition region, and recognize from (5–36) that for small y, $S \approx \pi y^2/\lambda x$, the width of the transition region is $|y| = \sqrt{\lambda x}$ (Figure 5–9). The presence of the transition region can be understood in terms of the Fresnel zone about the plane wave ray, which is described following (5–14). Consider a receiver point at the upper boundary of the transition region, as shown in Figure 5–9. The Fresnel zone for this point is also shown, and it is seen that the edge lies on its boundary. If the edge were to move up into the Fresnel zone, the receiver point would lie in the transition region of the edge, and the direct illumination would be distorted. For a receiver point at the lower boundary of the transition region, the edge would just block the Fresnel zone, thereby eliminating the direct contribution entirely. As an example, for x = 30 m and f = 900 MHz, as used in generating the curve in Figure 5–8, the boundaries of the transition region are at $y = \pm\sqrt{\lambda x}$ = ±3.16, which are seen to lie just past the first peak of the curve in the illuminated region and at the –16-dB point in the shadow region.

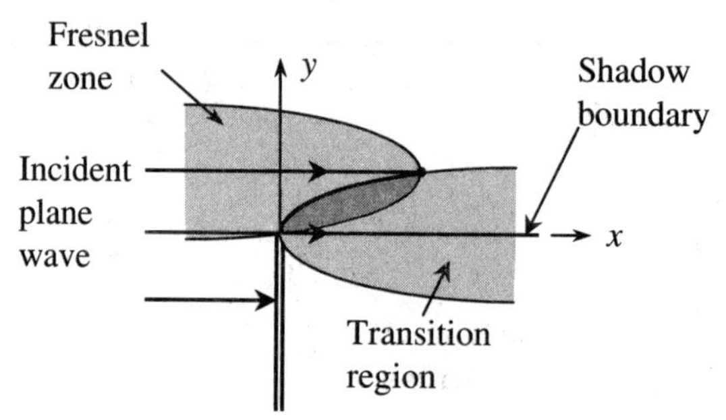

Figure 5-9 Transition region about the shadow boundary and its relation to the Fresnel zone for plane wave propagation to receiver points.

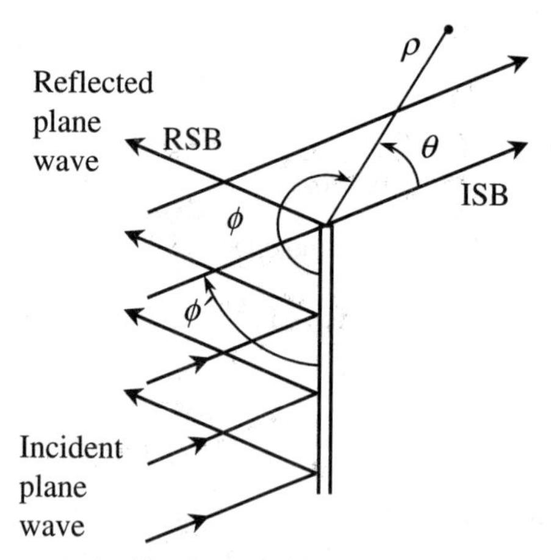

Figure 5-10 A plane wave incident on a half-plane illuminates the region above the incident wave shadow boundary (ISB) and can generate reflected waves in the region below the reflected wave shadow boundary (RSB).

5.3 Diffraction for other edges and for oblique incidence

The results obtained above for diffraction by an absorbing half-screen are representative of those found for screens having other boundary conditions and for diffraction by wedges with a finite apex angle. Rigorous analysis of various half-screens and wedges have shown that the total field consists of the geometrical optics fields and diffracted fields that are in the form of a cylindrical wave propagating away from the edge. The geometrical optical fields are those of the incident wave and the reflected wave in the case of nonabsorbing screens. These components may illuminate only a portion of space to one side of a shadow boundary, as indicated in Figure 5–10 for a half-screen. The diffraction coefficient, which gives the direction dependence of the diffracted fields, itself may depend on the type of screen, wedge angle, the angle of incidence ϕ', and the polarization. Near the shadow boundaries of the incident and reflected fields, the diffraction coefficient is multiplied by a transition function to make the total field bounded and continuous [4,8].

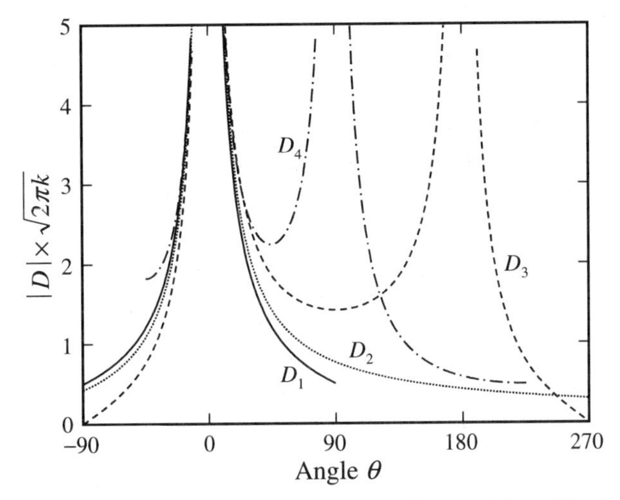

Figure 5-11 Variation with angle θ of the magnitude of the diffraction coefficients for (1) the Kirchhoff–Huygens analysis; (2) Felsen's absorbing screen; (3) the conducting half-plane (TE, $\phi' = 90°$), and; (4) the conducting 90° wedge (TM, $\phi' = 45°$). All coefficients are multiplied by the factor $\sqrt{2\pi k}$.

5.3a Absorbing screen

Felsen carried out a rigorous study of diffraction by wedges having absorbing faces [2]. When the internal wedge angle vanishes, the wedge becomes an absorbing half-screen, which is shown in Figure 5–10. His approach overcomes an inconsistency in the Kirchhoff–Huygens method that arises from the assumption that the total field just to the right of the screen is equal to the incident field above the edge, and zero below. [The diffracted field (5–43) is not zero in the plane of the screen ($\theta = -\pi/2$), which is inconsistent with the initial assumption.] For a plane wave incident at an angle ϕ', the results of the rigorous approach give the GTD diffraction coefficient

$$D = \frac{-1}{\sqrt{2\pi k}}\left[\frac{1}{\pi - |\phi - \phi'|} + \frac{1}{\pi + |\phi - \phi'|}\right] = \frac{-1}{\sqrt{2\pi k}}\left[\frac{1}{\theta} + \frac{1}{2\pi - \theta}\right] \tag{5–48}$$

where the angles ϕ, ϕ', and θ are in radians and are defined in Figure 5–10. Expression (5–48) applies for any polarization of the incident wave and for any interior wedge angle less than π.

When expressed in terms of $\theta = \pi - (\phi - \phi')$, the diffraction coefficient is independent of the angle ϕ' between the incident ray and the face of the wedge or screen. In addition, this result is valid even for $\theta > \pi/2$. The diffraction coefficient of (5–48) is compared to the Kirchhoff–Huygens coefficient (5–30) in Figure 5–11, where the angles have been expressed in degrees rather than radians. Note that in computing (5–48), θ must be in radians. The Kirchhoff–Huygens coefficient is plotted only up to 90° since the derivation is not valid for larger angles, while the Felsen coefficient is plotted up to 270°. It is seen from Figure 5–11 that the two coefficients are approximately equal when θ is small, where they both vary as $1/\theta$, but for larger values of $|\theta|$ they take on somewhat different values.

5.3b Conducting screen

If the screen is made of a thin conductor, the portion of the plane wave incident on the screen will be reflected, as in Figure 5–10. In this case the diffraction coefficient must be singular at the incident wave shadow boundary (ISB) and at the reflected wave shadow boundary (RSB). Furthermore, the diffraction coefficient will depend on the polarization of the incident wave. The GTD diffraction coefficient for a conducting screen is found to be [2,4]

$$D(\phi, \phi') = \frac{-1}{2\sqrt{2\pi k}}\left\{\frac{1}{\cos\frac{\phi-\phi'}{2}} + \frac{\Gamma_{E,H}}{\cos\frac{\phi+\phi'}{2}}\right\} \tag{5–49}$$

Here $\Gamma_E = -1$ is used for waves with $\mathbf{E}^{\mathrm{inc}}$ polarized parallel to the edge, and $\Gamma_H = +1$ is used for waves with $\mathbf{H}^{\mathrm{inc}}$ polarized parallel to the edge. The first term inside the brackets is singular at the shadow boundary of the incident wave where $\phi = \phi' + \pi$, while the second term is singular along the shadow boundary of the reflected wave where $\phi = \pi - \phi'$. When the receiver point is at a small angular displacement θ from either of the shadow boundaries, the corresponding term in (5–49) also varies as $1/\theta$, which is the same dependence observed in the GTD diffraction coefficients in (5–30) and (5–48).

The magnitude of the diffraction coefficient given by (5–49) is also plotted in Figure 5–11, with the angles expressed in degrees, for the TE polarization and assuming that $\phi' = 90°$ so that $\phi = 270° - \theta$. The coefficient (5–49) is seen to be close to (5–30) and (5–48) near the incident wave shadow boundary. It vanishes at $\theta = -90°$, which is the shadowed side of the conducting screen, and at the illuminated side of the conductor $\theta = 270°$, as called for by the boundary conditions on the tangential electric field. The reflected wave shadow boundary occurs at $\theta = 180°$, at which point (5–49) has the singular behavior seen in Figure 5–11.

5.3c Right-angle wedge

For wedges having finite wedge angle, the incident wave may illuminate one or both faces, depending on its direction of propagation, as shown in Figure 5–12 for a right-angle wedge. Thus, there can be an incident wave shadow boundary and a reflected wave shadow boundary, or two reflected wave shadow boundaries. If the wedge is conducting, the diffraction process will also depend on the polarization of the incident wave. For a right-angle conducting wedge [4] the GTD diffraction coefficient is

$$D(\phi, \phi') = D_1 + D_2 + \Gamma_{E,H}(D_3 + D_4) \tag{5–50}$$

where $\Gamma_{E,H}$ is as defined after (5–49) for $\mathbf{E}^{\mathrm{inc}}$ or $\mathbf{H}^{\mathrm{inc}}$ parallel to the edge, and

$$\begin{aligned} D_{1,2} &= \frac{-1}{3\sqrt{2\pi k}}\cot\frac{\pi \pm (\phi - \phi')}{3} \\ D_{3,4} &= \frac{-1}{3\sqrt{2\pi k}}\cot\frac{\pi \pm (\phi + \phi')}{3} \end{aligned} \tag{5–51}$$

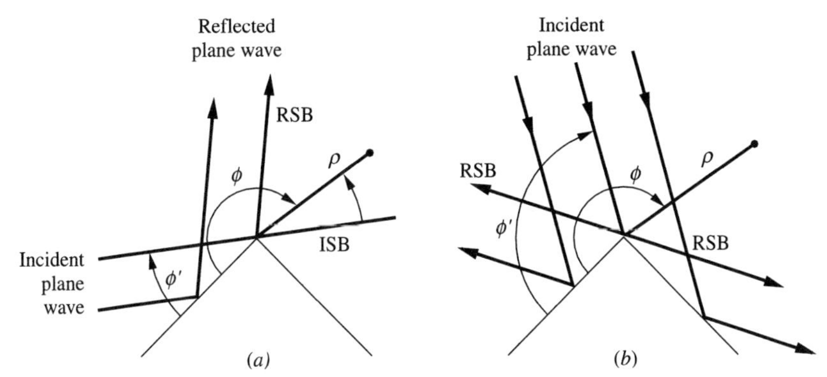

Figure 5-12 Depending on the direction of propagation, a plane wave incident on a 90° wedge will result in (a) shadow boundaries for the incident wave (ISB) and the plane wave reflected from the illuminated face (RSB), or (b) two shadow boundaries for the plane waves reflected from the two faces (RSB).

Depending on the direction of propagation of the incident wave, either D_1 or D_2 is singular along the incident wave shadow boundary, while D_3 or D_4 is singular along a reflected wave shadow boundary. As in previous cases, the diffraction coefficient varies as $1/\theta$ for small angular deviations from the shadow boundary. The UTD diffraction coefficient is found by multiplying each term in (5–50) by an appropriate transition function [4,8].

The magnitude of $D(\phi, \phi')$ from (5–50) is also plotted in Figure 5–11 for the TM polarization and assuming that the direction of incidence is like that shown in Figure 5–12a with $\phi' = 45°$. For this angle of incidence, $\phi = 225° - \theta$ and the diffraction coefficient is defined for $-45° < \theta < 225°$. Near the incident wave shadow boundary (5–50) is close to the other diffraction coefficients, and is singular on the reflected wave shadow boundary at $\theta = 90°$.

Diffraction by dielectric wedges is of interest for representing building corners. Transmission through the dielectric make such problems difficult to treat analytically. However, for absorbing dielectrics such as the corner of a brick building, transmission through the dielectric is weak, and the faces of the wedge can be modeled by means of a surface impedance. Solutions have been developed for this case, as, for example, in reference [9], although they do not appear to give significantly different results from those obtained using the diffraction coefficient for a conducting wedge [10]. A simple heuristic approach that is widely used is to replace the reflection coefficient of a perfect conductor in (5–49) and (5–50) with the plane wave reflection coefficients at the faces of the wedge [11]. Recall that if the wedge is assumed to have absorbing faces, Felsen's diffraction coefficient (5–48) is applicable.

5.3d Plane waves propagating oblique to the edge

If the direction of propagation of the incident plane wave makes an angle $\pi/2 - \psi$ with the edge, as indicated in Figure 5–13, all field components of the incident wave will have z dependence $\exp(-jkz \sin\psi)$. To satisfy the boundary conditions on the screen, the diffracted fields must have

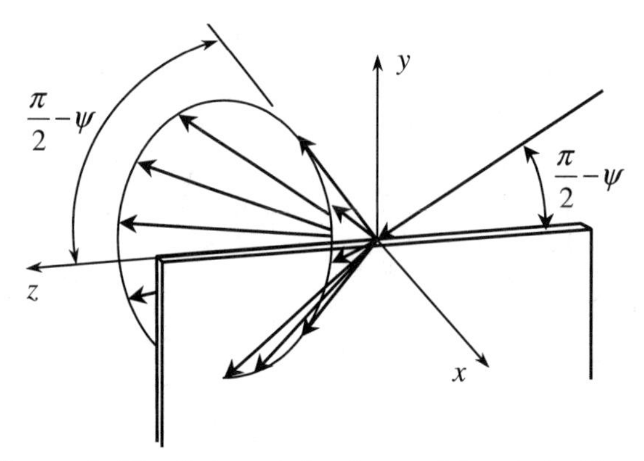

Figure 5-13 Cone of diffracted rays due to an obliquely incident ray making an angle $\pi/2 - \psi$ with the diffracting edge.

the same dependence on z. Thus, if we think of the plane wave fields as arising from a family of parallel rays, each ray that is incident on the edge will generate diffracted rays that originate from the point of incidence and lie on the surface of a cone, making an angle $\pi/2 - \psi$ with the edge, as in Figure 5–13. Also because of the z dependence, the wavenumber in the (x,y) plane is $k\cos\psi$. Except for the change in wavenumber, the electric and magnetic fields must satisfy the same differential equations and boundary conditions in the (x,y) plane as they would for normal incidence with $\psi = 0$. In the case of conducting boundary conditions, the incident wave must be decomposed into orthogonal polarizations, one having $\mathbf{E}^{\text{inc}}$ perpendicular to the edge, and the other having $\mathbf{H}^{\text{inc}}$ perpendicular to the edge [4, Chap. 6]. For wedges or screens with absorbing faces, this decomposition is not necessary since the diffraction coefficient is the same for both polarizations. Keeping in mind the polarization decomposition, the expressions previously developed for the dependence of the diffracted fields on ρ and θ remain valid with k replaced by $k\cos\psi$. Note, however, that ρ must still be interpreted as the radial or perpendicular distance from the edge to the receiver point, not the distance along a diffracted ray. Similarly, the angles ϕ, ϕ', and θ are to be measured in a plane perpendicular to the edge.

As an example, consider a conducting screen illuminated by a plane wave having unit amplitude, and propagating parallel to the (x,z) plane in Figure 5–13 at an angle ψ to the x axis. The angle of incidence, as measured from the screen face at $x = 0^-$ is $\phi' = \pi/2$ (see Figure 5–10). The diffraction angle $\phi = (3/2)\pi - \theta$, where $\theta = \arctan(y/x)$, so that in (5–49)

$$\cos\frac{\phi - \phi'}{2} = \cos\left(\frac{3\pi}{4} - \frac{\theta}{2} - \frac{\pi}{4}\right) = \sin\frac{\theta}{2} \qquad (5\text{–}52)$$

$$\cos\frac{\phi + \phi'}{2} = \cos\left(\frac{3\pi}{4} - \frac{\theta}{2} + \frac{\pi}{4}\right) = -\cos\frac{\theta}{2}$$

If the electric field of the incident wave is polarized along y, it is perpendicular to the edge, and we must use $\Gamma_H = +1$ in (5–49). Substituting these expressions into (5–49), along with replacing k by $k\cos\psi$, the GTD diffraction coefficient is $D(\theta)/\sqrt{\cos\psi}$, where

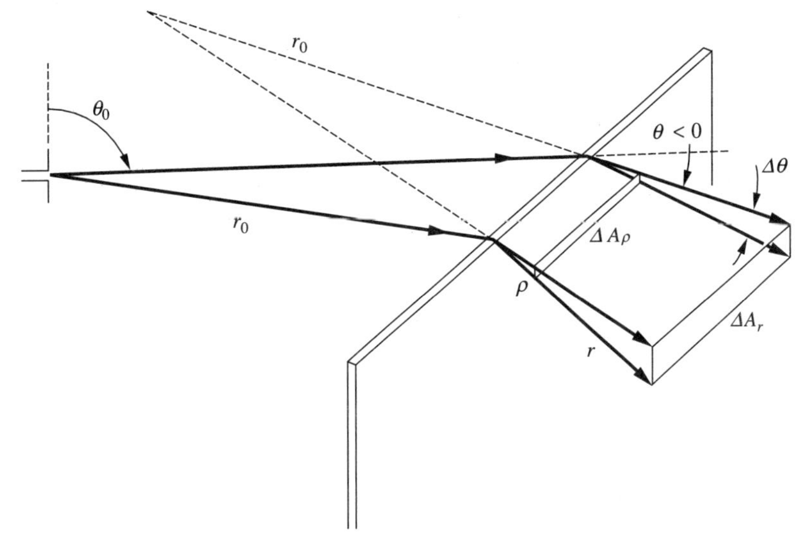

Figure 5-14 Rays used to describe the diffraction of a spherical wave by an edge.

$$D(\theta) = \frac{-1}{2\sqrt{2\pi k}}\left[\frac{1}{\sin(\theta/2)} - \frac{1}{\cos(\theta/2)}\right] \tag{5–53}$$

is the same as the coefficient for normal incidence ($\psi = 0$). The diffraction coefficient is singular at the incident wave shadow boundary $\theta = 0$ and at the reflected wave shadow boundary $\theta = \pi$. In terms of $D(\theta)$, the GTD expression for the diffracted field is given by

$$\mathbf{E}_D = \frac{e^{-jk(\rho\cos\psi + z\sin\psi)}}{\sqrt{\rho}} \frac{e^{-j\pi/4}D(\theta)}{\sqrt{\cos\psi}}(-\mathbf{x}_0\sin\theta + \mathbf{y}_0\cos\theta) \tag{5–54}$$

where $\rho = \sqrt{x^2 + y^2}$. The last factor in (5–54) gives the polarization of the electric field, which is perpendicular to the edge and is tangent to the cone of diffracted rays. If the incident wave had electric field polarized in the (x,z) plane, that is, along $\mathbf{x}_0\sin\psi - \mathbf{y}_0\cos\psi$, the magnetic field would be perpendicular to the edge. In this case $\Gamma_E = -1$ would be used in deriving (5–53), and the polarization of $\mathbf{E}_D$ would be perpendicular to the cone of diffracted rays and given by $\mathbf{x}_0\cos\theta\sin\psi + \mathbf{y}_0\sin\theta\sin\psi - \mathbf{z}_0\cos\psi$.

5.4 Diffraction of spherical waves

In Chapter 4 we have seen that antennas radiate electromagnetic fields in the form of spherical waves that propagate radially away from the source. The fields in the vicinity of a particular ray are like those of a plane wave. Thus when a ray encounters an edge, it will generate diffracted fields, which near the edge are like those generated by an incident plane wave. An example is shown in Figure 5–14, where an antenna illuminates an edge some distance away. The total field will be the sum of a geometrical optics (GO) component, which includes the incident and reflected waves, and a diffracted component generated at the edge. As a generalization of (5–43) to the case of spherical incident waves, we may write

$$\mathbf{E} = \mathbf{E}_{GO} + \mathbf{E}_D \tag{5–55}$$

Here the geometric optics field is composed of the incident spherical wave and reflected waves, which are also spherical and can be found from the image source, as in the case of ground reflection discussed in Chapter 4. The various spherical waves in (5–55) are understood to exist only within their appropriate illuminated regions.

5.4a Diffraction for rays incident at nearly right angles to the edge

The diffracted field E_D generated by the incident spherical wave can be found by considering two neighboring rays of nearly equal length r_0 from the antenna that are incident on the edge, as shown in Figure 5–14. Each ray generates diffracted rays lying on cones about the edge, as in Figure 5–13. Near the edge, where the radial distance ρ is small compared to r_0, the diffracted field will be the same as that produced by an incident plane wave. Hence the phasor field amplitude of the diffracted cylindrical wave will be

$$E_D(\rho, \theta) = \left[ZIf(\theta_0, \phi_0)\frac{e^{-jkr_0}}{r_0} \right]\left[e^{-j\pi/4}D(\theta)\frac{e^{-jk\rho}}{\sqrt{\rho}} \right] \tag{5–56}$$

for $\rho \ll r_0$. Here the terms in the first pair of brackets represent the field incident on the edge, while those in the second pair of brackets represent a cylindrical wave propagating away from the edge.

Near the edge the cylindrical wave in (5–56) satisfies the condition of conservation of power. However, at larger distances from the edge, the divergence of the ray cones excited by the two incident rays in Figure 5–14 must be taken into account to conserve power in a tube of rays. The divergence can be taken into account by considering the change in the cross-sectional area of a small tube of rays as they propagate away from the edge. To evaluate the area change most simply, assume that the two rays from the antenna in Figure 5–14 are nearly perpendicular to the edge so that in Figure 5–13 $\psi \approx 0$. A small tube of diffracted rays is shown in Figure 5–14 and is defined by two rays diffracted through slightly different angles θ and $\theta + \Delta\theta$ on each cone of diffracted rays. The upper rays on each cone appear to diverge from a point located at a distance r_0 behind the edge, as shown, as do the lower pair of rays. Thus the small area ΔA_ρ defined by the four rays will expand to the larger area ΔA_r at the distance r from the edge. Conservation of power in this small tube of rays requires that the ray fields vary inversely as the square root of the area. Thus in going from the distance ρ to the distance r, the magnitude of the field in (5–56) must be multiplied by the factor

$$\sqrt{\frac{\Delta A_\rho}{\Delta A_r}} = \sqrt{\frac{\rho}{r}\frac{r_0}{r_0 + r}} \tag{5–57}$$

The first ratio inside the square root on the right-hand side of (5–57) represents the vertical spreading of the rays in Figure 5–14, the second ratio represents the spreading in the plane parallel to the edge.

Multiplying (5–56) by (5–57), and accounting for the additional phase change as the field propagates from the distance ρ to the distance r, the diffracted field is found to be

$$E_D(r, \theta) = ZIf(\theta_0, \phi_0)e^{-j\pi/4}D(\theta)\frac{e^{-jk(r_0 + r)}}{\sqrt{r_0 r(r_0 + r)}} \qquad (5\text{–}58)$$

The product $r_0 r$ in the denominator in (5–58) can be thought of as due to the spreading of the rays in the vertical plane as they leave the antenna, and the spreading in the vertical plane as they leave the edge. The sum $r_0 + r$ is due to the ray spreading in the direction parallel to the edge. The rays used to construct (5–58) are said to be *astigmatic*, since they appear to diverge from different points in the plane perpendicular to the edge and in the plane parallel to the edge. The field in (5–58) is ray optical in nature, although it is not a spherical wave.

Near the shadow boundary, the GTD diffraction coefficient $D(\theta)$ must be multiplied by the transition function $F(S)$ given in (5–41). However, expression (5–36) for S must be modified to account for the phase-front curvature of the incident wave in the plane perpendicular to the edge. The result of a rigorous mathematical analysis [8] gives the following expression for this case:

$$S = 2k\frac{r_0 r}{r_0 + r}\sin^2(\theta/2) \qquad (5\text{–}59)$$

Since the transition function departs from unity only in the transition region, where θ is small, then using $\tan\theta \approx \theta$ in (5–36) and $\sin(\theta/2) \approx \theta/2$ in (5–59) leads to the same dependence on θ.

5.4b Diffraction for rays that are oblique to the edge

In Section 5–4d it was argued that the diffracted fields generated by rays incident on the edge at an angle $\pi/2 - \psi$, as shown in Figure 5–13, can be found by changing k into $k\cos\psi$ in the diffraction coefficient and in the transition function. For a spherical wave incident on the edge, the area ratio in (5–57) must also be modified to account for the fact that the distances r_0 and r are measured oblique to the edge, while ρ is measured perpendicular to the edge. This modification is accomplished by multiplying r_0 and r in (5–57) by $\cos\psi$, which is seen to have the net effect of dividing the area ratio by $\cos\psi$. For conducting boundary conditions, the polarization of the incident wave must be decomposed into two orthogonal components, one having electric perpendicular to the edge and the other having magnetic field perpendicular to the edge, as discussed in Section 5–3d. For wedges or screens with absorbing faces, the same diffraction coefficient applies for all polarizations of the incident wave.

Accounting for the $\cos\psi$ term in k and in the area ratio, the angle dependence of the diffracted field for oblique incidence can be written explicitly as

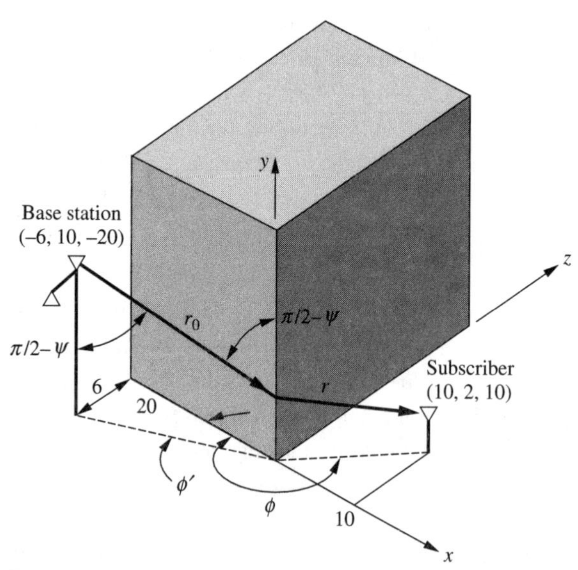

Figure 5-15 Path of a ray diffracted around a building corner from a lamppost-mounted base station antenna to a subscriber. All dimensions are in meters.

$$E_D(r, \theta) = ZIf(\theta_0, \phi_0)e^{-j\pi/4}\frac{D(\theta)}{\cos\psi}\frac{e^{-jk(r_0+r)}}{\sqrt{r_0 r(r_0+r)}} \tag{5–60}$$

where the diffraction coefficient $D(\theta)$ is given by the appropriate formula for normal incidence ($\psi = 0$). Note that r_0 and r are measured along the ray paths, but the angle θ is still measured in a plane perpendicular to the edge. The diffracted fields due to an incident plane wave can be recovered from (5–60) by taking r_0 to be much larger than r, and setting $A_0 = ZIFe^{-jkr_0}/r_0$.

An example of diffraction of oblique rays is shown in Figure 5–15, where the radiation at 1.8 GHz from a base station mounted at height $h_1 = 10$ m on a lamppost is diffracted around the corner of a building to a subscriber whose antenna is $h_2 = 2$ m above the ground. As measured in the plane perpendicular to the edge, the incident ray makes an angle $\phi' = \arctan(6/20) = 16.70°$ with the building face, while the diffracted ray makes an angle $\phi = 225°$. Assume that the base station antenna is vertically polarized, so that the magnetic field is parallel to the ground and hence perpendicular to the building edge. If the building is conducting, we must use $\Gamma_E = -1$ in (5–50), from which it is found that $D(\phi, \phi') = 0.987/\sqrt{2\pi k} = 0.0641$. The tangent of the angle $\pi/2 - \psi$ between these rays and the edge of the building is given by the path length in the horizontal plane $\sqrt{6^2+20^2} + 10\sqrt{2} = 35.02$ m divided by the vertical displacement of 8 m. Thus in degrees, $\psi = 90 - \arctan(35.02/8) = 12.87°$ and $\cos\psi = 0.975$. Finally, $r_0 = \sqrt{6^2+20^2}/\cos\psi = 21.42$ and $r = 10\sqrt{2}/\cos\psi = 14.50$. If the foregoing quantities are substituted into (5–60), then for an isotropic antenna with $|f(\phi, \phi')| = 1$, the magnitude of the field at the subscriber antenna is

$$|E_D| = |ZI|\frac{0.0641}{0.975\sqrt{21.42 \times 14.5 \times 35.92}} = 6.22 \times 10^{-4}|ZI| \text{ V/m}$$

The height y of the point of diffraction at the edge is given by $h_2 + 10\sqrt{2}\tan\psi$, or $y = 5.23$ m. It is seen in this example that cos ψ is close to unity, so that obliquity has only a small effect on the field. Thus, when the differences in height of the antennas is small compared to their horizontal separation, as in this example, diffraction around buildings can be approximated as occurring in the horizontal plane.

In the transition region (5–60) must be multiplied by the transition function $F(S)$, which is as defined in (5–41), with S given by [4, Chap. 6]

$$S = 2k\cos^2\psi\frac{r_0 r}{r_0 + r}\sin^2(\theta/2) \tag{5–61}$$

We argued previously that the width of the transition region is given by the condition $S = \pi$. As the angle ψ increases toward 90° in (5–61), $\cos\psi$ decreases to zero, requiring ever larger values of θ to achieve the condition $S = \pi$. In the limit as the incident ray approaches glancing incidence on the edge, the UTD field expressions must be used for all angles θ.

5.4c Path gain for wireless applications

In those regions of space that are illuminated by the geometrical optic fields, the diffracted fields are significant only near the shadow boundary. Exactly on the shadow boundary the diffracted field subtracts from the geometric optics field and the total field is one-half the geometrical optics field. Thus neglecting the diffracted field on the shadow boundary results in a 6-dB error. However, in the illuminated region outside the transition region, only a small error is made by neglecting the diffracted fields when computing the path loss.

In shadow regions the diffracted field provides the only contribution to the total field, and the path gain PG, which is the ratio of received power to transmitted power, can be computed using (4–5) and (4–33). With the help of expression (5–60) for the diffracted field, PG for isotropic antennas having $|f(\theta_0, \phi_0)| = 1$ is given by

$$\text{PG} = \left(\frac{\lambda}{4\pi}\right)^2\frac{|D(\theta)|^2}{\cos^2\psi}\frac{1}{r_0 r(r_0 + r)} \tag{5–62}$$

Recall that the λ^2 term in (5–62) represents the frequency dependence of the isotropic receiving antenna. However, PG has a additional dependence on frequency due to the fact that $|D(\theta)|^2$ is proportional to λ. In the example discussed after (5–60), $\lambda = 1/6$, and for an isotropic subscriber antenna the path gain is $(6.22\times 10^{-4})^2/(24\pi)^2$ or -101.7 dB.

As a basis for examining the effect of diffraction on the path gain in (5–62), let PG_o be the path gain between isotropic antennas that are separated by the same total distance $r_0 + r$ in free space. The effect of diffraction through an angle θ is then

$$\frac{\text{PG}}{\text{PG}_o} = \frac{|D(\theta)|^2}{\cos^2\psi}\frac{r_0 + r}{r_0 r} \tag{5–63}$$

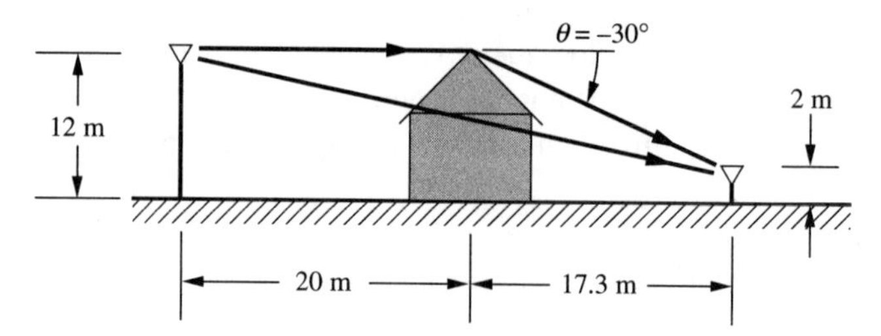

Figure 5-16 Communication paths between a base station and a mobile in the presence of a three-story townhouse.

An example is shown in Figure 5–16 for diffraction by a building of 900-MHz signals from a base station at the same height as the building. In this example the diffraction angle to the mobile is $\theta = -30°$ and we suppose that $\psi = 0$ and $r_0 = r$. Using the Felsen diffraction coefficient (5–48) and expressing (5–63) in decibels, the ratio is equivalent to –25.8 dB. Thus the path loss is 25.8 dB in excess of the free-space propagation loss. This is comparable to the excess wall loss incurred for the direct ray from the base station to the mobile that goes through the building if the exterior walls are made with brick or aluminized vapor barrier insulation and the windows are fitted with full screens. Somewhat more complex versions of this example have been used to study blockage of satellite paths by buildings [12] and terrestrial links by trucks [13]. For low antennas among high-rise buildings, signals propagate down the streets and around the sides of the buildings. While reflection assists the signals to turn corners into perpendicular streets, diffraction is the primary mechanism for illumination down the street and is associated with significant loss. This process is discussed in Chapter 8.

5.5 Diffraction by multiple edges

If the diffracted field generated by one edge is incident on a second edge, it will in turn produce diffracted fields that can be found by repeating the ray approach given above. Because the diffracted field of (5–58) or (5–60) has been constructed using ray arguments that are valid outside the transition region about the shadow boundary, cascading the geometrical optics diffraction events is valid when subsequent edges lie outside the transition region of the previous edge. When this is not the case, it is necessary to consider additional diffraction effects [14]. In this section we investigate diffraction by two edges for cases of special interest in wireless when the two edges are parallel, corresponding to a wave going over the top of a building or around the sides of a building, or when one edge is horizontal and one is vertical. To avoid the complexity associated with the polarization of the fields diffracted by the first edge as they encounter the second edge, we assume the wedges or screens to have absorbing faces, in which case the same diffraction coefficient applies to all components of the fields.

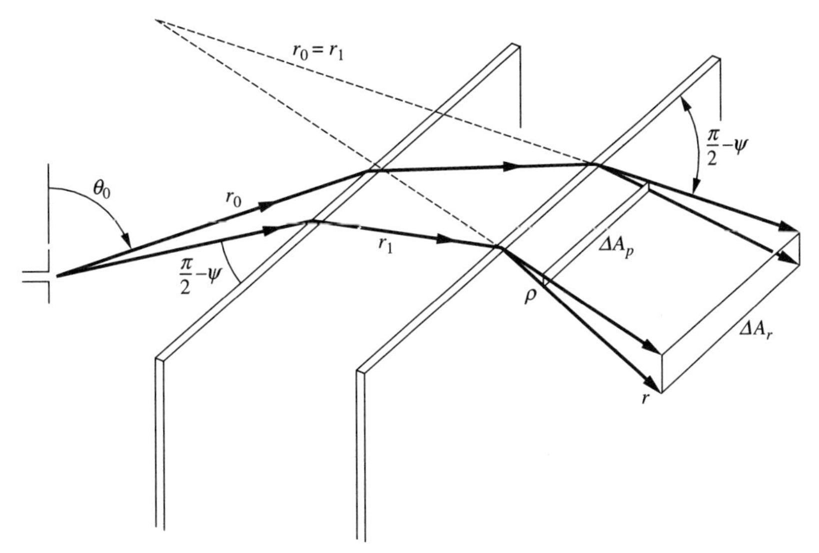

Figure 5-17 Rays used to describe the double diffraction of a spherical wave by two parallel edges.

5.5a Two parallel edges

Diffraction of a spherical wave over two parallel edges is shown in Figure 5–17 for rays making an angle $\pi/2 - \psi$ to the edges. The field E_{in} incident on the second edge is given by (5–60) with (r,θ) replaced by (r_1,θ_1), and it propagates at the same angle $\pi/2 - \psi$ to the second edge. Near the second edge, where the radial distance ρ is small compared to $r_0 + r_1$, the fields are like those of a cylindrical wave, and are given by

$$E_D(\rho, \theta) = E^{\text{inc}} e^{-j\pi/4} \frac{D_2(\theta)}{\sqrt{\cos\psi}} \frac{e^{-jk\rho\cos\psi}}{\sqrt{\rho}} \tag{5–64}$$

where $D_2(\theta)$ is the diffraction coefficient of the second edge. The factor $\cos\psi$ in (5–64) comes from the fact that the ray is incident oblique to the second edge, which causes the wavenumber transverse to the edge to be $k\cos\psi$. In (5–64) the diffraction angle θ is measured in a plane perpendicular to the edge. For propagation along the diffracted ray, the phase variation in (5–64) is replaced by $\exp(-jkr)$, which includes the phase change due to both the component of displacement radially away from the edge and the component parallel to the edge.

As the fields propagate away from the edge, the power in a tube of rays is spread over a larger area, as suggested in Figure 5–17. To conserve power, the field of (5–64) must be multiplied by the square root of the area ratio $\Delta A_\rho/\Delta A_r$. In the vertical plane the diverging rays come from the edge, whereas in the plane parallel to the edge, the diverging rays appear to come from a point located a distance $r_0 + r_1$ behind the edge. Thus the area ratio is given by

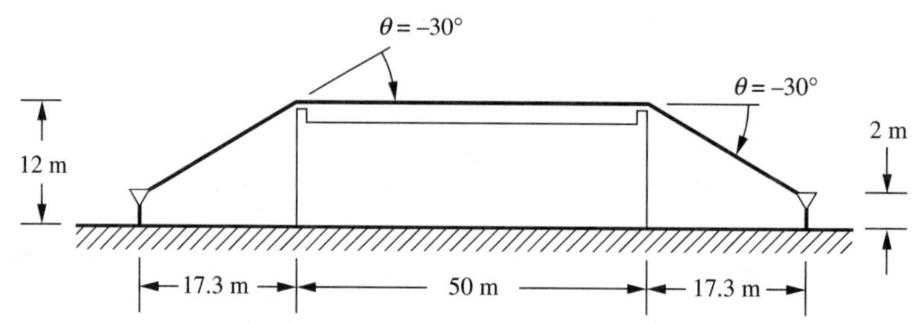

Figure 5-18 Mobile-to-mobile communication when the mobiles are located on either side of a large three-story building.

$$\sqrt{\frac{\Delta A_\rho}{\Delta A_r}} = \sqrt{\frac{\rho}{r\cos\psi}\frac{r_0+r_1}{r_0+r_1+r}} \tag{5–65}$$

where the perpendicular distance $r\cos\psi$ is used to find the spreading in the vertical plane.

Multiplying (5–64) by (5–65), substituting (5–60) for E^{inc}, and accounting for the phase change along the diffracted ray, the diffracted field is

$$E_D(r,\theta) = ZIf(\theta_0,\phi_0)e^{-j\pi/2}\frac{D_1(\theta_1)D_2(\theta)}{\cos^2\psi}\frac{e^{-jk(r_0+r_1+r)}}{\sqrt{r_0r_1r(r_0+r_1+r)}} \tag{5–66}$$

Expression (5–66) represents the GTD solution for the diffracted fields and is valid outside the transition region. Recalling the discussion in Section 4-b, the transition region expands as ψ approaches 90°, and in the limit includes the entire θ domain. Thus care must be taken in using (5–66) for large values of ψ. The foregoing process can be repeated for more edges, provided that each edge lies outside the transition region of the previous edge. In Chapter 6 we treat the case when each edge is in the transition region of the previous edge, which is important for modeling cellular and PCS propagation.

As was done in Section 5–4c, the path gain for isotropic antennas is found from (5–66) to be

$$\mathrm{PG} = \left(\frac{\lambda}{4\pi}\right)^2\frac{|D_1(\theta_1)D_2(\theta)|^2}{\cos^4\psi\ [r_0r_1r(r_0+r_1+r)]} \tag{5–67}$$

An example of double diffraction over a large three-story building for communication between two mobiles is shown in Figure 5–18 for $\psi = 0$. In this case the diffraction angles are $\theta_1 = \theta = -30°$ and the path segments are $r_0 = r = 20$ m and $r_1 = 50$ m. Using the Felsen diffraction coefficient given by (5–48), the path loss at 900 MHz found from (5–67) is 55 dB greater than if the two antennas were located 90 m apart in free space. This excess loss can be compared to the wall penetration loss for propagation directly through the building. Two exterior walls having brick or aluminized insulation will give about 20 dB excess loss. If the building is divided by interior walls spaced on average 4 m apart, and each having 3 dB loss, the 11 interior and two exterior

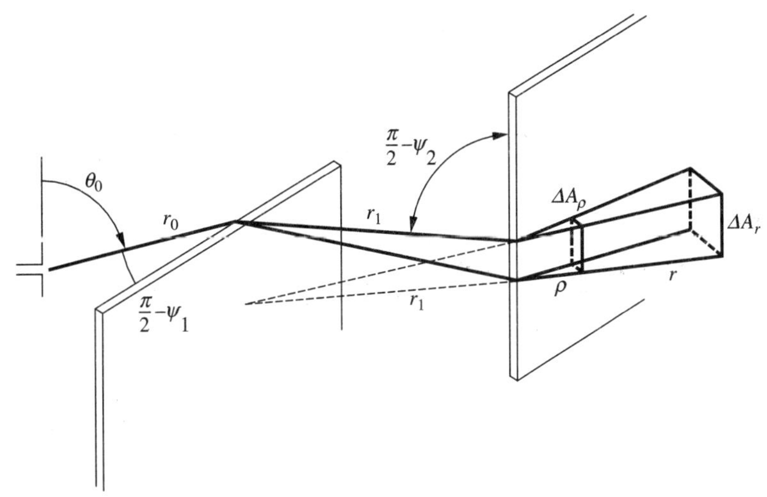

Figure 5-19 Rays used to describe the double diffraction of a spherical wave by a horizontal building edge and a vertical corner.

walls will give a total excess path loss of 53 dB. Although the wall loss values have been chosen somewhat arbitrarily, this example demonstrates that diffraction paths over or around buildings may have the same or even less excess loss than paths that go through buildings.

5.5b Two perpendicular edges

The geometry for diffraction over a horizontal edge, such as the roof of a building, and around a vertical building edge is shown in Figure 5–19. The field E^{inc} incident on the second edge is given by (5–60) with (r,θ) replaced by (r_1,θ_1). At radial distances ρ from the second edge that are small compared to r_1, the fields diffracted by it are cylindrical and given by

$$E_D(\rho, \theta) = E^{inc} e^{-j\pi/4} \frac{D_2(\theta)}{\sqrt{\cos\psi_2}} \frac{e^{-jk\rho\cos\psi_2}}{\sqrt{\rho}} \tag{5–68}$$

The factor $\cos\psi_2$ in (5–68) results from ray incidence oblique to the second edge, which causes the wavenumber transverse to the edge to be $k\cos\psi_2$. In (5–68) the diffraction angle θ is measured in a plane perpendicular to the edge. For propagation along the ray, the phase variation in (5–68) is replaced by $\exp(-jkr)$, which includes both the component of displacement radially away from the edge and the component parallel to the edge.

Subsequent ray spreading reduces the amplitude by the factor

$$\sqrt{\frac{\Delta A_\rho}{\Delta A_r}} = \sqrt{\frac{\rho}{r\cos\psi_2}\frac{r_1}{r_1 + r}} \tag{5–69}$$

Accounting for the ray spreading and the phase change along the ray, the doubly diffracted field is

$$E_D(r, \theta) = ZIf(\theta_0, \phi_0)e^{-j\pi/2}\frac{D_1(\theta_1)D_2(\theta)}{\cos\psi_1\ \cos\psi_2}\frac{e^{-jk(r_0 + r_1 + r)}}{\sqrt{r_0 r(r_0 + r_1)(r_1 + r)}} \tag{5–70}$$

Aside from the somewhat different dependence on the lengths of the ray segments, expression (5–70) is much like that of (5–66), so that both will give similar path loss for the same diffraction angles.

5.6 Summary

In this chapter we introduced the use of the Kirchhoff–Huygens approach to find the fields diffracted by an aperture. With it we have shown how the fields arriving at a point propagate in a limited region, called the Fresnel zone, about the ray path. A direct consequence of the local nature of the propagation is the need for ground clearances of the Fresnel zone on line-of-sight paths for point-to-point communications. Both the approach and the local properties we have found will be used in later chapters when developing models for propagation in built-up regions. Using the Kirchhoff–Huygens approach, we have investigated the problem of plane wave diffraction by a half-screen. This problem is a prototype for the diffraction of other high-frequency fields for which the wavelength is small compared to the distances. Its solution represents the diffracted field as a cylindrical wave propagating away from the edge with direction dependence given by the diffraction coefficient.

Studies of other diffracting objects using more rigorous mathematical approaches lead to the same representation for the diffracted field, with a difference only in the diffraction coefficient. When coupled with a description of propagation in terms of rays, the results for plane wave diffraction can be generalized to spherical waves radiated by antennas. These results allow evaluation of the excess loss due to diffraction, which was found to be significant in practical situations. The effects of diffraction can be cascaded for several diffracting edges, provided that each edge is not in the transition region about the shadow boundary of the previous edge. In Chapter 6 we consider the case when the edges are in the transition region.

Problems

5.1 Compute $|D(\theta)|/\sqrt{\rho}$ at the boundaries of the transition region in Figure 5–9, where $y = \pm\sqrt{\lambda\rho}$. Use the GTD diffraction coefficients of (5–30), (5–48), and (5–49), in the latter case assuming that $\phi' = 45°$.

5.2 For comparison, plot the magnitudes of the GTD and UTD forms of the diffraction coefficients in (5–30) and (5–44) over the range $|\theta| < \pi/2$ for $f = 300$ MHz and $\rho = 36$ m. Indicate the width of the transition region defined by the condition $S \le \pi$.

5.3 In Figure 5–16, assume that the base station height is 10 m rather than the 12 m shown. Compute the path gain in decibels, PG_{dB} for isotropic antennas at 3.5 GHz using the Felsen diffraction coefficient.

5.4 In Figure 5–18, assume that the antenna on the left is raised from 2 m to 60 m. For a frequency of 2.4 GHz and isotropic antennas, compute and plot the path gain PG_{dB} as a function of the antenna

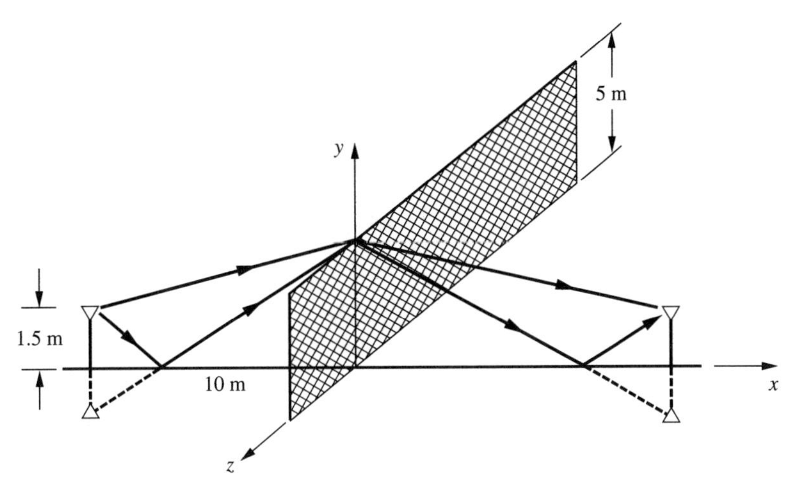

Figure P5-5

height. Use the Felsen diffraction coefficient for an absorbing screen to represent each edge of the building. Note that the antenna will pass two shadow boundaries so that the UTD forms of the two diffraction coefficients must be used.

5.5 Two personal communicators operating in the 900-MHz unlicensed band use vertically polarized, isotropic antennas that are 1.5 m above a flat earth, as shown in Figure P5–5. They are separated by a chain link fence, the one on the left being 10 m from the fence with its antenna at (–10, 1.5, 0), the antenna on the right being at (x, 1.5, 0) with $10 \le x \le 100$ m.

(a) Show that the Fresnel zone about the direct ray between the two antennas, in the absence of the fence, is completely blocked when the fence is present. As a result, diffraction over the top of the fence may be computed without accounting for the transition function.

(b) Accounting only for the ray going directly over the top of the fence, compute and plot the path gain PG_{dB} as a function of x in the range $10 \le x \le 100$ m. Use the diffraction coefficient (5–48) for a conducting edge.

5.6 If ground reflections are taken into account in part (b) of Problem 5–5, there are three rays in addition to the one going directly over the top. One of the additional rays appears to go from the first antenna to the image of the second, another from the image of the first antenna to the second, and the third from the image of the first to the image of the second. Repeat part (b) of Problem 5–5 including these three rays.

5.7 In Figure P5–5, the communicator on the right moves parallel to the fence at a distance of 10 m, so that the antenna is at (10,1.5,z), with $0 \le z \le 100$ m. Accounting only for the ray going directly over the top of the fence, compute and plot the path gain PG_{dB} as a function of z in this range. Neglect polarization coupling at the edge and use the diffraction coefficient (5–49) with $\Gamma_H = +1$.

5.8 In the geometry shown in Figure P5–8, a 1.8-GHz PCS subscriber antenna is 1 m above the flat earth and is located around the corner of a conducting building and behind a 5-m-high brick wall. The base station is mounted on a lamppost at a height of 10 m. One ray reaching the subscriber must diffract around the corner of the building and over the top of the brick wall, which can be treated as an absorber. The antenna is vertically polarized, so that the magnetic field of the incident ray is perpendicular to the building corner. A complicating factor in this problem is the need to find the value of y at which the ray is diffracted at the building corner and the value of x at which it is diffracted along

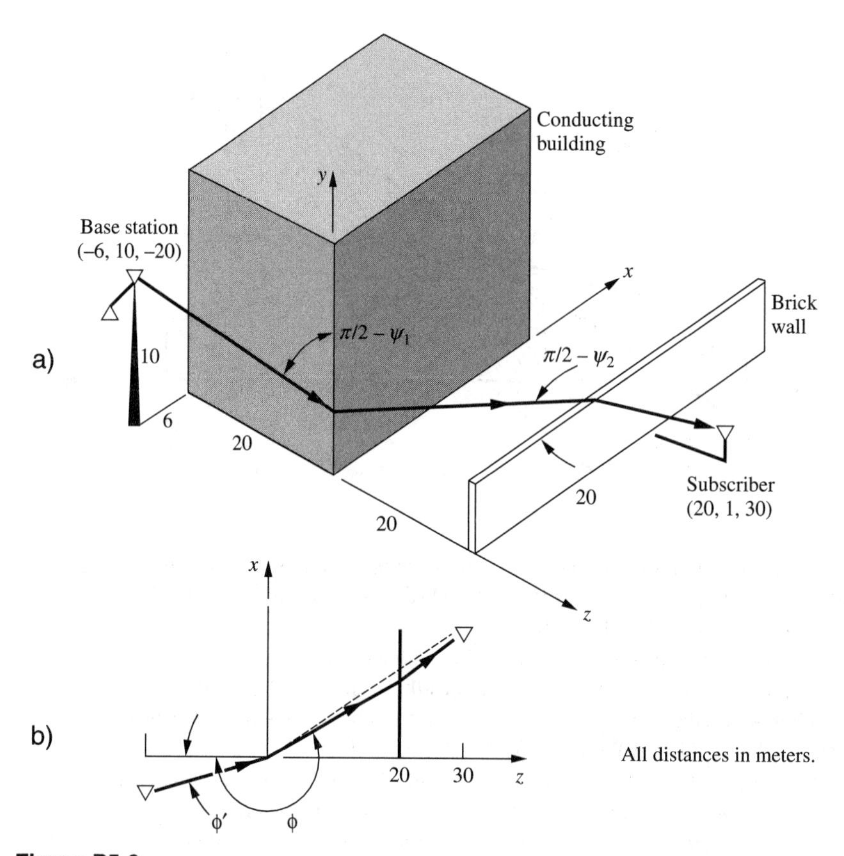

Figure P5-8

the wall. Equations whose solutions give these values can be found by requiring both the incident and diffracted rays to make the same angle with the edge in question. For example, at the building corner one has $\tan\psi_1 = (10 - y)/\sqrt{6^2 + 20^2} = (y - 5)/\sqrt{x^2 + 20^2}$.

(a) Find a second equation in x and y by considering $\tan\psi_2$. You will see that these equations are not easily solved for x and y.

(b) Because the ray diffracted at the brick wall lies on the surface of a cone of angle $\pi/2 - \psi_2$, as seen from above in Figure P5–8b, the projection of the diffracted ray appears to make a slight turn. An approximate solution for x is obtained using the straight line from the corner of the building to the subscriber. Find this value of x and substitute it into the formula above to find y. A correction to the value of x is obtained by substituting the value just obtained for y into the equation derived in part (a). Find the corrected value of x.

(c) Using either value of x and the appropriate GTD diffraction coefficients, compute the path gain PG_{dB} for isotropic antennas.

References

1. B. B. Baker and E. T. Copson, The Mathematical Theory of Huygens' Principle, 2nd ed., Oxford University Press, London, 1953.

2. L. B. Felsen and N. Marcuvitz, Radiation and Scattering of Waves, Prentice Hall, Upper Saddle River, N.J., 1973.
3. R. C. Hansen, ed., Geometrical Theory of Diffraction, IEEE Press, New York, 1981.
4. D. A. McNamara, C. W. I. Pistorius, and J. A. G. Malherbee, Introduction to the Uniform Geometrical Theory of Diffraction, Artech House, Norwood, Mass., 1990.
5. M. Born and E. Wolf, Principles of Optics, Pergamon Press, New York, Chap. 8, 1959.
6. C. A. Balanis, Advanced Engineering Electromagnetics, Wiley, New York, App. A, 1989.
7. M. Abramowitz and I. A. Stegun, eds., Handbook of Mathematical Functions, Dover Publications, New York, pp. 301–302, 1965.
8. R. F. Kouyoumjian and P. H. Pathak, A Uniform Geometrical Theory of Diffraction for an Edge in a Perfectly Conducting Surface, *Proc. IEEE*, vol. 62, pp. 1448–1461, 1974.
9. Y. Hwang, Y. P. Zhang, and R. F. Kouyoumjian, Ray-Optical Prediction of Radio-Wave Propagation in Tunnel Environments, Part 1: Theory, *IEEE Trans. Antennas Propagat.*, vol. 46, pp. 1328–1336, 1998.
10. Y. P. Zhang, Y. Hwang, and R. F. Kouyoumjian, "Ray-Optical Prediction of Radio-Wave Propagation in Tunnel Environments, Part 2: Analysis and Measurements, *IEEE Trans. Antennas Propagat.*, vol. 46, pp. 1337–1345, 1998.
11. R. J. Luebbers, Finite Conductivity Uniform GTD versus Knife Edge Diffraction in Prediction of Propagation Path Loss, *IEEE Trans. Antennas Propagat.*, vol. AP-32, pp. 70–76, 1984.
12. P. A. Tirkas, C. M. Wangsvick, and C. A. Balanis, Propagation Model for Building Blockage in Satellite Mobile Communication Systems, *IEEE Trans. Antennas Propagat.*, vol. 46, pp. 991–997, 1998.
13. A. J. Rustako, Jr., M. J. Gans, G. J. Owens, and R. S. Roman, Attenuation and Diffraction Effects from Truck Blockage of an 11-GHz Line-of-Sight Microcellular Mobile Radio Path, *IEEE Trans. Veh. Technol.*, vol. VT-40, pp. 211–215.
14. J. B. Andersen, UTD Multiple-Edge Transition Zone Diffraction, *IEEE Trans. Antennas Propagat.*, vol. 45, pp. 1093–1097, 1997.

CHAPTER 6

Propagation in the Presence of Buildings on Flat Terrain

In this chapter we integrate the propagation, reflection, and diffraction concepts of the previous chapters in order to understand and predict the path loss characteristics that have been observed over large portions of metropolitan regions. The goal here is to predict the range dependence for outdoor propagation and show how it depends on the system parameters, such as frequency and antenna height, and on the geometry of the builtup environment. In trying to understand and model propagation over builtup regions, it is first necessary to identify the features of the environment that are significant for radio propagation, especially the size and organization of the buildings.

To a visitor, a city is identified by its bridges, parks, tall buildings, and other unique features that are the settings for vacations and entertainment. However, these features occupy only a small portion of any metropolitan area. Away from the high-rise core, cities are composed of rows of buildings having nearly uniform height, with occasional high buildings. The buildings are taller and closer together near the core, becoming lower and more spread out near the edge of the city and into its suburbs, and finally turning into isolated houses and other buildings in the far suburbs and surrounding rural regions. Except for the distant suburbs and rural regions, land costs result in a close side-to-side spacing between neighboring houses along the access streets. In older regions of some cities the neighboring buildings are attached to each other with no gap between them; a design that has been picked up in modern townhouse developments. In many newer parts of cities the spacing between neighboring houses is less than the width of the houses. The houses are thus organized in rows along the streets, whose presence imposes a significant spacing between the fronts of the houses. Similarly, there is ample back-to-back spacing to accommodate a yard. As seen from the air, the separation of houses across the streets is about the same as the back-to-back separation across the yards, irrespective of the age or location of the housing development.

The layout of streets in the metropolitan region has much to do with the history of its founding and it development. Older cities that have grown slowly by themselves without planning or strong central authority have meandering streets and cul-de-sacs that enhance security. Some modern suburban developments have such street systems, with one of the stated goals being to make it difficult for vehicular traffic to move freely through them. Cities founded by generals or emperors are laid out with rectangular street grids for planning purposes and to facilitate central control. In some cities the orientation of the street grid was viewed as having mystical significance. The rapid development of cities in the United States left its own historical mark. For example, the Brooklyn and Queens sections of New York city grew from many small communities that had established local rectangular grids, each with its own orientation. As the population increased, the streets grids were extended until they collided, leaving a merged street pattern that is irregular when viewed overall, but rectangular on a local scale. In midwestern and western states of the United States, a rectangular road grid with 1-mile spacing was established to divide the land for European settlers and to provide access. Many cities in these states adopted the rectangular grid and further subdivided the land with parallel roads.

A simplified picture of the metropolitan area described above has a core of high-rise buildings surrounded by lower buildings that are of nearly uniform height over large areas. The street grid, or portions of it, forms a rectangular mesh that divides the land into blocks. The narrow width of these blocks ranges from 80 m or less up to more than 120 m. A row of buildings lines each side of the block and sometimes the ends of blocks as well. The length of a block is two or three times its width. The buildings are organized along the street grid with little or no side-to-side spacing and nearly equal front-to-front and back-to-back spacing. Propagation in the core can be modeled using ray tracing algorithms, as discussed in Chapter 8. Outside the core area, the base station antennas are near to or above the rooftops, and propagation takes place primarily over the buildings, as discussed in the remainder of this chapter. Because the Fresnel zone is narrow, when viewed from above the Fresnel zone appears to cross rows of buildings, with the spacing between rows being roughly equal, even when the street grid is not rectangular.

6.1 Modeling propagation over rows of low buildings

First consider the simple case depicted in Figure 6–1, where the propagation path is perpendicular to the rows of buildings, which are of nearly equal height and are located on flat terrain. In later sections of this chapter we consider propagation oblique to the street grid. Shadow fading due to irregular building height, as well as terrain and the effect of trees on the path loss, are discussed in the Chapter 7. Assuming the buildings to be of the same height, each row of buildings can be represented by a horizontal bar of height H_B above the ground, as shown in Figure 6–1 for buildings with peaked roofs. Since there are two such rows of buildings per city block, the rows are separated by a distance d that is about 40 to 60 m. The base station antenna is at a height h_{BS} above the ground and the subscriber (mobile) antenna at a height h_m, which is usually lower than the surrounding buildings.

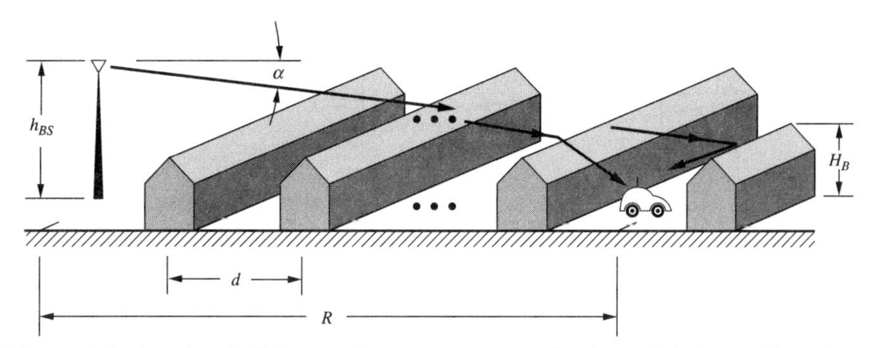

Figure 6-1 In a low building environment, propagation is modeled as taking place over the intervening rows of buildings, shown as uniform bars, with diffraction down to street level occurring at the rows near the subscriber.

Recall from Chapter 2 that the measured penetration loss into suburban houses is 4 to 7 dB at 800 MHz and higher at higher frequencies. Thus transmission loss through a row of houses (i.e., into and out of a house) will be 8 to 14 dB. More substantial buildings will give an even higher transmission loss through a row. There are also many rows of buildings per kilometer (a row spacing of 50 m, for example, corresponds to 20 rows per kilometer). Because of the high attenuation per row and the large number of rows per kilometer, the radio signal that arrives at a distant subscriber after passing through the buildings experiences a very large reduction in strength compared to free-space propagation. For example, for a distance of 0.5 km the signal would pass through 10 rows of houses giving a path loss of 80 to 140 dB in excess of free space. As suggested in Chapter 5, propagation paths in which the waves go over the buildings and are diffracted down to the subscriber may involve less excess path loss. Two such paths are indicated in Figure 6–1.

6.1a Components of the path gain

Taking into account the foregoing discussion, propagation from a elevated base station antenna to a subscriber is viewed as taking place over the tops of the intervening rows of buildings, with diffraction down to the subscriber occurring at the nearby rooftops [1,2]. With this viewpoint, the path gain is the product of three factors: (1) the free-space path gain PG_0, (2) the reduction PG_1 in the fields arriving at the buildings near the mobile due to diffraction past the previous rows of buildings, and (3) the reduction PG_2 in the fields as they diffract down to ground level. Thus when expressed in decibels, the path gain is

$$PG_{dB} = 10 \log PG_0 + 10 \log PG_1 + 10 \log PG_2 \tag{6–1}$$

while the path loss in decibels is the negative of (1). The free-space path gain was found in Chapter 4. For horizontal separations R that are large compared to the antenna height, the free-space path gain for isotropic antennas is

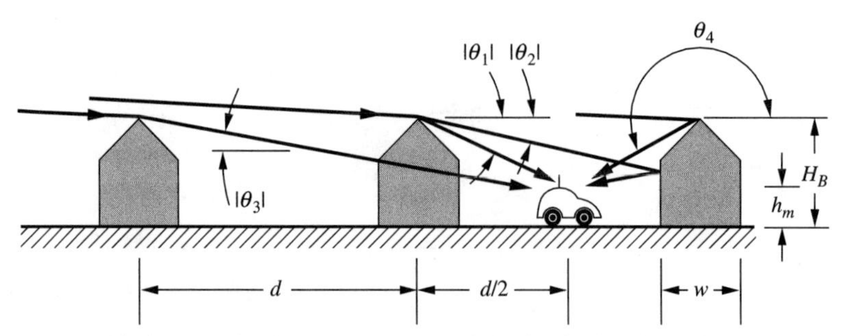

Figure 6-2 Various paths by which the rooftop fields can be diffracted down to street level for buildings that have peaked roofs running parallel to the rows.

$$\mathrm{PG}_0 = \left(\frac{\lambda}{4\pi R}\right)^2 \tag{6–2}$$

Expressions for PG_2 are obtained below by making use of the of the results derived in Chapter 5. However, PG_1 is more difficult to evaluate since the diffraction takes place past many rows, and each diffracting edge lies in the transition region of the previous edge. Methods for the evaluation of PG_1 are discussed in subsequent sections.

6.1b Modeling PG_2 by diffraction of the rooftop fields

Four of the possible paths by which the rooftop fields can be diffracted down to a street-level subscriber are shown in Figure 6–2 for buildings with peaked roofs. The most direct contribution is diffracted at the rooftop just before the subscriber. A second contribution is reflected back to the subscriber from the building across the street, while a third is diffracted from the rooftops two rows before the mobile and passes through the building just before the mobile. The fourth contribution comes from back diffraction at the rooftop just past the mobile. Although the exact geometry of these paths will depend on the roof shapes, for simplicity in comparing the various contributions and in drawing Figure 6–2, we have assumed all the roofs to be peaked, with the peaks running parallel to the street. Because our goal is to predict the range dependence and variability of the sector averages, we add the powers of the individual rays, rather than the fields, as discussed in Chapter 2.

For typical radio paths, the base station antenna height is small compared to the horizontal distance R, so that wave propagation over the rooftops is almost horizontal. In a later section we show that after propagating over a number of rows, the fields incident on the two rooftops nearest the mobile will have almost the same amplitude. The fields also have a variation with height near the rooftops that affects the diffracted fields. However, for simplicity in finding the path gain associated with diffraction down to street level from the two rooftops in Figure 6–2, we take the fields over both buildings to be those of horizontally propagating plane waves of unit amplitude. With the foregoing assumptions, PG_2 for isotropic antennas is given by

$$\mathrm{PG}_2 = \frac{1}{\rho_1}|D(\theta_1)|^2 + \frac{|\Gamma|^2}{\rho_2}|D(\theta_2)|^2 + \frac{|T|^2}{\rho_3}|D(\theta_3)|^2 + \frac{1}{\rho_4}|D(\theta_4)|^2 \tag{6–3}$$

where $|\Gamma|^2$ gives the reflection loss at the farther building, and $|T|^2$ gives the transmission loss through the building before the subscriber. Referring to Figure 6–2, we assume the subscriber to be located midway between the rows, and let w be the depth of the building perpendicular to the street. The distances ρ_i in (6–3) from the diffracting edges to the subscriber are then given by

$$\begin{aligned}
\rho_1 &= \rho_4 = \sqrt{(H_B - h_m)^2 + (0.5d)^2} \\
\rho_2 &= \sqrt{(H_B - h_m)^2 + (1.5d - w)^2} \\
\rho_3 &= \sqrt{(H_B - h_m)^2 + (1.5d)^2}
\end{aligned} \tag{6–4}$$

while the angles θ_ι are

$$\theta_i = -\arcsin\frac{H_B - h_m}{\rho_i} \qquad \text{for } i = 1, 2, 3 \tag{6–5}$$

$$\theta_4 = \pi + |\theta_1|$$

The boundary conditions at the peaked roof may be conducting or absorbing. However, for the small bending angles associated with paths 1, 2, and 3, the diffraction coefficients for either boundary condition are nearly the same. For simplicity we use the Felsen coefficient (5–48) for an absorbing screen, whose magnitude squared for the four rays can be approximated by

$$\begin{aligned}
|D(\theta_i)|^2 &= \frac{1}{2\pi k}\left(\frac{1}{\theta_i} + \frac{1}{2\pi - \theta_i}\right)^2 \approx \frac{1}{2\pi k}\frac{1}{|\theta_i|^2} \qquad \text{for } i = 1, 2, 3 \\
|D(\theta_4)|^2 &= \frac{1}{2\pi k}\left(\frac{1}{\pi + |\theta_1|} + \frac{1}{\pi - |\theta_1|}\right)^2 \approx \frac{1}{2\pi k}\left(\frac{2}{\pi}\right)^2
\end{aligned} \tag{6–6}$$

The error introduced by the approximation indicated in (6–6) is less than 14% even for angles approaching 45°. Using these approximations in (6–3) and recalling that $\rho_4 = \rho_1$,

$$\mathrm{PG}_2 = \frac{1}{2\pi k}\left[\frac{1}{\rho_1|\theta_1|^2} + \frac{|\Gamma|^2}{\rho_2|\theta_2|^2} + \frac{|T|^2}{\rho_3|\theta_3|^2} + \frac{1}{\rho_1(\pi/2)^2}\right] \tag{6–7}$$

Since $|\theta_1|$ is substantially less than $\pi/2$, the last term in (6–7) is smaller than the first term and will be neglected. If w is a small fraction of d, the distance/angle pairs (ρ_2,θ_2) and (ρ_3,θ_3) are nearly equal, as seen from (6–4) and (6–5), and the relative sizes of the second and third terms in (6–7) are given by the reflection and transmission coefficients. For the cases considered in Chapter 3, a typical number for $|\Gamma|^2$ is 0.1, corresponding to a 10-dB loss. Transmission loss $|T|^2$ through houses is on the order of 10 dB or more. Thus the third term in (6–7) is about the same as the second term, or less than the second term, depending on the building construction. Finally, to compare the second and third terms in (6–7) with the first term, we further approximate the angles $|\theta_i|$ for $i = 1,2,3$ by $\sin|\theta_i|$. Thus, neglecting the last term in (6–7), and substituting for $\sin|\theta_i|$ from (6–5), it is easily shown that (6–7) becomes

$$PG_2 = \frac{1}{2\pi k}\frac{1}{(H_B - h_m)^2}\left[\rho_1 + \rho_2|\Gamma|^2 + \rho_3|T|^2\right] \tag{6–8}$$

From Figure 6–2 it is seen that ρ_2 and ρ_3 are less than $3\rho_1$. Thus if $|\Gamma|^2 \approx 0.1$, and $\rho_2 \approx 3\rho_1$, the first term in (6–8) will be several times larger than the second term. For narrow streets and/or taller buildings, both of which cause the angle $|\theta_1|$ to be large, or for more reflective buildings, the first and second terms are more nearly equal.

There are other paths by which the rooftop fields reach ground level, in addition to the four shown in Figure 6–2. However, they involve diffraction through larger angles, passage through more buildings, or additional reflections or diffractions. All these paths are expected to give smaller contributions to the total signal than path 1. An approximate way to account for all such paths is to assume that their total contribution is about the same as that of path 1. In that case the total average power can be found by doubling the contribution along path 1, so that

$$PG_2 = \frac{2}{\rho_1}|D(\theta_1)|^2 \approx \frac{1}{\pi k \rho_1}\frac{1}{|\theta_1|^2} \approx \frac{\lambda\rho_1}{2\pi^2(H_B - h_m)^2} \tag{6–9}$$

Equation (6–9) gives various degrees of approximation to the height gain of the subscriber antenna, which is seen to depend on the building height and the street width.

When measuring the height gain of the subscriber antenna, it is common to aggregate measurements from all over a metropolitan area [3], which is equivalent to taking an average of (6–9) over a building height distribution. This averaging is avoided if the height gain measurements are made in a smaller city having uniform height buildings. One such set of height gain measurements was made in Reading, England, where buildings are fairly uniform in height, with an average height of about 12.5 m [4]. Measurements were made using a distant 191.25 MHz TV transmitter as the signal source with the receiving antenna mounted on a van with a telescoping antenna mast. At each location of the van, the received signal strength was measured at $h_m =$ 3, 5, 7, 10 m. The median value of the received signal in decibels at each height, relative to the value at a height of 3 m [5], is plotted in Figure 6–3. For comparison, we have also plotted as a continuous curve the height gain computed from (6–9) using the Felsen diffraction coefficient. The very good agreement between measured and computed height gain indicates the validity of the mechanism of diffraction of rooftop fields down to the subscriber. Figure 6–3 also shows agreement with simulations that account for random variations in building height, as discussed later. Ikegami et al also substantiated this mechanism by comparing predictions with measurements along a street in Tokyo [1].

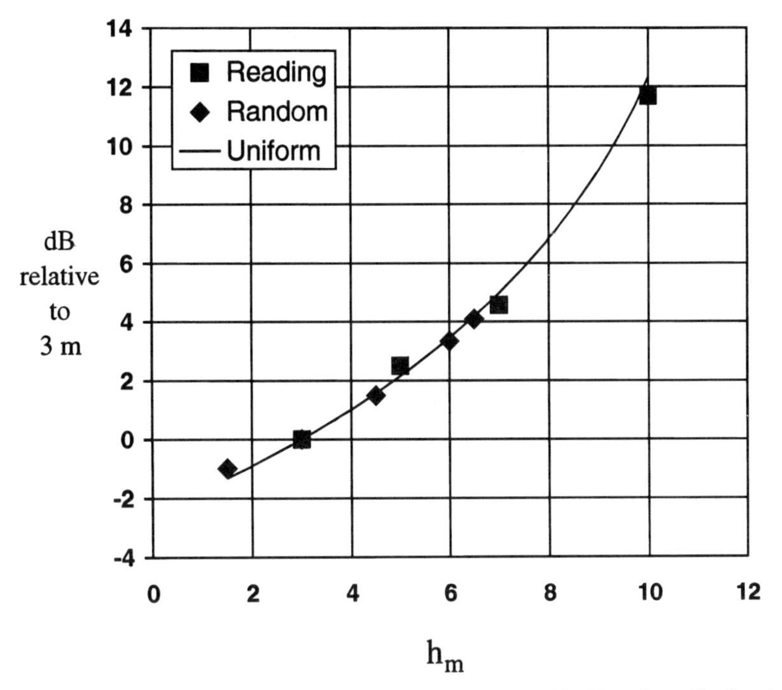

Figure 6-3 Height gain of the median signal measured in Reading, England where the buildings are of nearly equal height $H_B \approx 12.5$ m. For comparison, the theoretical height gain labeled "Uniform" is shown. In both cases the height gain has been normalized to 0 dB for $h_m = 3$ m. Points marked "Random" are discussed in Chapter 7. This figure is taken from [5] (©1992 IEEE).

6.2 Approaches to computing the reduction PG_1 of the rooftop fields

To complete the description of propagation to subscribers in portions of a metropolitan region outside the high-rise core, it is necessary to compute the reduction in the field arriving at the last rooftop before the mobile due to propagation over the previous rows of buildings. In typical cellular installations, the base station height is small compared to the range for coverage, so that the angle α in Figure 6–1 is small. For example, if the base station is 21 m (seven stories) above the surrounding buildings, then at a distance of 1 km, the angle $\alpha \approx 1.2°$. Because the wave propagation is nearly horizontal, the fundamental wave mechanism will involve multiple diffraction past successive rows of buildings. Waves diffracted down to street level may be reflected back up to the rooftops and rejoin the waves propagating over the buildings after a second diffraction. Because this process involves two diffraction events through significant angles, and one or more reflections, the resulting contributions will be small and are ignored here.

Besides the amplitude reduction as a result of diffraction in the vertical plane by the rows of buildings, the fields radiated by the base station antenna will experience a reduction due to the spreading of the rays in the horizontal plane. Since the buildings are uniform transverse to the direction of propagation, the ray spreading in the horizontal plane is independent of diffraction

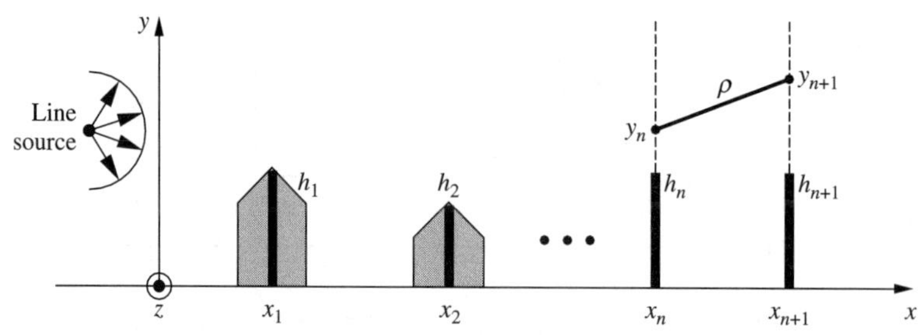

Figure 6-4 Replacing the rows of buildings by absorbing screens, the field everywhere can be found by recursively computing the field in the plane above the $n + 1$ row from the field above the n row using physical optics.

in the vertical plane. As a result, the effects of the rows of buildings on the spherical wave radiated by a base station antenna will be the same as those found for a cylindrical wave radiated by a line source parallel to the rows. To view this statement in a different way, suppose that the reduction of the field at the top of the buildings is expressed as a factor multiplying the free-space field in the absence of buildings. Then the same factor will be found for cylindrical waves as is found for spherical waves. Thus we may reduce the three-dimensional problem to a simpler two-dimensional line source problem, as indicated in Figure 6–4.

The propagation process described above involves only diffraction through small angles in the vertical plane, which does not depend strongly on the cross-sectional shape of the diffracting obstacle. For simplicity, we can therefore replace the rows of buildings by diffracting screens, as shown in Figure 6–4. Since transmission through the buildings is ignored for UHF frequencies, as are ground reflections, the screens are assumed to be absorbing and to be semi-infinite. Diffraction past an array of absorbing screens is a classic problem in electromagnetics. When all the screens have the same height and are spaced by the same distance d, simple closed-form solutions are found for the field incident on each edge for the special cases of (1) plane wave propagation in the x direction, and (2) excitation by a line source located at the same height as the edges ($y_0 = 0$ in Figure 6–4) and at a distance d before the first edge [6]. For plane wave incidence at a finite angle α, or for other locations of the source, an alternative analysis must be carried out.

Several analytic and numerical approaches have been used to find the rooftop fields after diffraction past many absorbing screens. One approach makes use of the repeated application of the physical optics approximation that was invoked in Chapter 5 to find the fields diffracted by a single absorbing screen [7–10]. A variation on the physical optics method uses Fourier transforms to represent the fields between two screens as a spectrum of plane waves in order to propagate the field from one screen to the next [11]. Alternatively, the parabolic equation method has been used to find the multiply diffracted field [12]. Finally, uniform theory of diffraction (UTD) methods have been applied for limited numbers of screens [13–18]. Because it is simple to understand, amenable to both analytic and numerical study, robust in dealing with variations in screen height and separation, and flexible in incorporating effects of vegetation, we use the physical optics approach of Chapter 5.

6.2a Physical optics approach to computing field reduction

In the physical optics approach, the field in the region above the $n = 1$ screen in Figure 6–4 is taken to be the incident field as if the base station antenna were radiating in free space. The field above the first screen is then used as the equivalent source of the field above the next screen, and so on. Let (y_n,z_n) stand for the (y,z) coordinates in the plane above row n. Then the field $H(x_{n+1},y_{n+1},z_{n+1})$ in the plane of the $n + 1$ screen is found from the field $H(x_n,y_n,z_n)$ in the plane above row n. For vertical polarization and for propagation that is essentially horizontal, equation (5–1) gives

$$H(x_{n+1}, y_{n+1}, z_{n+1}) = \int_{-\infty}^{\infty}\int_{h_n}^{\infty} \Lambda(\alpha_{n+1}, 0)H(x_n, y_n, z_n)\frac{jke^{-jkr}}{4\pi r}dy_n dz_n \tag{6–10}$$

Here r is the distance from the secondary source point (y_n,z_n) in the plane $x = x_n$ to the receiver point (y_{n+1},z_{n+1}) in the plane $x = x_{n+1}$. For propagation that is essentially horizontal

$$\Lambda(\alpha_{n+1}, 0) = \cos\alpha_{n+1} + 1 \approx 2 \tag{6–11}$$

The lower limit of the y_n integration is at the top of the buildings, which is the height h_n above a reference. The integration in (6–10) must be repeated many times to find the fields diffracted over the many rows of buildings covered in a typical cellular link.

As discussed above, the reduction in the rooftop fields can be found from the solution of a two-dimensional line source problem, for which the fields in (6–10) have no variation along z. In this case the integration over z_n can be carried out using the approximations leading to (5–21), which reduces (6–10) to the single integration

$$H(x_{n+1}, y_{n+1}) \approx e^{j\pi/4}\sqrt{\frac{k}{2\pi}}\int_{h_n}^{\infty} H(x_n, y_n)\frac{e^{-jk\rho}}{\sqrt{\rho}}dy_n \tag{6–12}$$

The distance

$$\rho = \sqrt{(x_{n+1} - x_n)^2 + (y_{n+1} - y_n)^2} \tag{6–13}$$

is shown in Figure 6–4. Repeated numerical evaluation of the integral in (6–12) is a viable approach to finding the fields. When the buildings are of uniform height and, the row spacing is uniform, the field at the top of the buildings can be found in terms of Borsma's functions, as discussed in subsequent sections.

6.2b Solutions for uniform row spacing and building height

Consider uniformly spaced rows of buildings with $x_{n+1} - x_n = d$ that are of equal height $h_n = 0$ above the reference plane, as shown in Figure 6–5. Because the propagation is essentially horizontal, the primary contribution to the integral in (6–12) comes from values of y_n near to y_{n+1}. Thus ρ can be replaced by d in the denominator of (6–12), and in the exponent by the Fresnel approximation

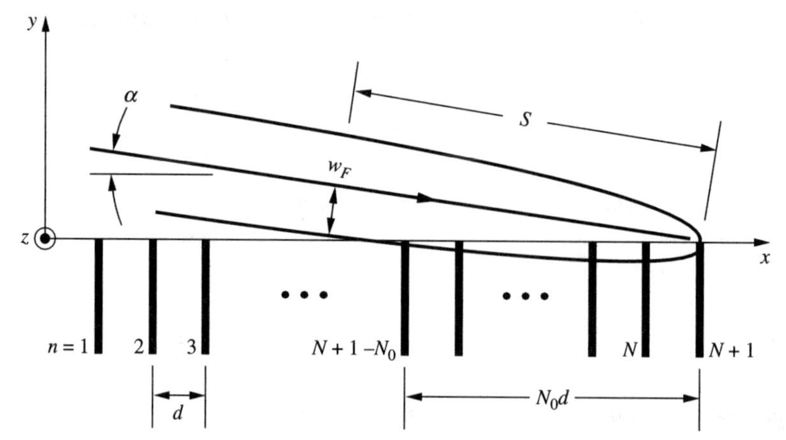

Figure 6-5 The settling number N_0 is determined by the number of screens that penetrate the Fresnel zone about the plane wave ray reaching the top of the $n + 1$ screen.

$$\rho \approx d + \frac{(y_{n+1} - y_n)^2}{2d} \tag{6–14}$$

Since $\sqrt{k/2\pi} = 1/\sqrt{\lambda}$, cascading the integrals (6–12) past N screens gives

$$H(x_{n+1}, y_{n+1}) = \frac{e^{jN\pi/4} e^{-jkNd}}{(\lambda d)^{N/2}} \int_0^{\infty} dy_1 \int_0^{\infty} dy_2 \ldots \int_0^{\infty} dy_N H(d, y_1) \times \exp\left[-j\frac{k}{2d}\sum_{n=1}^{N} (y_{n+1} - y_n)^2\right] \tag{6–15}$$

where y_n refers to the location of a field or integration point in the plane $x_n = nd$. In the following sections we consider two different forms for the field illuminating the first screen at $x_1 = d$, for which we can obtain solutions to (6–15) at the top of the $N + 1$ screen, where $y_{N+1} = 0$. These solutions are in terms of functions studied by Borsma [8,19]. The first is a plane wave, which gives the field reduction for elevated antennas used in macrocellular systems that cover distances from 1 to about 20 km or more, and the second is a line source that may be located at or below the rooftops, as in microcellular systems covering distance up to about 1 km.

6.3 Plane wave incidence for macrocell predictions

In macrocellular applications, the fields radiated by an elevated base station antenna will cross many rows of buildings before reaching the subscriber. As a result, end effects associated with the first few rows will be eliminated, and fields at the rooftop will settle to a value relative to free space that depends only on the angle α indicated in Figure 6–5. The settled value can therefore be found by considering an incident plane wave propagating past a series of rows, as will be seen from the computed results. Assuming that an incident plane wave of unit amplitude propagates downward at an angle α to the horizontal, the field in the plane $x_1 = d$ is

$$H(d, y_1) = \exp(-jkd\cos\alpha)\exp(jky_1\sin\alpha) \tag{6–16}$$

For small angles α we may use the approximation $\cos\alpha \approx 1$ in (6–16), so that when it is substituted into (6–15), the field for $y_{N+1} = 0$ is given by

$$H(x_{N+1}, 0) = \frac{e^{jN\pi/4}e^{-jk(N+1)d}}{(\lambda d)^{N/2}}\int_0^\infty dy_1\int_0^\infty dy_2\ldots\int_0^\infty dy_N\exp(jky_1\sin\alpha) \times \exp\left[-j\frac{k}{2d}\left(y_1^2 - 2\sum_{n=1}^{N-1} y_{n+1}y_n + 2\sum_{n=2}^{N} y_n^2\right)\right] \tag{6–17}$$

The next step is to change the variables of integration from y_n to v_n using the transformations

$$v_n = y_n\sqrt{\frac{jk}{2d}} \tag{6–18}$$

$$dy_n = dv_n\sqrt{\frac{2d}{jk}} = dv_n e^{-j\pi/4}\sqrt{\frac{\lambda d}{\pi}}$$

and introducing the dimensionless parameter g_p, defined by

$$g_p = \sin\alpha\sqrt{\frac{d}{\lambda}} \tag{6–19}$$

Substituting (6–18) and (6–19) in (60–17), the multiple integral can be written in the form

$$H(x_{N+1}, 0) = \frac{e^{-jk(N+1)d}}{\pi^{N/2}}\int_0^\infty dv_1\int_0^\infty dv_2\ldots\int_0^\infty dv_N\exp(2\sqrt{j\pi}g_p v_1) \times \exp\left(-v_1^2 + 2\sum_{n=1}^{N-1} v_{n+1}v_n - 2\sum_{n=2}^{N} v_n^2\right) \tag{6–20}$$

Aside from the phase term $\exp[-jk(N+1)d]$ in (6–20), the dependence of $H(x_{N+1},0)$ on frequency, row spacing d, and angle of incidence α is through the single parameter g_p defined in (6–19).

6.3a Solution in terms of Borsma's functions

The final step in the evaluation of (6–2) is to expand the first exponential of the integrand in the Taylor series

$$\exp(2\sqrt{j\pi}g_p v_1) = \sum_{q=0}^{\infty}\frac{1}{q!}(2\sqrt{j\pi}g_p)^q v_1^q \tag{6–21}$$

Substituting (6–21) into (6–20) and interchanging the order of integration and summation gives the field at the rooftops as

$$H(x_{N+1}, 0) = e^{-jk(N+1)d} \sum_{q=0}^{\infty} \frac{1}{q!}(2\sqrt{j\pi} g_p)^q I_{N,q}(1) \tag{6–22}$$

In (6–22), $I_{N,q}(1)$ is one of Borsma's functions $I_{N,q}(\beta)$ for $\beta = 1$. These functions are defined by [19]

$$I_{N,q}(\beta) = \frac{1}{\pi^{N/2}} \int_0^{\infty} dv_1 \int_0^{\infty} dv_2 \ldots \int_0^{\infty} dv_N \left[v_1^q \exp\left(-\beta v_1^2 + 2\sum_{n=1}^{N-1} v_{n+1} v_n - 2\sum_{n=2}^{N} v_n^2 \right) \right] \tag{6–23}$$

Borsma has shown that the functions defined by (6–23) can be evaluated using the following recursion relation for q ≥ 2:

$$I_{N,q}(\beta) = \frac{N(q-1)}{2(N+1)^{\beta-1}} I_{N,q-2}(\beta) + \frac{1}{2\sqrt{\pi}(N+1)^{\beta-1}} \sum_{n=\beta-1}^{N-1} \frac{I_{n,q-1}(\beta)}{\sqrt{N-n}} \tag{6–24}$$

For $\beta = 1$ the starting functions in the recursion relation are

$$I_{0,q}(1) = \begin{cases} 1 & \text{for } q = 0 \\ 0 & \text{for } q > 0 \end{cases} \tag{6–25}$$

$$I_{N,q}(1) = \frac{(1/2)_N}{N!} \tag{6–26}$$

$$I_{N,1}(1) = \frac{1}{2\sqrt{\pi}} \sum_{n=0}^{N-1} \frac{(1/2)_n}{n!\sqrt{N-n}} \tag{6–27}$$

In (6–26) and (6–27) the term $(1/2)_n$ denotes Pochhammer's symbol [20], defined by

$$\begin{aligned} (a)_0 &= 1 \\ (a)_1 &= a \\ (a)_n &= a(a+1)\ldots(a+n-1) \end{aligned} \tag{6–28}$$

Using the recursion relation (6–24), and the initial terms (6–25) to (6–27), a matrix of values of $I_{N,q}(1)$ can be computed efficiently and then used in the summation (6–22) to find the field incident on the rooftop of the $N + 1$ row of buildings, where $y_{N+1} = 0$.

A simple closed-form result is obtained when the incident plane wave is propagating parallel to the line of the rooftops, so that the angle $\alpha = 0$. In this case g_p in (6–19) vanishes and only the first term with $q = 0$ contributes to the summation (6–20). Thus

$$H(x_{N+1}, 0) = e^{-jk(N+1)d} \frac{(1/2)_N}{N!} \approx e^{-jk(N+1)d} \frac{1}{\sqrt{\pi N + 1}} \tag{6–29}$$

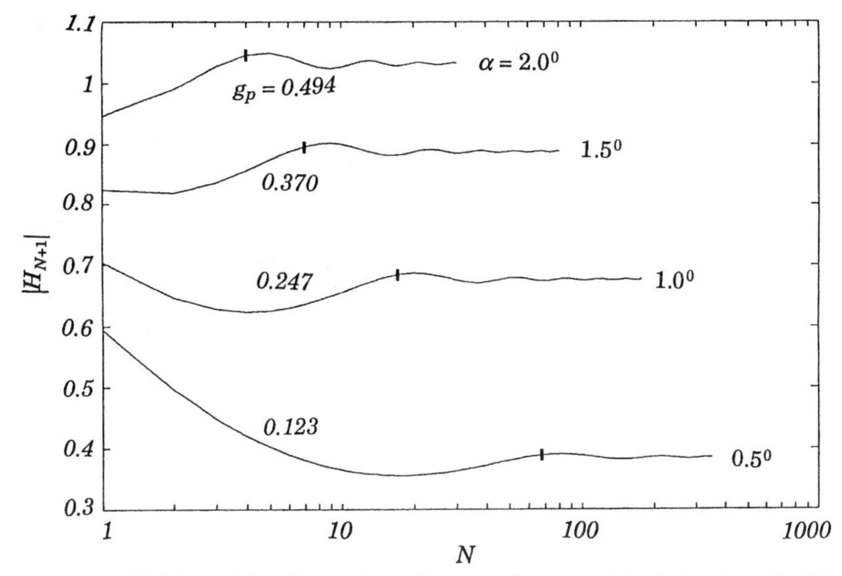

Figure 6-6 Field reaching the rooftop of successive rows $N + 1$ due to an incident plane wave of unit amplitude for various values of the parameter $g_p \approx \alpha\sqrt{d/\lambda}$. The angles listed are for $d = 200\lambda$. Although a continuous curve has been drawn, only integer values of N have physical significance. Values of N_0 are indicated by the vertical bars [8] (©1992 IEEE).

The approximation in (6–29) is obtained by using the asymptotic approximation to Pochhammer's symbol [20]. For propagation parallel to the line of rooftops, each rooftop lies on the shadow boundary of the previous rooftop. Thus the second rooftop lies on the shadow boundary of the first, and the field incident on the second rooftop has an amplitude that is one-half the plane wave amplitude incident on the first row, as discussed in Chapter 5. Substituting $N = 1$ into the exact form in (6–29), the field amplitude is 1/2, as expected. Even the approximate form in (6–29) gives a field value very near 1/2 for $N = 1$. With diffraction past more rows, the field amplitude is seen from (6–29) to decrease monotonically, but only weakly. If it were valid to simply multiply the diffraction coefficients for the various rows of buildings, the field amplitude incident on the $N + 1$ row for $\alpha = 0$ would be $1/2^N$, which is a much stronger rate of decrease. However, since each diffracting edge lies in the transition region of the previous edge, a simple multiplication of the diffraction coefficients is not valid.

When the plane wave is incident from above the line of rooftops, as in Figure 6–5, a different dependence of the field amplitude on N emerges. Figure 6–6 shows the dependence on N of the magnitude of the field computed from (6–20) for various values of the parameter g_p. While the various plots in Figure 6–6 have been drawn as continuous curves, only the points corresponding to integer values of N have physical significance. The curves are also labeled by the angle of incidence α for a row spacing $d = 200\lambda$. After passing a number of rows, the field incident on successive rooftops is seen to oscillate with decreasing amplitude about a finite settled value. For high angles of incidence, the settled value may be slightly greater than unity, due to a reinforcement of the incident field by that diffracted from the preceding row. However, for small angles of incidence the settled value decreases monotonically with angle. The initial variation of the field can be viewed as an end effect associated with the transition from free-space propagation to diffraction past the rows. The influence of this transition decreases as the fields propagate past more rows and the field settles to a steady value.

The number of rows N_o that must be crossed to achieve settling of the field is found by considering those rows whose rooftops lie inside the first Fresnel zone about the ray to a particular rooftop for large N, as indicated in Figure 6–5. Since the source of the incident plane wave is at infinity, the half-width of the Fresnel zone at a distance s from the end of the ray is

$$w_F = \sqrt{\lambda s} \tag{6–30}$$

The perpendicular distance from a rooftop to the ray is $s \tan\alpha$, so that the rooftops for which $w_F > s \tan\alpha$ will lie in the Fresnel zone, while those for which $w_F < s \tan\alpha$ will lie outside the Fresnel zone. The distance $N_0 d$ to the last rooftop just inside the Fresnel zone is the value of $s \approx N_0 d$ such that $w_F = N_0 d \tan\alpha$. From this condition and (6–30) it is found for small angles α that

$$N_0 \approx \frac{\lambda}{d \tan\alpha^2} \approx \frac{1}{g_p^2} \tag{6–31}$$

where g_p is the parameter defined in (6–19). This value of N_0 is indicated by the vertical bar on each curve in Figure 6–6. It is seen that N_0 identifies the end of the transition region, and the start of the settled field. As an example, for a row spacing of $d = 60$ m and a frequency of 900 MHz, $d/\lambda = 180$ and the settling number for an angle $\alpha = 0.5°$ is $N_0 = 73$, corresponding to a distance $N_0 d = 4.4$ km. After treating the case of an incident cylindrical wave in Section 6–4, it is found that the settled field value can be used at much shorter distances.

6.3b Using the settled field to find the path loss

Let $Q(g_p)$ be the magnitude of the settled value of the field, as taken from plots such as those shown in Figure 6–6. The dependence of the settled field on the parameter g_p is plotted in Figure 6–7 using logarithmic scales. For small values of g_p the slope of $Q(g_p)$ is near unity, so that $Q(g_p)$ varies approximately linearly with g_p. For larger values of g_p the slope decreases and $Q(g_p)$ approaches a maximum value slightly greater than 1. A simple polynomial expression has been fit to the variation of $Q(g_p)$ shown in Figure 6–7 over the range $g_p < 1$. This polynomial is given by [21]

$$Q(g_p) = 3.502 g_p - 3.327 g_p^2 + 0.962 g_p^3 \tag{6–32}$$

Since $Q(g_p)$ gives the reduction in the field at the rooftop before the subscriber due to propagation over the previous rows of buildings, the path gain factor PG_1 in equation (6–1) is the square of $Q(g_p)$, or

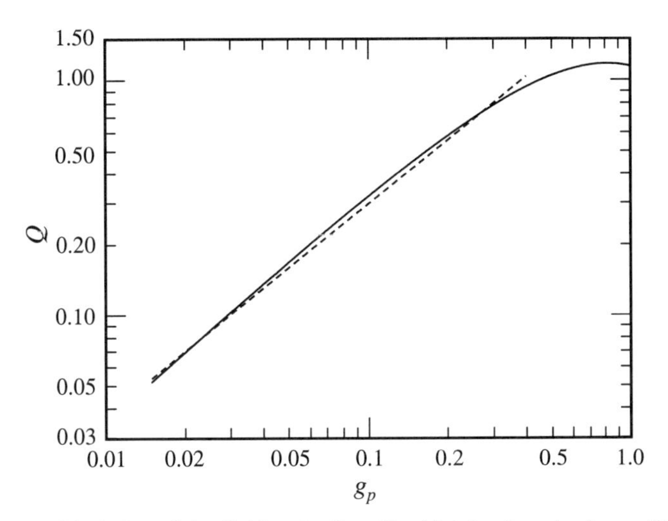

Figure 6-7 Variation of the field reduction Q, which is given by the settled field, with the dimensionless parameter g_p [2]. The dashed line corresponds to the simple approximation (6–34) (©1988 IEEE).

$$\mathrm{PG}_1 = [Q(g_p)]^2 \tag{6–33}$$

Expression (6–33) is the last of the terms needed to compute the overall path gain or path loss.

As an example of field reduction due to diffraction over the buildings, suppose that $h_{BS} - H_B = 22$ m, corresponding to the base station antenna seven stories above the surrounding buildings, and $d = 60$ m. Then for 900-MHz signals, $g_p = 0.3$ for a distance $R = 1$ km from the base station, while $g_p = 0.03$ for a distance $R = 10$ km. For these parameters, the field reduction due to propagation past the rows of buildings is found from (6–32) to be $Q(0.3) = 0.777$ at $R = 1$ km, whereas at $R = 10$ km the reduction is $Q(0.03) = 0.102$. The ratio of the foregoing values of $Q(g_p)$ at $g_p = 0.3$ and, 0.03 is 7.6, which suggest a simple approximation for its dependence over the range of g_p of interest for macrocellular systems. Because $10^{0.9} = 7.9$, the values cited above lead to the following approximation for $Q(g_p)$:

$$Q(g_p) \approx 0.1\left(\frac{g_p}{0.03}\right)^{0.9} \tag{6–34}$$

Approximation (6–34) is plotted as the dashed line in Figure 6–7 and is seen to be close to the actual curve for $g_p < 0.4$.

The overall path gain defined in (6–1) can now be computed using (6–2), (6–9), and (6–33), with the help of (6–19) and (6–34). The result for isotropic antennas is

$$\mathrm{PG}_{\mathrm{dB}} = 10\log\left(\frac{\lambda}{4\pi R}\right)^2 + 10\log\left(\frac{\lambda\rho_1}{2\pi^2(H_B - h_m)^2}\right) + 10\log\left[2.347\left(\frac{h_{\mathrm{BS}} - H_B}{R}\sqrt{\frac{d}{\lambda}}\right)^{0.9}\right]^2 \tag{6–35}$$

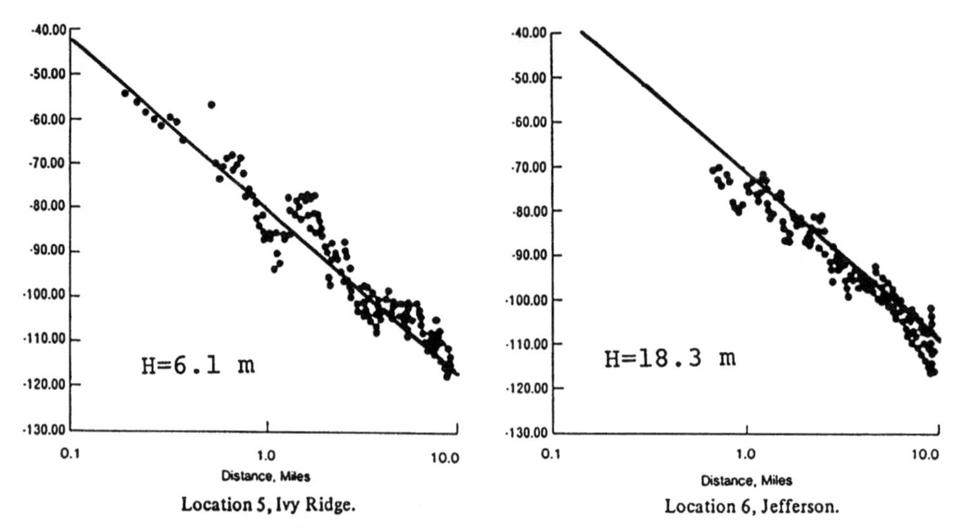

Figure 6-8 Comparison of the path loss predictions with measurements of the small-area averages made in Philadelphia [22] for two transmitter heights $H = h_{\text{BS}} - H_B$ (©1978 IEEE).

The predictions given by (6–35) are compared in Figure 6–8 with 820-MHz measurements of the small-area average received power made in areas of Philadelphia outside the high-rise core [22]. The dots in Figure 6–8 represent the measured power, and the straight lines are computed from (6–35) for the two different values of $H = h_{\text{BS}} - H_B$. As found from a map of the measurement area, the average row separation is $d = 35$ m. The building height is taken as $H_B = 7.7$ m, the subscriber antenna height as $h_m = 1.5$ m, and the predictions were adjusted to account for net antenna gains and radiated power, which totaled 23.3 dBm. The predictions are seen to be in good agreement with the measurements, thereby supporting the assumption that propagation takes place over the tops of the rows of buildings between the base station and mobile.

To compare (6–35) with the Hata model, let R_k be the distance in kilometers and f_M be the frequency in megahertz. In terms of these variables, the path loss L in decibels, which is the negative of (6–35), can be expressed as

$$L = 89.5 - 10\log\left[\frac{\rho_1 d^{0.9}}{(H_B - h_m)^2}\right] + 21\log f_M - 18\log(h_{\text{BS}} - H_B) + 38\log R_k \qquad (6\text{–}36)$$

If h_{BS} is set to 30 m in the Hata model for an urban area, which is given by equation (2–9), the coefficient of $\log R_k$ is 35.2 instead of the theoretical value 38 given in (6–36). The coefficient of $\log f_M$ in the Hata model is 26.2, as compared to the theoretical value of 21. The base station antenna height enters the Hata model as $13.8\log h_{\text{BS}}$ rather than the term $18\log(h_{\text{BS}} - H_B)$ in (6–36). Thus the term in the Hata model for base station height gain uses a smaller coefficient multiplying the log of a larger argument so that the two expressions are nearly equal. For example, if $h_{BS} = 30$ m and $H_B = 10$ m, the Hata base station height gain is 20.4 dB, whereas the theoretical gain is 23.4 dB.

The constant term of 69.6 dB in the Hata model accounts for the building environment and should be compared with the first two terms in (6–36). If we assume that $d = 60$ m, $H_B = 10$ m, and $h_m = 1.5$ m, then $\rho_1 = 31.2$ m and the first two terms of (6–36) sum to the value 77.1 dB, which is about 7.5 dB more than the Hata model. To summarize, for the foregoing building geometry and the antenna heights $h_m = 1.5$ m and $h_{BS} = 30$ m, the theoretical and Hata models for the path loss are

$$\text{Theory:} \quad L = 53.7 + 21 \log f_M + 38 \log R_k$$

$$\text{Hata:} \quad L = 49.2 + 26.2 \log f_M + 35.2 \log R_k$$

Note that at $R_k = 10$ km and $f_M = 900$ MHz, the theoretical path loss is 153.4 dB, whereas the Hata model gives 161.8 dB. Changing the building geometry slightly in the theoretical model can easily account for the discrepancy between the two models. In any case, the theoretical dependence on distance and frequency is close to the measurement-based model, and the overall path loss is in agreement. This observation lends further credence to the view that propagation takes place over the buildings, with diffraction of the rooftop fields down to the mobile.

6.4 Cylindrical wave incidence for microcell predictions

When the base station is lowered to the level of the surrounding rooftops to limit cell radius to distances on the order of 1 km, the reduction in the rooftop fields will depend on the number of rows that are crossed. Modeling the process of diffraction over a few rows to find the reduction in the rooftop fields, as compared to free-space propagation, requires that we consider a cylindrical wave rather than a plane wave, as in Section 6–3. The cylindrical wave is generated by a line source parallel to the rows of buildings, as shown in Figure 6–9. To find a solution to this problem in terms of Borsma's functions, the line source is assumed to be located at a distance d before the first row. If the height of the base station above or below the rooftops is given by the variable $y_0 = h_{BS} - H_B$, the z-directed magnetic field of the cylindrical wave is

$$H(x,y) = \frac{e^{-jk\rho}}{\sqrt{\rho}} \qquad (6–37)$$

where $\rho = \sqrt{x^2 + (y - y_0)^2}$.

Expression (6–37) is now used in equation (6–15) as the field $H(d,y_1)$ incident on the first row of buildings. Consistent with the approximations used to derive (6–15), ρ in the denominator of (6–37) is replaced by d, and the Fresnel approximation (6–14) is used for ρ in the exponent. With these approximations, the field incident on the rooftop $y_{N+1} = 0$ of the $N + 1$ row is

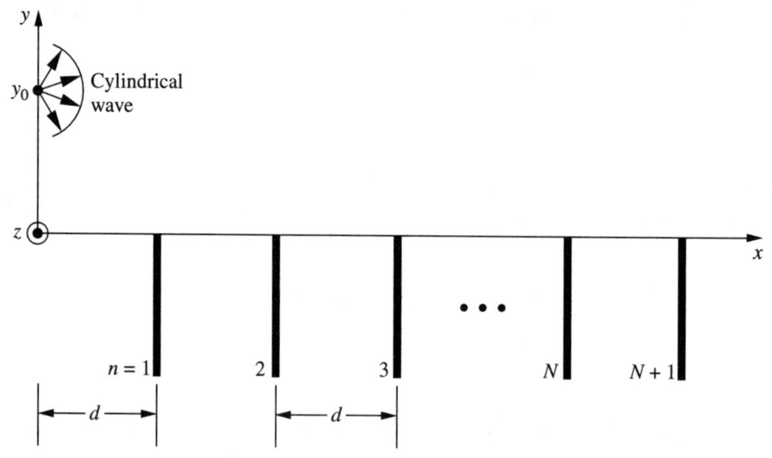

Figure 6-9 A line source above absorbing screens of uniform height and spacing is used to find the reduction of the rooftop fields for base station antenna height near that of the rooftops.

$$H(x_{N+1}, 0) = \frac{e^{jN\pi/4} e^{-jk(N+1)d}}{(\lambda d)^{N/2}\sqrt{d}} \int_0^{\infty} dy_1 \int_0^{\infty} dy_2 \ldots \int_0^{\infty} dy_N \times \exp\left[-j\frac{k}{2d}\left(y_0^2 - 2y_0 y_1 + 2y_1^2 - 2\sum_{n=1}^{N-1} y_{n+1} y_n + 2\sum_{n=2}^{N} y_n^2\right)\right] \tag{6–38}$$

Defining the dimensionless parameter

$$g_c = \frac{y_0}{\sqrt{\lambda d}} \tag{6–39}$$

and using the change of variable defined in (6–18), the multiple integration of (6–38) can be written as

$$H(x_{N+1}, 0) = \frac{e^{-jk(N+1)d} e^{-jky_0^2/2d}}{\pi^{N/2}\sqrt{d}} \int_0^{\infty} dv_1 \int_0^{\infty} dv_2 \ldots \int_0^{\infty} dv_N \exp(2\sqrt{j\pi} g_c v_1) \times \exp\left(-2v_1^2 + 2\sum_{n=1}^{N-1} v_{n+1} v_n - 2\sum_{n=2}^{N} v_n^2\right) \tag{6–40}$$

6.4a Solution in terms of Borsma's functions

From (6–40) it is seen that aside from the phase terms and amplitude before the integrals, the frequency and physical dimensions enter only through the dimensionless parameter g_c defined in (6–39). The multiple integrals in (6–40) for the incident cylindrical wave is of the same form as that in (6–20) for an incident plane wave, except that g_c is used in place of g_p and there is a factor of 2 multiplying v_1^2 in the exponent. Thus expanding the exponential term containing g_c in a Taylor series like that in (6–22), and integrating term by term [9], it is found for an incident cylindrical wave that

$$H(x_{N+1}, 0) = \frac{e^{-jk(N+1)d} e^{(-jky_0^2)/2d}}{\sqrt{d}} \sum_{q=0}^{\infty} \frac{1}{q!}(2\sqrt{j\pi}g_c)^q I_{N,q}(2) \tag{6–41}$$

In (6–41), the terms $I_{N,q}(2)$ are Borsma's functions defined in (6–23) for $\beta = 2$. These functions can be found from the recursion relation (6–24) using the initial terms [19]

$$I_{N,0}(2) = \frac{1}{(N+1)^{3/2}} \tag{6–42}$$

$$I_{N,1}(2) = \frac{1}{4\sqrt{\pi}} \sum_{n=1}^{N} \frac{1}{n^{3/2}(N+1-n)^{3/2}} \tag{6–43}$$

If the line source is at the same height as the rooftops, so that $y_0 = 0$, g_c vanishes and only the first term in the sum (46–1) contributes. Thus using (6–42) for this case yields

$$H(x_{N+1}, 0) = \frac{e^{-jk(N+1)d}}{\sqrt{(N+1)d}} \frac{1}{N+1} \tag{6–44}$$

The first fraction in (6–44) is the cylindrical wave (6–37) that would have reached the location of the $N+1$ rooftop under free-space propagation conditions. Thus the reduction of the fields due to propagation over the rows of buildings is given by the factor $1/(N+1)$. As argued previously, this same factor will also give the reduction of the spherical wave fields radiated by an antenna located at the rooftop height. By way of comparison, for a plane wave incident with $a = 0$, the field reduction factor is, $1/\sqrt{\pi N + 1}$ which decreases less rapidly with N.

As found for the case when $y_0 = 0$, the reduction of the rooftop fields will depend on the number of rows of buildings that the ray crosses. We therefore define the field reduction factor $Q_{N+1}(g_c)$ for the field reaching the $N+1$ row by normalizing $H(x_{N+1},0)$ to the cylindrical wave field (6–37) reaching the location of the $N+1$ rooftop under free-space conditions. Thus

$$Q_{N+1}(g_c) \equiv \left| \frac{H(x_{N+1}, 0)}{e^{-jk\rho}/\sqrt{\rho}} \right| \tag{6–45}$$

where $\rho = \sqrt{(N+1)^2 d^2 + y_0^2}$. Approximating ρ in the amplitude of the cylindrical wave by $(N+1)d$, from (6–41) it is seen that

$$Q_{N+1}(g_c) = \sqrt{N+1} \left| \sum_{q=0}^{\infty} \frac{1}{q!}(2\sqrt{j\pi}g_c)^q I_{N,q}(2) \right| \tag{6–46}$$

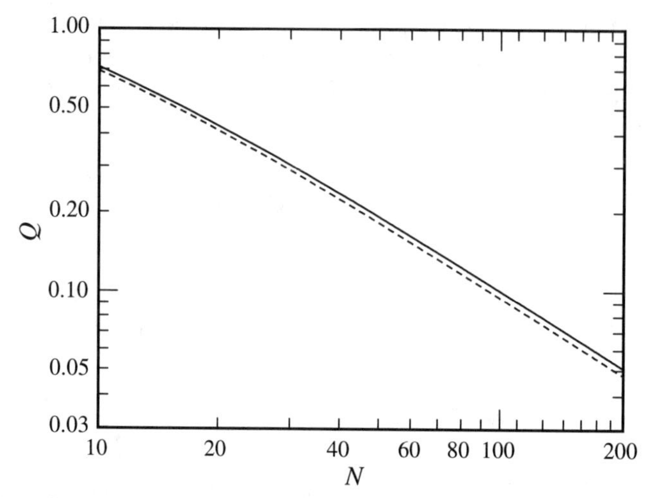

Figure 6-10 Comparison of the cylindrical wave field reduction factor $Q_{N+1}(g_c)$ (solid curve) of argument $g_c = y_0/\sqrt{\lambda d}$ with the plane wave factor $Q(g_p)$ (dashed curve) of argument $g_p = g_c/(N+1)$ for $g_c = 2.80$. Only integer values of N have physical significance.

For $|g_c|$ large, many terms must be used when computing the sum in (6–46). This can cause overflow problems when evaluating $(g_c)^q$ and $q!$, which can be avoided by using Sterling's formula [20] for $q!$ when $q > 150$.

Consider, for example, the antenna height $y_0 = 12.5$ m with $\lambda = 1/3$ m and d = 60 m, so that $g_c = 2.80$. The variation with N of $Q_{N+1}(g_c)$ is plotted in Figure 6–10 for these parameters. For comparison, consider the plane wave reduction $Q(g_p)$ for values of g_p found from the local angle of incidence $\alpha \approx y_0/(N+1)d$. Substituting this expression for α into (6–19) gives

$$g_p = \sin\alpha\sqrt{\frac{d}{\lambda}} \approx \frac{y_0}{(N+1)d}\sqrt{\frac{d}{\lambda}} = \frac{g_c}{N+1} \tag{6–47}$$

In Figure 6–10 we have also plotted the plane wave reduction factor $Q(g_p)$ obtained from (6–32) for the values of g_p given by (6–47). The agreement seen in Figure 6–10 between the plane wave reduction factor and the cylindrical wave reduction factor establish the validity of the plane wave approach to finding the field reduction for elevated base station antennas.

To gain a tighter bound on the conditions under which the plane wave reduction $Q(g_p)$ may be used, and conversely, when it is necessary to use the cylindrical wave reduction $Q_M(g_c)$, consider the Fresnel zone about the ray from the line source to the edge $N + 1$ shown in Figure 6–9. For N large, the width of the Fresnel zone at the $n = 1$ row is $w_F = \sqrt{\lambda d}$. If $y_0 > \sqrt{\lambda d}$ (i.e., $g_c > 1$), the first rooftop lies outside the Fresnel zone and the end effect associated with the first row will be negligible. In this case the plane wave reduction $Q(g_p)$ may be used, as in the case of Figure 6–10 for which $g_c = 2.80$. Conversely, when the base station is close to the rooftops or below, so that $g_c < 1$, the cylindrical wave reduction $Q_M(g_c)$ must be used. We will come back to

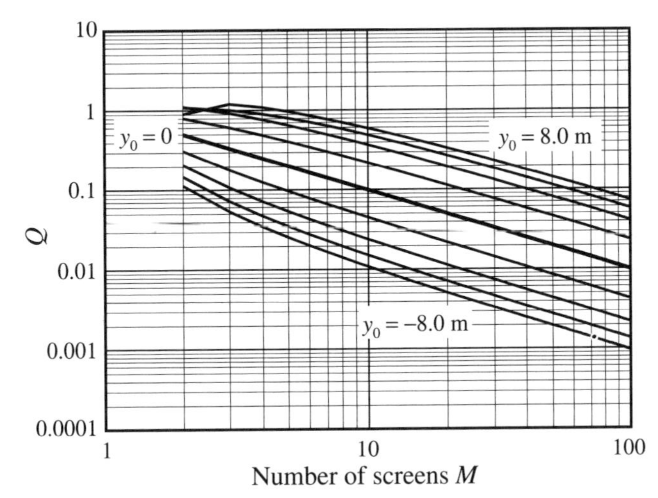

Figure 6-11 Variation with M of the cylindrical wave reduction factor $Q_M(g_c)$ for the field reaching the rooftop of row M. The curves are for y_0 varying in steps of 2 m when d = 50 m and λ = 1/6 m (g_c varying in steps of 0.693). Only integer values of M have physical significance.

this issue after discussing approximations for very low antennas. Note that the plane wave reduction $Q(g_p)$ can be used even though the plane wave settling number N_0 is greater than the number of rows N crossed by the wave. As an example, for the parameters used in Figure 6–10, and for $N = 100$, it is found from (6–47) that $g_p = 0.0275$, so that from (6–31) the settling number is $N_0 = 1320$. In other words, the cylindrical wave reaching the row $N + 1 = 101$ does not experience the end effect of the first row, so that $Q(g_p)$ may be used for the reduction factor. However, the field due to a plane wave incident at the same angle would exhibit end effects because of its wider Fresnel zone.

The variation of $Q_M(g_c)$ with $M = N + 1$ for low base station antenna heights is shown in Figure 6–11 using logarithmic scales. The curves have been plotted for values of y_0 ranging from –8 m to +8 m in steps to 2 m and assuming that d = 50 m and f = 1.8 GHz. From the definition (6–39) of g_c, these plots may be viewed as being for values of g_c ranging from –2.77 to +2.77 in steps of 0.693. When $y_0 = 0$, the field reduction varies as $1/M = 1/(N + 1)$, as discussed previously. If the base station is above the buildings, for small M there is no reduction in the field, while for larger values of M it is seen from Figure 6–11 that $Q_M(g_c)$ decreases as $1/M$. When the base station antenna is below the rooftops, $Q_M(g_c)$ initially varies more rapidly than $1/M$, but decreases ultimately as $1/M$. Alternatively, treating M as a parameter, we can plot $Q_M(g_c)$ as a function of g_c or of the relative base station antenna height $y_0 \equiv h_{BS} - H_B$ using (6–39). As an example, $Q_{20}(g_c)$ for $M = 20$ is plotted in Figure 6–12 versus y_0 for a frequency of 1.8 GHz and a row spacing d = 50 m (see curve labeled $Q_{20,\text{perpend}}$). The curves in Figure 6–12 that are labeled $Q_{10,\text{obliqe}}$ and Q_{exp} are discussed in subsequent sections.

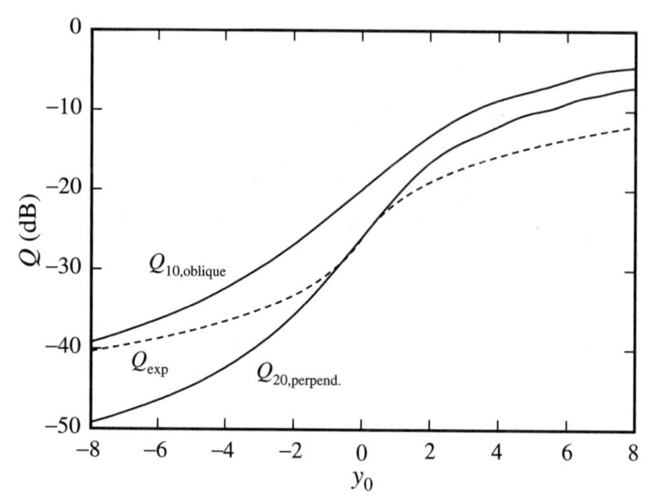

Figure 6-12 Variation with y_0 of the reduction factor $Q_M(g_c)$ for 1.8-GHz radiation at $R = 1$ km and $d = 50$ m. For propagation perpendicular to the rows, $M = 20$ rows are crossed and $g_c = y_0/\sqrt{\lambda d}$. For propagation oblique to the rows at $\psi = 60°$ (see Section 6–4d), $M = 10$ and $g_c = y_0/\sqrt{2\lambda d}$. The curve labeled "exp" is obtained from the fit to measurements in San Francisco.

6.4b Path loss for low base station antennas

Recall that the total path loss L consists of the free-space loss, the loss due to diffraction of the rooftop fields down to the mobile and the field reduction due to propagation over the rooftops. Thus using (6–9), it is found for isotropic antennas that

$$L = -10\log\left(\frac{\lambda}{4\pi R}\right)^2 - 10\log\left[\frac{\lambda\rho_1}{2\pi^2(H_B - h_m)^2}\right] - 20\log Q_M(g_c) \qquad (6\text{–}48)$$

This theoretical expression for the path loss can be compared to the regression fit to the path loss L measured in residential sections of San Francisco, as given by (2–19). For those measurements $H_B - h_m \approx 8$ m and $d = 50$ m, so that $\rho_1 = 26.2$ m and (6–48) becomes

$$L = 114.5 + 30\log f_G + 20\log R_k - 20\log Q_M(g_c) \qquad (6\text{–}49)$$

To complete the comparison with the measurements, we define an experimental value Q_{exp} of the reduction in the rooftop fields by using the regression fit given by (2–19) for the path loss L in (6–49), and remove from it the free-space path loss and the diffraction loss down to the mobile, as given by the first three terms on the right in (6–49). In this way it is found that for a frequency of 1.8 GHz and at a distance $R = 1$ km, corresponding to $M = 20$,

$$20\log Q_{\text{exp}} = -26.1 + 14.9\,\text{sgn}(y_0)\log(1 + |y_0|) \qquad (6\text{–}50)$$

The values of Q_{exp} in decibels obtained from (6–50) are also plotted in Figure 6–12. As compared to Q_{exp}, the theoretical curve for $Q_{20}(g_c)$ is seen to be optimistic by a few decibels for

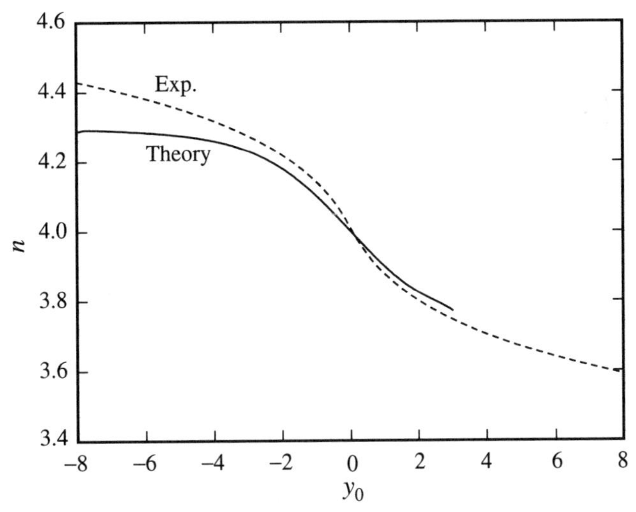

Figure 6-13 Comparison of the antenna height dependence of the range exponent n obtained from the fit to measurements in San Francisco with the theoretical value obtained from the logarithmic derivative at $M = 10$.

high base station antennas, but pessimistic by nearly 10 dB for base station antennas well below the rooftops. Measurements made in Europe also indicate that the theory is somewhat pessimistic for low base station antennas [23]. Approximately 3 dB of the discrepancy for $y_0 < 0$ can be explained by the fact that the evaluation of $Q_M(g_c)$ was made for the geometry of Figure 6–9, which incorporates only one path from the transmitter to the rooftops. Accounting for multiple paths, such as those described in connection with diffraction of the rooftop fields down to street level, will in effect increase $Q_M(g_c)$ by about 3 dB.

The range exponent n predicted by the theory can also be compared to the San Francisco measurements for low base station antennas. From (6–49), n consists of the value 2 arising from free-space portion of the path loss and a contribution from the variation of $Q_M(g_c)$ with M. This later contribution can be approximated by the logarithmic derivative $S_M(g_c)$, defined by

$$S_M(g_c) = \frac{\log Q_{M+1}(g_c) - \log Q_M(g_c)}{\log(M+1) - \log M} \tag{6–51}$$

which is a function of the antenna height through the parameter g_c. The range exponent is then given by

$$n = 2 + 2S_M(g_c) \tag{6–52}$$

Evaluating the logarithmic derivative (6–51) for $M = 10$, corresponding to the middle of the range over which measurements were made in San Francisco, the variation of n with antenna height y_0 is plotted in Figure 6–13. By comparison, the range index n_{exp} of the regression fit to the measurements is one-tenth of the coefficient of $\log R_k$ in expression (2–19), or

$$n_{\text{exp}} = 4.01 - 0.44 \operatorname{sgn} y_0 \log(1 + |y_0|) \tag{6–53}$$

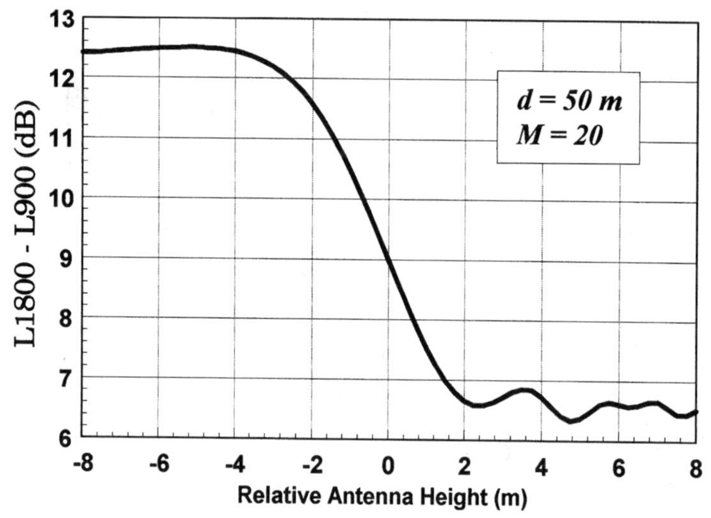

Figure 6-14 Predicted dependence of the difference in path loss at 1800 and 900 MHz on the base station antenna height [21] (©1994 IEEE).

This expression is also plotted in Figure 6–13, from which it is seen that the theory is in overall agreement with the measurements.

The difference between the path loss $L(1800)$ at 1800 MHz and the path loss $L(900)$ at 900 MHz is of interest for comparing coverage in cellular and PCS systems. The free-space loss for isotropic antennas increases by 6 dB when the frequency is doubled. In the presence of buildings, the loss due to diffraction of the rooftop fields down to street level has a frequency dependence that leads to an additional 3 dB per doubling of frequency. Finally, the reduction of the rooftop fields $Q_M(g_c)$ has a frequency dependence through the parameter g_c. The difference in path loss at 1800 and 900 MHz due to $Q_M(g_c)$ will also depend on the height of the base station antenna relative to the surrounding buildings, being negative for high antennas and positive for low antennas. Taken together the three contributions in (6–49) lead to the difference $L(1800) - L(900)$ that is plotted in Figure 6–14 as a function of the relative antenna height y_0 for a distance $R = 1$ km [21]. For base stations below the buildings the difference is greater than 12 dB, whereas for high antennas it approaches 6 dB. Measurements indicate the same inverse relation between the path loss difference and the antenna height, with the differences ranging from 11 to 8 dB [24,25].

6.4c Path loss for mobile-to-mobile propagation

When both ends of the radio link are located at street level, as in mobile-to-mobile communications, y_0 is negative and may have a large magnitude. In this case the sum in (6–46) is poorly convergent. A simple alternative representation for this case is found by recognizing that the diffracted fields generated at the rooftop next to the transmitting mobile are in the form of a cylindrical wave emanating from an equivalent line source whose strength is product of the incident

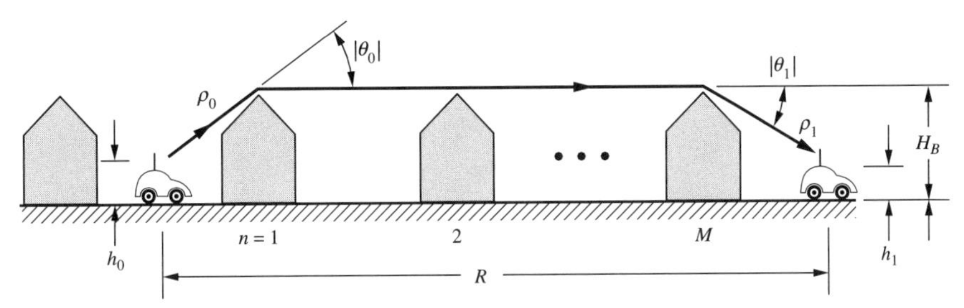

Figure 6-15 Geometry for computing path loss between mobiles at street level.

field and the diffraction coefficient. Referring to Figure 6–15, the cylindrical wave generated at the rooftop next to the transmitting mobile on the left travels across the rooftops before being diffracted down to the receiving mobile on the right. The effect of intervening rooftops on this cylindrical wave is the same as that for a line source at the level of the rooftops. Because the cylindrical wave passes $M - 1$ rows, the effect of the intervening rooftops is given by $Q_{M-1}(0) = 1/(M-1)$.

Considering only the single ray diffracted at the rooftop to the right of the transmitting antenna in Figure 6–15, the foregoing approximation for very low transmitting antennas leads to the following formula for the effective value Q_e of the field reduction at the rooftop of row M:

$$Q_e = \frac{1}{\sqrt{\rho_0}}|D_T(\theta_0)|\frac{1}{M-1} \tag{6–54}$$

The relation between this expression for the field reduction, and the cylindrical and plane wave expressions is demonstrated in Figure 6–16 for 900-MHz signals incident on the row $M = 20$ for a row spacing of $d = 50$ m. When computing Q_e from (6–54), the UTD form of Felsen's diffraction coefficient for an absorbing screen has been used and the transmitter was assumed to be at a distance d to the left of the diffracting building in Figure 6–15. For low antennas ($y_0 < -\sqrt{\lambda d}$), when the Fresnel zone is completely blocked by the first row, it is seen from Figure 6–15 that Q_e is close to $Q_M(g_c)$. As the transmitting antenna is raised to rooftop height, Q_e diverges from $Q_M(g_c)$. Later in the chapter, these differences are discussed in terms of higher-order corrections to diffraction by an edge. Figure 6–16 also shows how the cylindrical wave $Q_M(g_c)$ merges with the plane wave factor $Q(g_p)$ for $y_0 > \sqrt{\lambda d}$, as discussed previously.

Using the same arguments that lead to expression (6–9) for the effective diffraction coefficient at both ends of the link, the path loss for the link in Figure 6–15 is

$$L = -10\log\left(\frac{\lambda}{4\pi R}\right)^2 - 10\log\left[\frac{\lambda\rho_0}{2\pi^2(H_B-h_0)^2}\right] - 10\log\left[\frac{\lambda\rho_1}{2\pi^2(H_B-h_1)^2}\right] + 20\log(M-1) \tag{6–55}$$

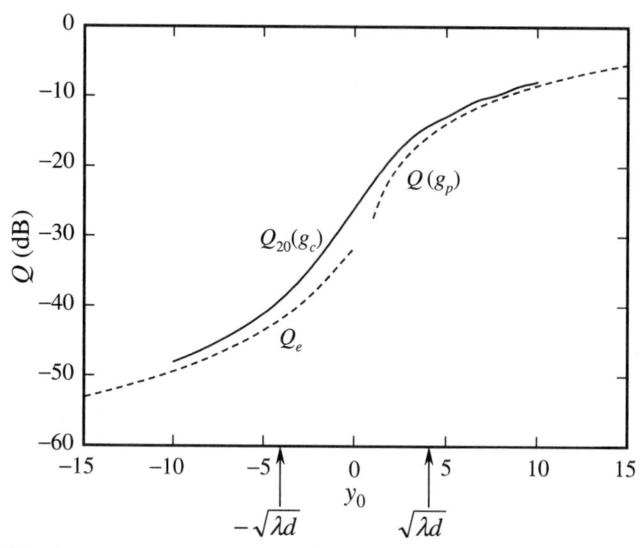

Figure 6-16 Comparison of the cylindrical wave reduction $Q_M(g_c)$ with that of the plane wave reduction $Q(g_p)$ for $g_p = g_c/M$ and with the approximate reduction Q_e for low antennas. The computations are for $\lambda = 1/3$ m, $d = 50$ m, and $M = 20$.

As an example, suppose that both mobiles are of the same height $h_0 = h_1$ and are at the same distance $d/2$ from the center of the nearest row of buildings. In this case $R = Md$ and the diffraction loss is the same at both ends of the link. If we use the approximation $\rho_0 = \rho_1 \approx d/2$ in (6–55), after some manipulation the path loss becomes

$$L = 20 \log (16\pi^3) + 20 \log [M(M-1)] + 40 \log\left(\frac{H_B - h_1}{\lambda}\right) \tag{6–56}$$

If $H_B - h_1 = 10$ m and $\lambda = 1$ m (300 MHz), the path loss is $94 + 20 \log [M(M-1)]$ dB. When going past $M = 10$ rows, the total path loss is therefore 133 dB.

6.4d Propagation oblique to rows of buildings

When the propagation path is oblique to the rows of buildings, as shown in top view in Figure 6–17, the diffraction formulas for PG_1 and PG_2 need to be modified. If the horizontal distance is large compared to the base station antenna height relative to the buildings, the wave incident on the rooftops has wavenumber along the rows that is given by $k \sin \psi$, where ψ is the angle shown in Figure 6–17. As discussed in Chapter 5, the wavenumber in the plane transverse to the rows is therefore $k \cos \psi$, and the wavelength in the plane is $\lambda/\cos \psi$. Hence the solution for oblique propagation is obtained by replacing k and λ appearing in PG_1 and PG_2 by $k \cos \psi$ and $\lambda/\cos \psi$, respectively. This modification to (6–9) results in

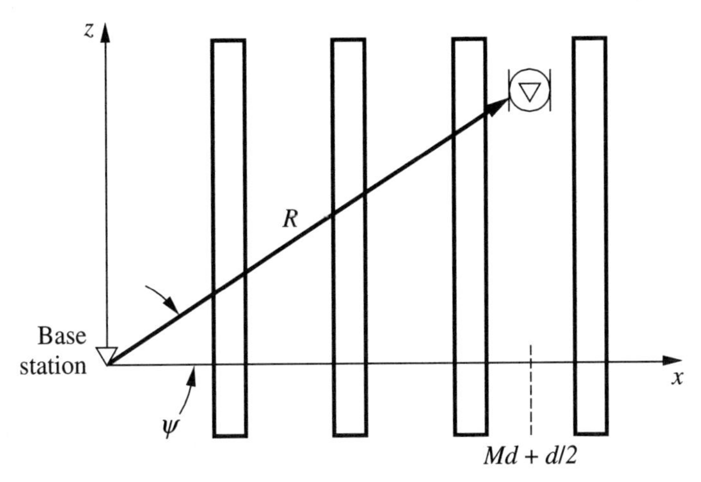

Figure 6-17 Path geometry as seen from above for propagation oblique to the street grid.

$$PG_2 = \frac{1}{\pi k \rho_1 \cos\psi} \frac{1}{|\theta_1|^2} \approx \frac{\lambda \rho_1}{2\pi^2 \cos\psi \, (H_B - h_m)^2} \tag{6–57}$$

where $\rho_1 = \sqrt{(H_B - h_m)^2 + (d/2)^2}$. As noted in Chapter 5, this expression is only valid provided that the subscriber antenna lies outside the transition region about the shadow boundary. This condition requires that $H_B - h_m$ be greater than the half-width of the transition region, which at the center point $d/2$ between the rows of buildings is $\sqrt{\lambda d/(2\cos\psi)}$. For ψ large enough, the transition zone widens to encompass the subscriber antenna, and (6–57) must be replaced by the UTD diffraction coefficient. Thus while PG_2 in (6–57) increases with increasing angle ψ, in the limit as $\psi \to 90°$, $PG_2 \to 1/4$.

The wavelength enters $PG_1 = Q_M^2$ through the parameter g_c, which becomes

$$g_c = y_0 \sqrt{\frac{\cos\psi}{\lambda d}} \tag{6–58}$$

Note that the number of rows M crossed by the wave is the integer part of $(R/d)\cos\psi$, so that for a fixed distance R, the number of rows decreases with ψ. The decrease of g_c and M with increasing ψ have opposite effects on PG_1. Overall, PG_1 is found to increase with increasing ψ. As an example, if $\psi = 60°$, then $g_c = y_0/\sqrt{2\lambda d}$, and $M = 10$ corresponds to a distance of 1 km for $d =$ 50 m. The height dependence of Q_{10} is plotted in decibels in Figure 6–12 for $\psi = 60°$ and is labeled $Q_{10,\text{oblique}}$. It is seen from Figure 6–12 that at 1 km, PG_1 is indeed larger than it is at $\psi =$ 0.

The wavelength enters the free-space path loss for isotropic antennas through the effective area of the receiving antenna, which is independent of the direction of propagation, and expression (2) for PG_0 therefore remains unchanged. Because both PG_1 and PG_2 increase with ψ, and

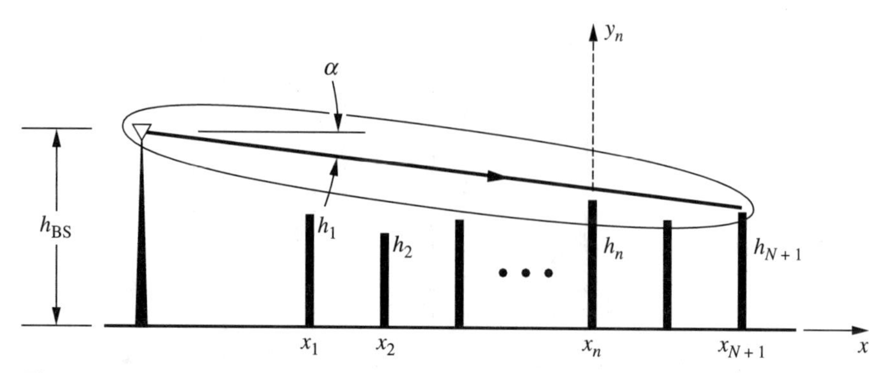

Figure 6-18 Use of the Fresnel zone to determine the size of the aperture needed for numerical evaluation of the physical optics integrals.

PG_0 is unchanged, the theoretical model indicates that the path loss at a given horizontal distance R is less on oblique paths than it is on paths that are perpendicular to the street grid. Few measurements have been made of the effect of street orientation on path loss. The measurements made in San Francisco, and their regression fit, which is discussed in Chapter 2, show little dependence on orientation. The cause of this discrepancy between the theory and measurements is not known.

6.5 Numerical evaluation of fields for variable building height and row spacing

In Sections 6–3 and 6–4, solutions for the reduction in the rooftop fields due to propagation past the rows of buildings were expressed in terms of Borsma's functions for uniform building height and uniform row spacing. If these conditions do not hold, as in Figure 6–18, the field reduction can be found from repeated numerical evaluation of the integral in (6–12). To carry out the numerical integration, it is necessary to terminate the integral at a finite upper limit and to replace the integration by a discrete summation. An abrupt termination of the integral is equivalent to placing an absorbing screen above the termination point, and would therefore artificially generate diffracted waves that are not present in the actual problem. To avoid this spurious diffraction contribution, the termination must be done in a continuous way. One approach is to integrate numerically up to some limit and then integrate to infinity analytically using a simplified approximate integrand [10]. This approach cancels the endpoint contribution from the numerical integration by the endpoint contribution from the analytic approximation. Alternatively, the windowing approach used in Chapter 5 to study the local properties of radiation can be used to suppress the endpoint contribution by making the field go smoothly to zero before the integration is terminated.

6.5a Windowing to terminate the integration

The windowing approach is implemented by multiplying the integrand in (6–12) by a function $W(y)$ that has the definition

$$W(y) = \begin{cases} 1 & \text{for } 0 \le y \le y_w \\ w_T(y) & \text{for } y_w < y \le y_w + T \\ 0 & \text{for } y_w + T < y \end{cases} \tag{6–59}$$

Here $w_T(y)$ is a continuously varying function that goes from unity to zero over the interval $y_w < y \le y_w + T$. Various functions will do for $w_T(y)$, one of which is the Kaiser–Bessel function $K(\xi)$, defined by

$$K(\xi) = 0.40208 + 0.49858\cos\xi + 0.09811\cos(2\xi) + 0.00123\cos(3\xi) \tag{6–60}$$

where

$$\xi = \pi\frac{y - y_w}{T} \tag{6–61}$$

The window size needed for accurate computations is related to the Fresnel zone. Figure 6–18 shows the Fresnel zone about the ray from the transmitter to the rooftop of the farthest row $N + 1$ of buildings at a distance R. The window width $y_w^{(n)}$ at the row n must include at least the Fresnel zone. Experience with numerical computations for various building profiles indicates that stable numerical results are obtained when the width is given by

$$y_w^{(n)} = y_{N+1} + (x_{N+1} - x_n)\tan\alpha + 3\sqrt{\lambda R} \tag{6–62}$$

Here α is the angle between the ray from the base station antenna to the $N + 1$ row, and the horizontal. When the ray in Figure 6–18 from the base station antenna to the $N + 1$ row is blocked by intervening buildings, the terms $y_{N+1} + (x_{N+1} - x_n)\tan x$ in (6–62) are replaced by a line defined such that all the intervening rooftops lie on or below the line.

The width $T^{(n)}$ of the smoothing function at row n is found in terms of the Fresnel zone in going from row n to row $n + 1$. Stable numerical results are found using

$$T^{(n)} = 15\sqrt{\lambda(x_{n+1} - x_n)} \tag{6–63}$$

As an example, assume that the buildings are of uniform height $h_n = H_B$, that the base station antenna is at the rooftop height $h_{BS} = H_B$ so that $\alpha = 0$, and that the row spacing is uniform with $d = 60$ m. For a frequency of 1 GHz and the overall path length of $R = 3$ km, it is found from (6–62) that $y_w^{(n)} = 90$ m or 300λ, and from (6–63) that $T^{(n)} = 63.6$ m or 212λ. Thus the overall integration aperture is 153.6 m or 512λ.

Using the window function (6–59) in the integrand in (6–12) for the field above the $n + 1$ screen

$$H(x_{n+1}, y) \approx e^{j\pi/4}\sqrt{\frac{k}{2\pi}}\int_{h_n}^{y_w + T} A_n(y, y')e^{-jk\rho_n(y, y')}dy' \tag{6–64}$$

where

$$A_n(y, y') = W(y')\frac{H(x_n, y')}{\sqrt{\rho_n(y, y')}} \tag{6–65}$$

In (6–64) and (6–65), $\rho_n(y, y') = \sqrt{d_n^2 + (y - y')^2}$ where $d_n = x_{n+1} - x_n$ is the distance between row n and row $n + 1$. Since the wave propagation is nearly horizontal, the phase of $H(x_n,y')$ will be only slowly varying with y' and can be viewed as part of the complex amplitude.

6.5b Discretization of the integration

Because of the large size of the integration aperture, when descretizing the integration in (6–64) it is desirable to approximate the integrand in such a way that the step size Δy is a significant fraction of a wavelength, or larger. This can be accomplished by approximating the amplitude $A_n(y')$ and phase $k\rho$ over the discrete interval of the integration, and integrating the approximation in closed form. Simple algebraic expressions result if linear approximations are used for the phase and amplitude, as is shown below. If quadratic approximations are used, the closed-form expressions contain Fresnel integrals [26]. The linear approximations for the phase and amplitude in the interval between $m\,\Delta y$ and $(m + 1)\,\Delta y$ make use of their values at the endpoints of the interval. It has been found that stable numerical results are obtained when Δy is about $\lambda/3$ to λ / 2. Since the integration aperture may be several hundred λ, as discussed in Section 6–5a, there are on the order of 1000 integration steps.

Because the integration is repeated many times, it is necessary to compute the integral at the discrete field points $y = p\,\Delta y$ above the $n + 1$ row. In writing the linear approximation, we make use of the simplifying notation

$$\begin{aligned} A_n[p, m] &\equiv A_n(p\Delta y, m\Delta y) \\ \rho_n[p, m] &\equiv \sqrt{d_n^2 + (p\Delta y - m\Delta y)^2} = \sqrt{d_n^2 + (p - m)^2(\Delta y)^2} \end{aligned} \tag{6–66}$$

Using this notation, the linear approximations are

$$\begin{aligned} A_n[p\Delta y, y'] &= A_n[p, m] + \frac{A_n[p, m+1] - A_n[p, m]}{\Delta y}(y' - m\Delta y) \\ \rho_n[p\Delta y, y'] &= \rho_n[p, m] + \frac{\rho_n[p, m+1] - \rho_n[p, m]}{\Delta y}(y' - m\Delta y) \end{aligned} \tag{6–67}$$

The approximations in (6–67) may now be used in the integration over the interval from m Δy to $(m + 1)\,\Delta y$, which can be shown to give

$$\int_{m\Delta y}^{(m+1)\Delta y} A_n(p\Delta y, y')e^{-jk\rho_n(p\Delta y, y')}dy' \tag{6–68}$$

$$= \Delta y \frac{A_n[p, m+1]e^{-jk\rho_n[p, m+1]} - A_n[p, m]e^{-jk\rho_n[p, m]}}{-jk(\rho_n[p, m+1] - \rho_n[p, m])}$$

$$+ \Delta y \frac{(A_n[p, m+1] - A_n[p, m])(e^{-jk\rho_n[p, m+1]} - e^{-jk\rho_n[p, m]})}{k^2(\rho_n[p, m+1] - \rho_n[p, m])^2}$$

The field $H(x_{n+1}, p\,\Delta y)$ at the height $p\,\Delta y$ above the $n + 1$ row can be found by summing the integrals (6–68) over the integration steps. The lower limit of the sum is taken to be I_n, which is the integer part of $(h_n/\Delta y)$, where h_n is the height of row n, and the upper limit is L_n, which from (6–62) and (6–63) is seen to be the integer part of $(y_w^{(n)} + T^{(n)})/\Delta y$. In terms of these limits, the summation for the field gives

$$H(x_{n+1}, p\Delta y) = \frac{\Delta y e^{j\pi/4}}{\sqrt{2\pi k}} \sum_{m=I_n}^{L_n} \left\{ \frac{A_n[p, m+1]e^{-jk\rho_n[p, m+1]} - A_n[p, m]e^{-jk\rho_n[p, m]}}{-j(\rho_n[p, m+1] - \rho_n[p, m])} + \frac{(A_n[p, m+1] - A_n[p, m])(e^{-jk\rho_n[p, m+1]} - e^{-jk\rho_n[p, m]})}{k(\rho_n[p, m+1] - \rho_n[p, m])^2} \right\} \tag{6–69}$$

To find the field at points $q\Delta y$ above the $n + 2$ row, we define $A_{n+1}[q,p]$ at points $p\Delta y$ above the $n + 1$ row. With the help of (6–65) and (6–69), it is seen that

$$A_{n+1}[q, p] = W(p\Delta y)\frac{H(x_{n+1}, p\Delta y)}{\sqrt{\rho_{n+1}[q, p]}} \tag{6–70}$$

where $\rho_{n+1}[q,p]$ is as defined in (6–66) with d_{n+1} replacing d_n. Taken together, (6–69) and (6–70) represent recursion relations that can be used to find the field incident on the $N + 1$ row when the field incident on the first row is specified. Two applications of the numerical integration method are discussed below. Further applications are treated in Chapter 7.

6.5c Height dependence of the settled field

One advantage of the numerical integration method, even for buildings of uniform height, is that it can give the field incident at any point in the plane of the $M = N + 1$ row, not just the top of the row. The height dependence of the field incident on the 120 row due to a plane wave of unit amplitude incident on the first row is plotted in Figure 6–19. The buildings are assumed to be of uniform height and the row separation is taken to be $d = 50$ m. The plane wave is incident at a glancing angle $\alpha = 1°$ and has a frequency of 900 MHz. For these parameters, the settling number from (6–31) is $N_0 = 21$, which is smaller than the value of M for which Figure 6–19 is plot-

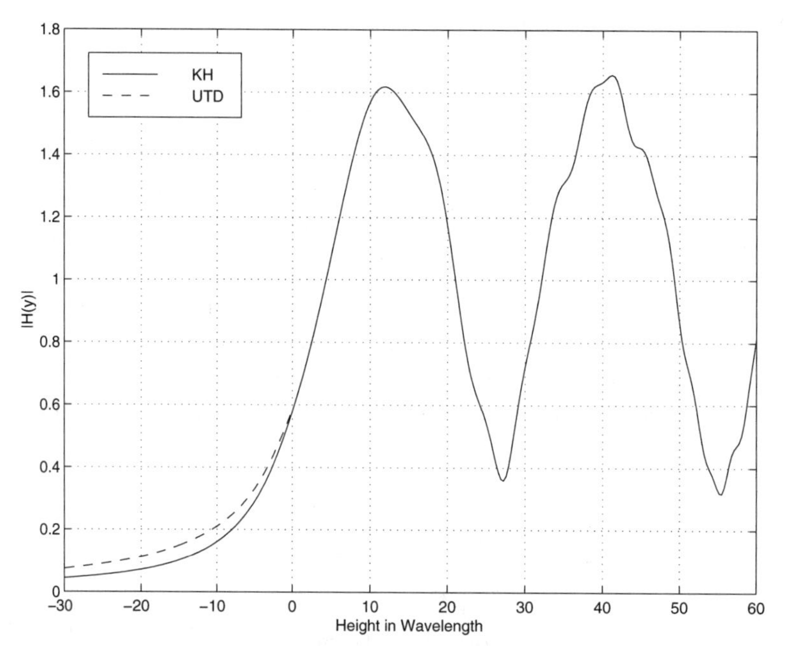

Figure 6-19 Height variation of the field incident on the $M = 120$ row for a 900-MHz plane wave incidence at $\alpha = 1°$ when the row spacing is $d = 50$ m. The simple diffraction approximation is also shown for $y < 0$.

ted. In this plot y is the height above the rooftops, so that the value of the field at $y = 0$ is the settled value $Q(g_p)$. For the parameters used in the calculation, $g_p = 0.214$, and from (6–32) it is found that $Q(g_p) = 0.61$, which agrees with the field amplitude at $y = 0$ in Figure 6–19.

Above the buildings, the field variation is similar to that of a standing wave resulting from the summation of the incident wave and a wave reflected from the plane of the rooftops. Like the waves reflected by a grating, this wave is in fact the sum of the waves diffracted from the rooftops, whose phase variation causes them to add constructively in the specular direction and other grating lobe directions. Because the diffraction coefficients decrease with angle, the reflected field is greatest in the specular direction. Since the wavenumber of the incident plane wave in the y direction is $k \sin \alpha$, the distance between interference minima is $(2\pi/k \sin \alpha)/2$, or $\lambda/(2 \sin \alpha)$. For the parameters assumed, it is 9.5 m, or 28.5λ, which is seen to be in good agreement with the actual distance between the minima in Figure 6–19. For an incident wave of unit amplitude, the minimum of the standing-wave pattern is $1 - |\Gamma|$, while the maximum is $1 + |\Gamma|$. In Figure 6–19 it is seen that $|\Gamma| \approx 0.62$.

For values of y between the rooftops ($y = 0$) and the first peak of the standing wave at $\lambda/(2 \sin \alpha)$, the field amplitude is a distorted version of a standing wave. For a pure standing wave with $\Gamma = -1$, the y dependence of the field is given by the difference of the terms

$\exp[\pm jky \sin\alpha]$. Well below the peak where $y \ll \lambda/(4\sin\alpha)$, the variation can be approximated by

$$e^{jky\sin\alpha} - e^{-jky\sin\alpha} = 2j\sin(ky\sin\alpha) \approx 2jky\sin\alpha \tag{6–71}$$

In the range $0 \le y \ll \lambda/(4\sin\alpha)$, the standing-wave variation in (6–71) can be combined with the settled amplitude at $y = 0$ by adding the squares of the value $Q(g_p)$ of the field at $y = 0$, and the pure standing-wave variation in (6–71). The resulting expression is

$$|H(x_n, y)| \approx \sqrt{(2ky\sin\alpha)^2 + Q^2(g_p)} = \sqrt{\left(\frac{4\pi y}{\lambda}\sin\alpha\right)^2 + Q^2(g_p)} \tag{6–72}$$

Equation (6–72) gives the height gain for roof mounted antennas, such as those used for TV or for wireless local loops. For example, consider a 200-MHz TV signal radiated by an antenna at an elevation $h_{BS} - H_B = 150$ m and propagating over rows of buildings separated by $d = 60$ m. At a distance of 10 km, $\sin\alpha$ is 0.015, and $\lambda/(4\sin\alpha) = 16.7$ m. The dimensionless parameter g_p in (6–18) has value 0.095, and from (6–32), $Q(g_p) = 0.303$. To increase the received field by a factor of $\sqrt{2}$, or 3 dB, from its value for $y = 0$, it is necessary to raise the antenna to $y = \lambda Q(g_p)/(4\pi\sin\alpha)$, or $y = 2.4$ m.

The field variation in Figure 6–19 below the rooftop ($y < 0$) is seen to be similar to that of the field in the shadow region for diffraction by a single edge, as shown in Figure 5–8. One significant difference is the amplitude at the shadow boundary. When a geometrical optics field is diffracted by a single edge, the field on the shadow boundary is one-half the value of the field that would be there in the absence of the edge. For an incident plane wave of unit amplitude, the field on the shadow boundary has a magnitude of 1/2, as seen in Figure 5–8. However, after diffraction past many rows the field incident on any one rooftop, which is near the shadow boundary of the previous rooftop, has the same value $Q(g_p)$ as that incident on the previous rooftop. This effect results from the height dependence of the field amplitude incident on the previous row, which results in higher-order diffraction effects [15,16].

Since the diffracted field $e^{-j\pi/4}D_T(\theta)/\sqrt{\rho}$ given in (5–43) has value 1/2 on the shadow boundary, the higher order diffraction effects can be approximately accounted for by doubling D_T. Thus for comparison in Figure 6–19, we have plotted $2Q(g_p)|D_T(\theta)|/\sqrt{\rho}$, which is seen to somewhat overestimate the field amplitude compared to the numerically generated variation. At 30λ, or 10 m, below the rooftops, the diffraction approximation is almost double the numerically generated value, and therefore closer to the value previously obtained without including the extra factor of 2. Thus the higher-order diffraction effects are significant near rooftop level but less significant at street level. In particular, the higher-order effects are responsible for the discrepancy between Q_e and $Q_M(g_c)$ for $y < 0$ in Figure 6–16.

6.5d Influence of roof shape

In the foregoing discussions, the roofs were assumed to be peaked, with the peak running parallel to the row, as in Figure 6–20c. If the peaks run perpendicular to the row, the actual fields can

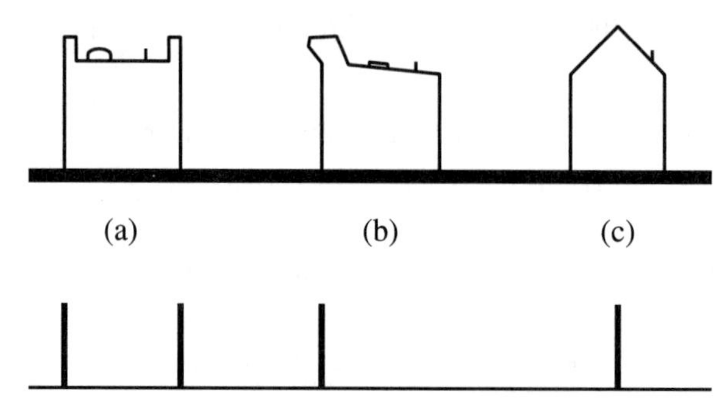

Figure 6-20 Building profiles showing two styles of flat roof and a peaked roof, and their modeling by absorbing screens [27] (©1999 IEEE).

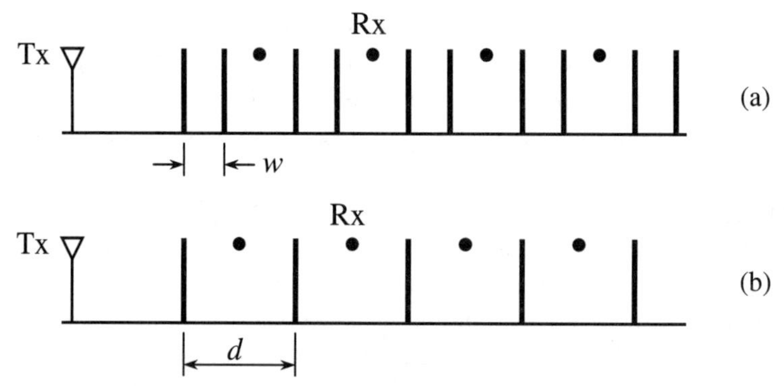

Figure 6-21 Transmitter, field point, and screen locations for comparing the reduction in the rooftop fields for buildings having flat roofs with parapets to buildings having peaked roofs running parallel to the rows [27] (©1999 IEEE).

only be found from a two-dimensional numerical integration. Such integrations are computationally intensive and cannot be repeated for multiple rows. It is assumed that diffraction past rows of buildings with such roof will not be significantly different than when the peaks run parallel to the rows. Depending on their construction, buildings with flat roofs may be represented by one or two diffracting screens, as shown in Figure 6–20. Some buildings have a decorative cornice at the front, as in Figure 6–20b, that is higher than the rest of the roof. In this case the building can be represented by a single diffracting screen. The location of the screen at the front of the building rather than the center will change the diffraction loss down to the subscriber but will not have a significant effect on the field propagating over the rows of buildings.

Larger buildings with flat roofs sometimes have parapets around the roofs, as suggested in Figure 6–20a. The presence of the two parapets will cause a greater reduction of the rooftop fields than will the single diffracting edges in Figure 6–20b and c. To quantify this effect, computations were made for diffraction past absorbing screens positioned as shown Figure 6–21a and c. In the first case of Figure 6–21a, the screens representing the two parapets of a building are separated by the building width w, while the distance to the face of the building in the next

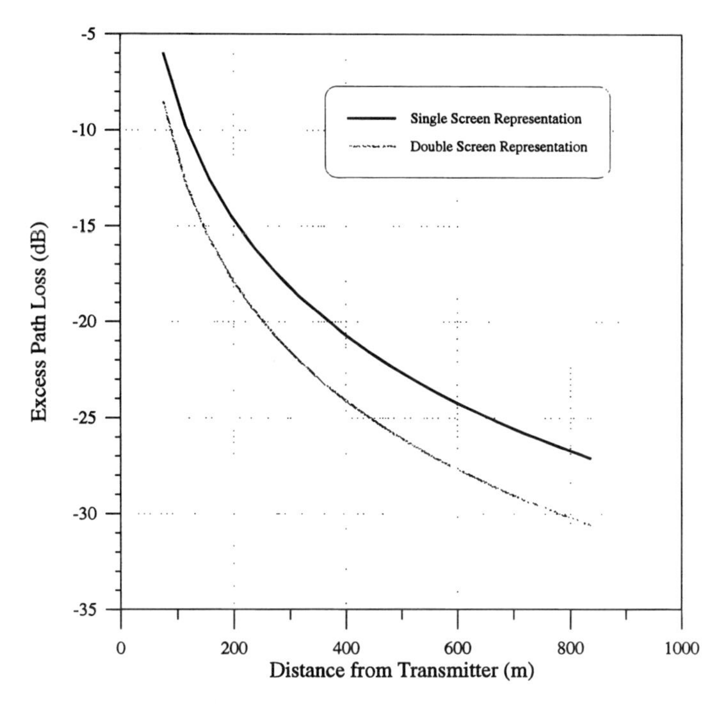

Figure 6-22 Field reduction in excess of free-space loss for propagation over buildings having flat roofs with parapets and buildings having peaked roofs running parallel to the rows [27] (©1999 IEEE).

row is $d - w$. The second case of Figure 6–21c is for buildings with peaked roofs having the same row separation d. In both cases the field computed at the points labeled Rx, which are midway between the buildings and at rooftop level, are used in plotting the path loss in excess of free space in Figure 6–22. The plots of Figure 6–22 are for a base station antenna at rooftop height, $w = 10$ m, $d = 40$ m, and a frequency of 900 MHz. After two rows of buildings, the offset of the parapet buildings from the peaked roof buildings is nearly constant at 3.3 dB. This implies that for both types of roofs, the range exponent for the distance variation will be the same, but the signal level will be offset by a constant 3.3 dB, in addition to the difference in the diffraction loss down to the subscriber. Other numerical examples have shown that the offset decreases with increasing base station height above the buildings [27].

In going from an integration point in the plane above the first parapet of a building to a field point in the plane above the other parapet, we have included only the direct ray. One could also include an additional roof reflected ray. If the parapets are low, the roof reflections can affect the field diffracted over a single building by a few decibels [28]. If the height of the parapet is greater than the maximum width of the Fresnel zone $(1/2)\sqrt{\lambda w}$, reflections from the roof have little effect on the fields diffracted over the roof [28]. At 900 MHz, the Fresnel width for a building $w = 10$ m wide is 0.9 m. For larger buildings, the Fresnel width at 900 MHz will be greater than the height of the parapets. For buildings without parapets, and for buildings with other types of roofs, it is conjectured that the roof shape will result in a difference in path loss of a few decibels, but will not affect the range exponent.

6.6 Summary

Away from the high-rise core, the buildings of many cities are low and of roughly uniform height, with occasional tall buildings. Except in distant suburbs, the buildings are arranged in rows by the street grid, with small gaps between neighbors in a row. Because of the significant loss in going through a building, propagation must take place over the buildings. Various paths involving diffraction bring the rooftop fields down to street level. With this picture, the path loss can be divided into three components: (1) the free-space path loss, which is that associated with horizontal distance R between the base station and subscriber; (2) the reduction of the rooftop fields due to propagation over the rows of buildings; and (3) the loss due to diffraction of the rooftop fields down to street level. The free-space loss was discussed in Chapter 4. Although there are many possible diffraction paths by which the rooftop fields can reach street level, the dominant path is the one from the rooftop before the subscriber. By doubling its power to account for the contributions from all other paths, a simple approximation is obtained for the loss associated with the diffraction down to street level.

Computing the reduction of the rooftop fields due to propagation past a row of buildings is a significantly more complex problem. We have adopted the physical optics approach in which the fields above one row of buildings is found by integrating over the fields in the plane above the previous row. Iterating this procedure from row to row allows the fields to be calculated for many rows. When the buildings are of uniform height and the row spacing is uniform, the fields at the tops of the buildings can be found using Borsma's functions. For elevated base stations and $R > 1$ km, the field reduction is found in terms of the function $Q(g_p)$ whose argument $g_p \approx [(h_{BS} - H_B)/R]\sqrt{d/\lambda}$ contains all of the significant path parameters. Simple approximations for $Q(g_p)$ yield closed-form expressions for path loss that show good agreement with measurements. For microcellular coverage over distances $R < 1$ km, base station antennas may be located near the building height, so that $(h_{BS} - H_B) \leq \sqrt{\lambda d}$, and the field reduction is dependent on the number of rows M that lie between the base station and subscriber. If the base station is located at a distance d before the first row, the field reduction as $Q_M(g_c)$ is again given by Borsma's functions, whose argument $g_c = (h_{BS} - H_B)/\sqrt{\lambda d}$ contains the path parameters. In this case there is no closed-form approximation for $Q(g_p)$ from which to compute the path loss, except when $h_{BS} = H_B$ and $Q_M(0) = 1/M$. For very low base station antennas, with $(h_{BS} - H_B) \leq -\sqrt{\lambda d}$, as in mobile-to-mobile communications, a simple closed-form approximation Q_e can again be found.

When the building height and row separations are not uniform, or if the field point is not at the rooftop, the physical optics integrals must be evaluated numerically. As one example, this method was used to find the height variation of the field incident on the plane of one row, after passing many other rows. Above the rooftops, the fields were found to have a standing-wave variation, whose height gain may be useful in locating antennas for fixed wireless applications.

At points below the rooftops, the variation of the field is found to be like that predicted by the expression for diffraction loss down to the mobile, except for a multiplying factor of 2 (6 dB). The numerical results show up the higher-order diffraction effects that result from the height variation of the field incident on the row before the subscriber. Numerical integration was also used to examine the effect of the roof shape. Limited computations suggest that roof shape will offset the path loss by a few decibels, but will not affect the range exponent. In Chapter 7, the numerical integration method is used to examine other problems of interest in wireless systems.

Problems

6.1 In Figure 6–2, assume that $H_B = 12$ m, $d = 50$ m, and $w = 12$ m. Also assume that $|\Gamma|^2 = |T|^2 = 0.1$ and that $f = 900$ MHz. Compute and plot PG_2 versus subscriber antenna height in the range $1 \le h_m \le 9$ m using:

(a) Expression (6–3) together with the Felsen diffraction coefficient of (6–6).

(b) The first and last approximations in (6–9).

6.2 The transmitting antenna for an FM station at 100 MHz is located at a height $h_{BS} = 160$ m above ground, and services a residential area with houses that are $H_B = 10$ m high and have row separation $d = 60$ m. A receiving antenna is located at rooftop level on a house at a distance $R = 12$ km from the transmitter. Find the path loss between the antennas.

6.3 Suppose that $h_{BS} = 30$ m, $H_B = 12$ m, $d = 50$ m, $h_m = 1.5$ m, and $f = 1.5$ GHz. We wish to find the variation of the path loss L with distance R in the range $1 \le R \le 10$ km.

(a) Using expression (6–36), plot L versus R.

(b) Repeat part (a) using the Hata formula (2–9).

6.4 For the geometry shown in Figure 6–9, assume that $d = 60$ m and $N + 1 = 50$, so that $R = 3$ km and $f = 900$ MHz.

(a) For what base station antenna height y_0 will the Fresnel zone about the ray from the base station to the last edge at (3000, 0) just touch the first edge, which is closest to the base station? What is the value of the parameter g_p for this value of y_0?

(b) If $y_0 = 15$ m, on graph paper make a sketch showing the Fresnel zone about the ray from the base station antenna to the last edge (use different scales for horizontal and vertical dimensions).

(c) On the same graph paper sketch the Fresnel zone to the last edge for a plane wave incident at the angle $\alpha = \arctan(15/3000)$.

6.5 In Figure 6–15, assume that the mobiles are at a horizontal distance d from the edges of the nearest buildings. Using the approximation $\rho_0 = \rho_1 \approx d$ in (6–55), compute the path loss between isotropic antennas at 300 MHz if $H_B = 12$ m, $h_m = 2$ m, and $d = 60$ m.

6.6 Referring to Figure 6–17, suppose that the subscriber travels along a street at a distance $x = 630$ m from the base station. Assume that $h_{BS} = H_B = 9$ m, so that $Q_M(g_c) = 1/M$, and that $h_m = 1.6$ m, $d = 60$ m, and $f = 1.8$ GHz. Compute and plot the path loss L for isotropic antennas as a function of z over the range $0 \le z \le 1.2$ km. For comparison, plot the path loss obtained from expression (2–19) for this case.

6.7 Repeat Problem 6–6 when the base station antenna is raised to a height $h_{BS} = 15$ m. For this case $y_0 > \sqrt{\lambda d/\cos\psi}$ and the plane wave factor $Q(g_p)$ can be used to find the reduction in the rooftop fields.

6.8 Frequencies near 3.5 GHz have been proposed for fixed wireless applications, which would replace the wires from the local telephone center to individual houses. Suppose that a base station antenna located $y_0 = 5$ m above the rooftops is used to cover houses out to a distance of 1 km. Assume that the propagation is perpendicular to the rows of buildings, and that $d = 50$ m.

(a) Near the edge of the coverage area, where the signal is weakest, find the height y of the subscriber antenna that gives a signal increase of 6 dB as compared to rooftop height $y = 0$.

(b) If the base station antenna has a gain of 18 dB and the subscriber antenna has a gain of 7 dB, find the path loss at the edge of the coverage area for the value of y found in part (a).

References

1. F. Ikegami, S. Yoshida, T. Takeuchi, and M. Umehira, Propagation Factors Controlling Mean Field Strength on Urban Streets, *IEEE Trans. Antennas Propagat.*, vol. AP-32, pp. 822–829, 1980.
2. J. Walfish and H. L. Bertoni, A Theoretical Model of UHF Propagation in Urban Environments, *IEEE Trans. on Antennas Propagat.*, vol. AP-36, pp. 1788–1796, 1988.
3. Y. Okumura, E. Ohmori, T. Kawano, and K. Fukuda, Field Strength and Its Variability in VHF and UHF Land-Mobile Radio Service, *Rec. Electron. Commun. Lab.*, vol. 16, pp. 825–873, 1968.
4. M. J. Brooking and R. Larsen, Results of Height Gain Measurements Taken in Different Environments, *Tech. Rep. MTR 84/42*, GEC Research Laboratories, Marconi Research Centre, Chelmsford, Essex, England, 1985.
5. L. R. Maciel and H. L. Bertoni, Theoretical Prediction of Slow Fading Statistics in Urban Environments, *Proc. IEEE ICUPC'92 Conference*, pp. 01.01.1–4, 1992.
6. S. W. Lee, Path Integrals for Solving Some Electromagnetic Edge Diffraction Problems, *J. Math. Phys.*, vol. 19, pp. 1414–1422, 1978.
7. L. E. Vogler, An Attenuation Function for Multiple Knife-Edge Diffraction, *Radio Sci.*, vol. 19, pp. 1541–1546, 1982.
8. H. H. Xia and H. L. Bertoni, Diffraction of Cylindrical and Plane Waves by an Array of Absorbing Half Screens, *IEEE Trans. Antennas Propagat.*, vol. 40, pp. 170–177, 1992.
9. L. R. Maciel, H. L. Bertoni, and H. H. Xia, Unified Approach to Prediction of Propagation over Buildings for All Ranges of Base Station Antenna Height, *IEEE Trans. Veh. Technol.*, vol. 42, pp. 41–45, 1993.
10. J. H. Whitteker, Numerical Evaluation of One-Dimensional Diffraction Integrals, *IEEE Trans. Antennas Propagat.*, vol. 45, pp. 1058–1061, 1997.
11. J-E. Berg and H. Holmquist, An FFT Multiple Half-Screen Diffraction Model, *Proc. IEEE Vehicular Technology Conference*, pp. 195–199, 1994.
12. M. F. Levy, Diffraction Studies in Urban Environments with Wide-Angle Parabolic Equation Method, *Electron. Lett.*, vol. 28, pp. 1491–1492, 1992.
13. F. Ikegami, T. Takeuchi, and S. Yoshida, Theoretical Prediction of Mean Field Strength for Urban Mobile Radio, *IEEE Trans. Antennas Propagat.*, vol. 39, pp. 299–302, 1991.
14. J. B. Andersen, Transition Zone Diffraction by Multiple Edges, *IEE Proc. Microwave Antennas Propag.*, vol. 141, pp. 382–384, 1994.
15. J. B. Andersen, UTD Multiple-Edge Transition Zone Diffraction, *IEEE Trans. Antennas Propagat.*, vol. 45, pp. 1093–1097, 1997.
16. M. J. Neve and G. B. Rowe, Contributions towards the Development of a UTD-Based Model for Cellular Radio Propagation Prediction, *IEE Proc. Microwave Antennas Propag*,. vol. 141, pp. 407–414, 1994.
17. W. Zhang, A More Rigorous UTD-Base Expression for Multiple Diffractions by Buildings, *IEE Proc.-Microwave Antennas Propag.*, vol. 142, pp. 481–484, 1995.

18. L. Juan-Llacer and N. Cardona, UTD Solution for the Multiple Building Diffraction Attenuation Function for Mobile Radiowave Propagation, *Electron. Lett.*, vol. 33, pp. 92–93, 1997.
19. J. Borsma, On Certain Multiple Integrals Occurring in a Waveguide Scattering Problem, *SIAM J. Math. Anal.*, vol. 9, pp. 377–393, 1978.
20. M. Abramowitz and I. A. Stegun, eds., Handbook of Mathematical Functions, Dover Publications, New York, pp. 256–257, 1965.
21. H. L. Bertoni, W. Honcharenko, L. R. Maciel, and H.H. Xia, UHF Propagation Prediction for Wireless Personal Communications, *Proc. IEEE*, vol. 82, pp. 1333–1359, 1994.
22. G. D. Ott and A. Plitkins, Urban Path-Loss Characteristics at 820 MHz, *IEEE Trans. Veh. Technol.*, vol. VT-27, pp. 189–197, 1978.
23. E. Damosso, ed., COST Action 231: Digital Mobile Radio towards Future Generation Systems, *European Commission Final Report EUR 18957*, Luxembourg, Belgium, Chap. 4, 1999.
24. P. E. Mogensen, P. Eggers, C. Jensen, and J. B. Andersen, Urban Area Radio Propagation Measurements at 955 and 1845 MHz for Small and Micro Cells, *Proc. IEEE GLOBECOM'91*, pp. 1297–1302, 1991.
25. L. Melin, M. Ronnlund, and R. Angbratt, Radio Wave Propagation: A Comparison Between 900 and 1800 MHz, *Proc. IEEE Vehicular Technology Conference*, pp. 250–252, 1993.
26. T. Hansen and R. A. Shore, Incremental Length Diffraction Coefficients for the Shadow Boundary of a Convex Cylinder, *IEEE Trans. Antennas Propagat.*, vol. 46, pp. 1458–1466, 1998.
27. L. Piazzi and H. l. Bertoni, On Screen Placement for Building Representation in Urban Environments Considering 2D Multiple Diffraction Problems, *Proc. IEEE Vehicular Technology Conference*, 1999.
28. C. J. Haslett, Modeling and Measurements of the Diffraction of Microwaves by Buildings, *IEE Proc. Microwave Antennas Propag.*, vol. 141, pp. 397–401, 1994.

CHAPTER 7

Shadow Fading and the Effects of Terrain and Trees

Although the effects of randomness that cause shadow fading and the effects of terrain and trees are rather different in nature, each plays a significant role in modifying the predictions of path loss discussed in Chapter 6. Their study is also linked in part by the use of numerical evaluation of the physical optics integrals to describe them quantitatively. Shadow fading is studied by considering random variations of building height, position, and construction. Terrain effects involve diffraction over hills, which is discussed here for bare hills and for hills covered with buildings. In the latter case, the height of the buildings varies systematically as a result of the terrain they are built on. Path loss to subscribers located in a forest and on roads or in clearings in a forest are found using refraction and diffraction concepts. To model the effect of trees planted next to houses, we again make use of the physical optics approximation by including partial transmission through the trees.

7.1 Shadow fading statistics

As a subscriber moves along a street, she or he passes buildings of different size, position, and construction, as suggested in Figure 7–1, as well as gaps between buildings, especially those occurring at street intersections. As a result of these differences, the small-area average of the received signal will show a variation that is referred to as *shadow fading*. The source of the fading can be understood from the narrow width of the Fresnel zone, as measured along the row. As an example of the horizontal width of the Fresnel zone in this context, assume that the distance from the subscriber to the nearest row of intervening buildings is $d/2$ and that the base station is at a distance $R \gg d$. Then the full horizontal width of the Fresnel zone at the roof of the nearest intervening row is equal to $2\sqrt{\lambda d/2}$. For 900-MHz signals and for $d = 60$ m, the Fresnel zone width is 6.2 m, which is the width of a house or other small building.

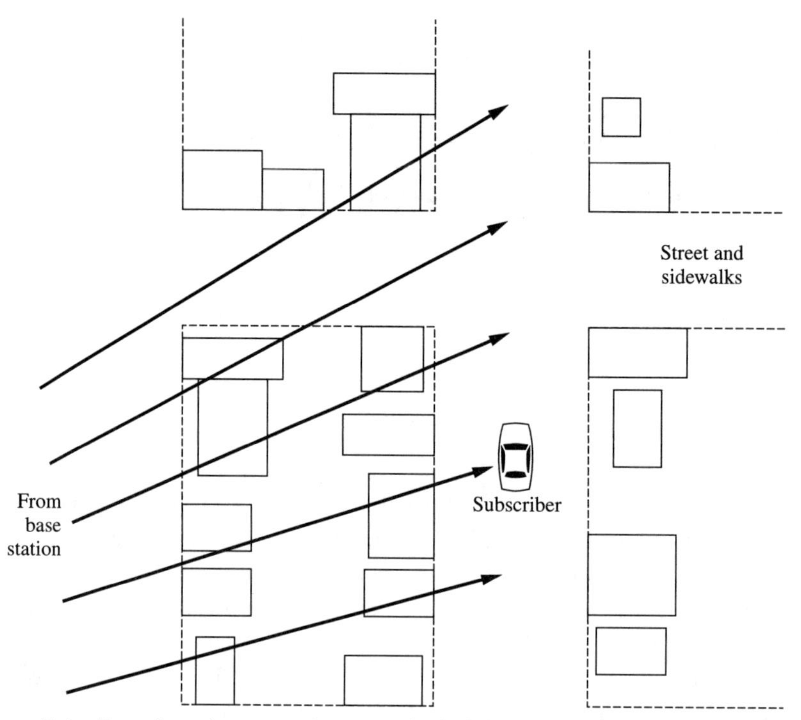

Figure 7-1 Top view of propagation over the buildings to a street-level subscriber.

One source of shadow fading is the difference in the position and composition of the diffracting roof edge, which leads to random variations in the diffraction loss down to the mobile. A second independent source of shadow fading is the variation of the field reaching the rooftop next to the subscriber as a result of the variation in the height of the buildings in the previous rows. These same mechanisms will also cause random variations in the small-area average at the same location along parallel streets, as shown in side view in Figure 7–2. The two sources of variation work in cascade, so that the small-area average signal is the product of two factors with random variations. When the signal is expressed in decibels, the logs of the two factors add to give the overall variation. Because a sum of random variables tends to a normal or Gaussian distribution, for a wide class of individual distribution functions, the distribution of the variations of the small area average signal in decibels will approximate that of a normal distribution. In this case the distribution function of the signal in watts is lognormal. Since the lognormal distribution is obtained independent of the of the distribution function, it will hold in all cities, as has been observed.

7.1a Variation of the rooftop fields

The variation in the rooftop fields, resulting from random variations in the heights of the buildings in previous rows, can be simulated using the numerical integration of the physical optics

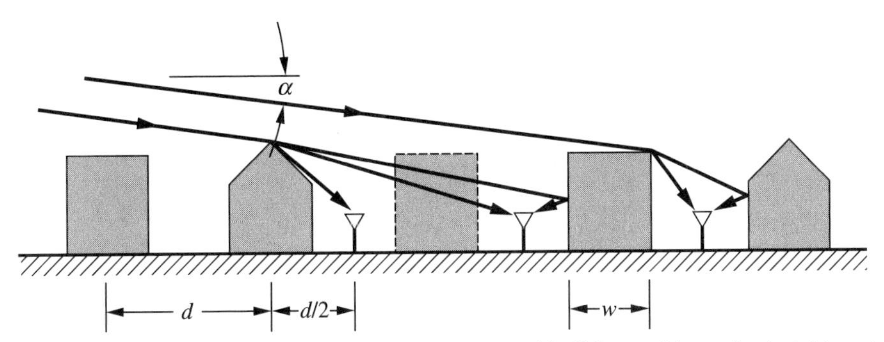

Figure 7-2 Side view of propagation over rows of buildings of irregular height and shape to street-level subscribers.

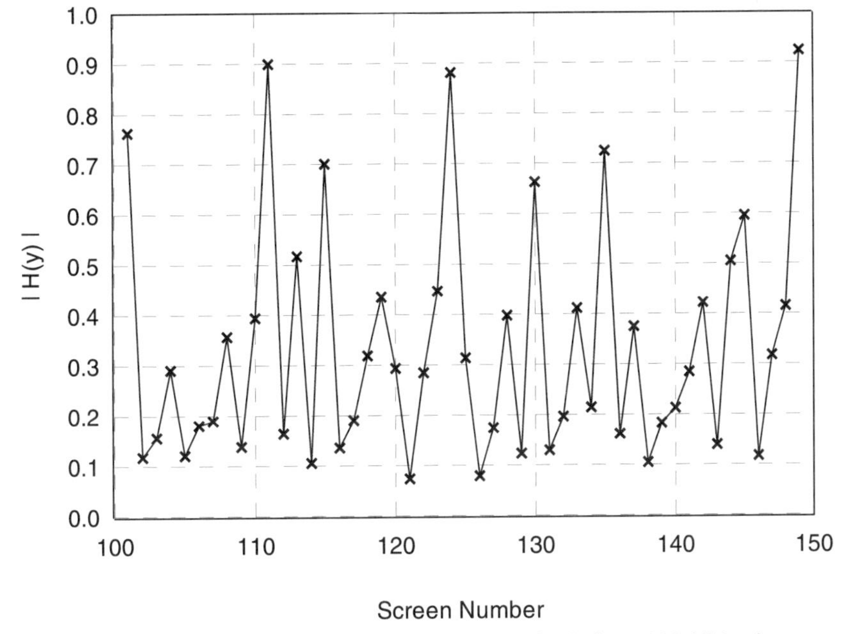

Figure 7-3 Variation from row to row of the rooftop fields for a 900-MHz plane wave of unit amplitude incident at an angle $\alpha = 0.5°$ on rows of buildings whose height H_B is uniformly distributed from 8 to 14 m for $d = 50$ m.

approximation discussed in Section 6–5. A random set of building heights is first generated based on the discretization interval Δy. To do so, let I be the integer part of $\Delta H_B/\Delta y$, where ΔH_B is the range of building heights to be considered, and generate a series of random integers I_n between 0 and I for $n = 1, 2, 3, \ldots$. The heights of the buildings is then taken to be $h_n = I_n \Delta y$, and the random integers I_n are used for the lower limits in the summation of (6–68). For example, if the building heights are distributed uniformly over the range from 5 to 11 m (1 to 3 stories), and

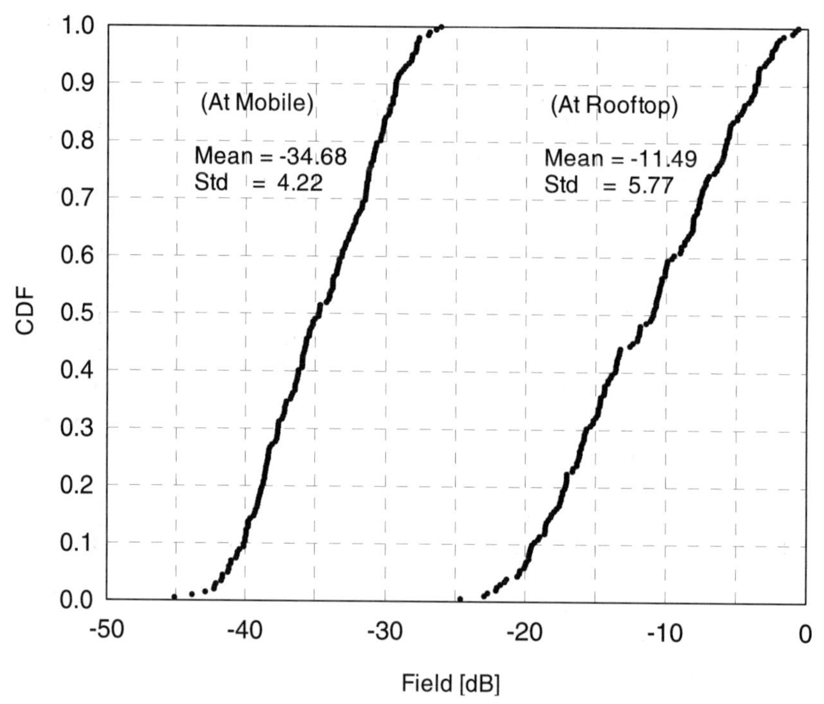

Figure 7-4 Cumulative distribution functions for the decibel values of the rooftop fields in Figure 7–3, and for the fields diffracted down to a subscriber (h_m = 1.8 m) accounting only for the differences in building height.

$\Delta y = \lambda/3.33 = 1/10$ m at 900 MHz, then $I = 60$ and the values of I_n are random integers chosen to be uniformly distributed between 0 and 60.

Having chosen the building heights, the fields above the succeeding rooftops are computed by repeated application of (6–68) and (6–69), taking the field incident on the $n = 1$ row to be that of a plane wave or of a cylindrical wave radiated by a line source. An example of the rooftop field computed in this manner is shown in Figure 7–3 versus row number, where the x's represent the rooftop field and the connecting lines are included to help identify the sequential order. These results have been computed for a 900-MHz plane wave of unit amplitude incident at an angle $\alpha = 0.5^{\circ}$ to the horizontal and for a row spacing $d = 50$ m. Building heights are assumed to be uniformly distributed over a range $\Delta H_B = 6$ m. It is seen from Figure 7–3 that the rooftop fields vary by about a factor of 10, from 0.09 to just above 0.9, corresponding to a 20-dB variation.

To form a cumulative distribution function (CDF) of the field variation, the foregoing computations are repeated so as to generate a population of 200 or more samples for screens beyond the settling number N_0 discussed in Chapter 5, which is $N_0 = 90$ for the parameters chosen for Figure 7–3. Figure 7–4 shows the CDF of the rooftop fields plotted in Figure 7–3. The distribution of the field amplitude expressed in dB is plotted on a linear scale, against which a uniform distribution plots as a straight line. The plotted CDF is seen to be close to linear, indicating that the distribution is close to uniform. For a unit amplitude incident wave, the mean value of the distribution is about –11.5 dB, or 0.27 on the field amplitude scale. By comparison, for the parameters used to compute Figure 7–4, $g_p = 0.1069$ and $Q = 0.34$, corresponding to –9.4 dB. Thus the mean value of the rooftop fields in the case of random building height will be somewhat smaller than the settled field for uniform height buildings [1]. The standard deviation of the variation in the rooftop fields is 5.8 dB.

Not accounting for any other sources of variability, when these rooftop fields are diffracted down to street level the standard deviation of the field in decibels will be smaller. This reduction results from the fact that the highest values of rooftop fields occur at the tallest buildings, for which diffraction down to street level has the greatest loss. Conversely, the rooftop fields have the smallest values at low buildings, for which diffraction loss down to street level is least. For buildings ranging in height from 8 to 14 m, and the parameters used above, the cumulative distribution function of the field at a 1.8-m-high subscriber antenna is shown in Figure 7–4. The mean signal at the subscriber is –34.7 dB, compared to the value of –32.4 dB that would be obtained if all the buildings were of the same height of 11 m. The standard deviation of the distribution is 4.2 dB, which is about three-fourths that of the rooftop fields. Again it is seen that the CDF is nearly linear when plotted on a linear scale, indicating that it is close to a uniform distribution and definitely unlike the normal distribution that is typically observed at street level.

7.1b Combined variations for street-level signal

To simulate the statistical properties of the variation in the small-area average seen at street level, the variation of the rooftop fields described above is combined with the variations of diffraction loss down to street level, several of which are shown in Figure 7–2. The exact distance from the diffracting roof edge to the subscriber and the diffraction angle will depend on the roof shape, which is shown as flat or peaked in Figure 7–2. The diffraction coefficient will also depend on the boundary conditions, which may be taken as conducting in the case of buildings with aluminum siding or aluminized vapor barrier insulation, or absorbing in the case of brick buildings. Street intersections and other gaps between buildings will have a major effect on the diffraction loss. As seen in Figure 7–2, the diffraction angles will be significantly smaller, and the received signal larger, when a building is missing on the side of the subscriber nearest the base station. If the gap is on the side of the subscriber farther from the base station, it is assumed that the reflected ray is missing.

An example of simulations that included the foregoing sources of variability were carried out by an independent, random assignment of the foregoing properties to the buildings in each row [2]. Half the buildings were assumed to have triangular roofs, and half to have flat roofs; for half, the absorbing knife-edge diffraction coefficient was used, while a conducting knife edge was assumed for the other half. At 10% of the rows, the building was removed to compute diffraction down to street level, and the power reflection coefficient from buildings was taken as $|\Gamma|^2 = 0.10$. The CDF obtained from such simulations is shown in Figure 7–5 for buildings hav-

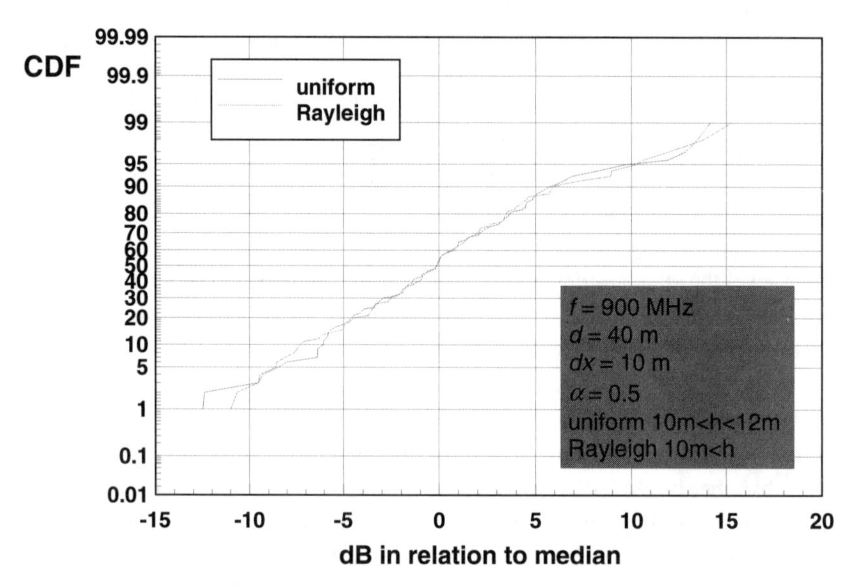

Figure 7-5 Cumulative distribution function about the median for the decibel values of the small-area average signal received by a subscriber at street level, accounting for the variation in building height as well as differences in construction and gaps between them [2] (©1992 IEEE).

ing both uniform and Rayleigh distributions of height, using the nonlinear Gaussian scale for the vertical axis. It is seen that for both height distributions, the CDF is nearly a straight line indicative of a normal distribution. In computing Figure 7–5, $f = 900$ MHz, $\alpha = 0.5°$, $d = 40$ m, and the building width is $w = 10$ m. The standard deviation of the street-level signal obtained from the simulations [2] is plotted in Figure 7–6 versus the range of building heights ΔH_B for uniform distributions of height. Because of the other random variations, there is a finite standard deviation of the signal even when $\Delta H_B = 0$, which is seen to increase with ΔH_B and with frequency.

It would be difficult to observe the predicted dependence on ΔH_B, since one is not likely to find cities in which the buildings are the same except for the value of ΔH_B. However, the increase in the standard deviation with frequency has been observed in measurements that were made simultaneously at frequencies in the 900-MHz and 1.8-GHz bands [3,4]. In Aalborg, Denmark [3], where the buildings range from about three to five stories, it was found that the shadow fading was highly correlated at the two frequencies, as the theory suggests. However, the standard deviations of 7.5 and 8 dB at the two frequencies are higher than those given by the simulations, and the difference of 0.5 dB between them is less than that of the simulations. Measurements in Darmstadt, Germany [4] gave standard deviations of 5.9 and 6.4 dB at the two frequencies, which are closer to the values obtained in the simulation, but the difference between them is still only 0.5 dB. To achieve a closer correspondence between simulations and measurements, it is necessary to know the actual distribution of building heights, and more careful mod-

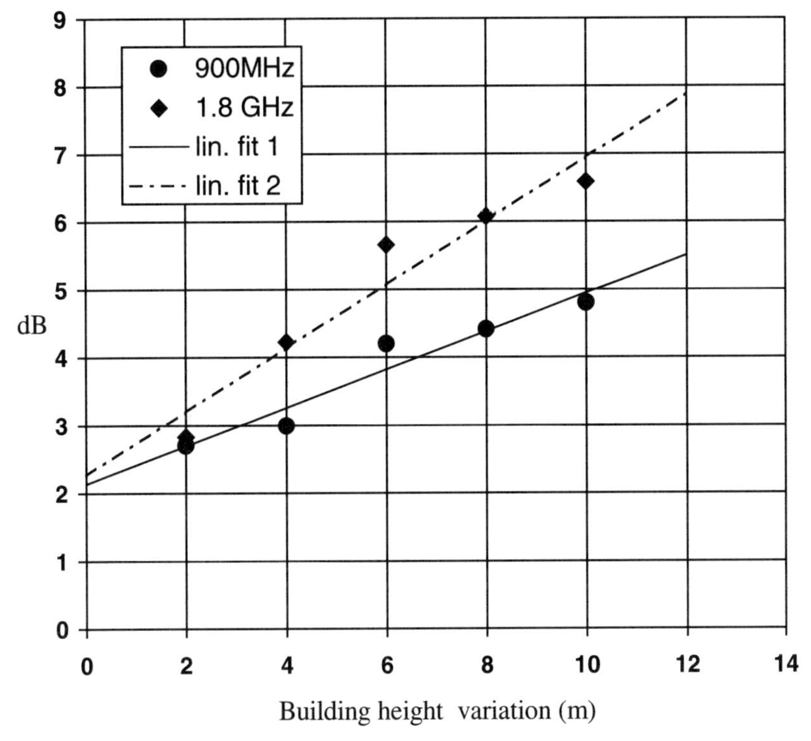

Figure 7-6 Standard deviation of the distribution of simulated small-area averages as a function of the building height variation ΔH_B for 900 MHz and 1.8 GHz [2](©1992 IEEE).

eling is required for street intersections and other gaps between the buildings. Simulations of the average height gain of the subscriber antenna, accounting for random building variations [2], are plotted in Figure 6–3. The height gain of the average received signal is found to be the same as obtained for buildings of uniform height and to be consistent with measurements.

7.2 Modeling terrain effects

Statistical and deterministic approaches have been used to account for terrain effects on radiowave propagation. As part of their extensive measurement program, Okumura et al [5] made measurement on rolling terrain. Their measurements applied to relatively long paths, and terrain roughness was defined by the difference in height Δh between the 10 and 90% points in CDF of terrain height over the last 10 km of the path from the base station to the subscriber location. They plotted the reduction of the small-area averages, as compared to flat terrain, for mobiles located at points having the median height of the terrain variation versus the roughness Δh. They gave no indication of the horizontal distance between undulations, which will also affect the signal. The average reduction at 922 MHz for $\Delta h = 50$ m was about 3 dB, although the scatter in the

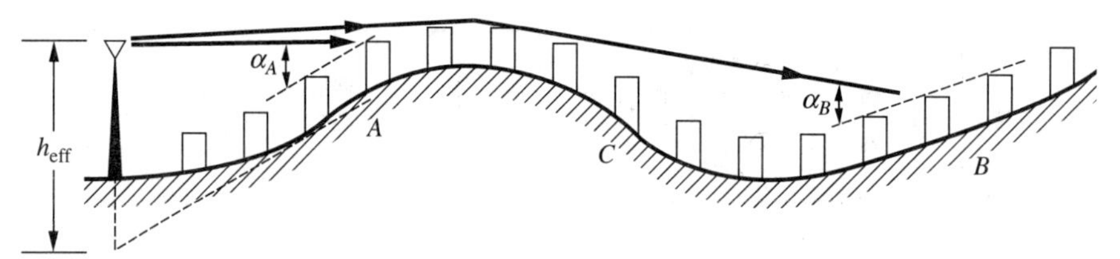

Figure 7-7 Propagation to the rooftops of buildings on rolling terrain.

measurements was about 10 dB. An additional correction was given for mobiles at the tops or bottoms of the terrain profile. Again for $\Delta h = 50$ m, this correction was about 9 dB. In general, terrain roughness reduces the average signal and increases its variability.

An extensive literature deals with deterministic propagation over terrain in the absence of buildings. Approaches based on the geometrical diffraction theory account for the tops of hills that block the LOS path (see, e.g., [6–10]). Full-wave approaches, such as the parabolic equation method [11,12] and the method of moments [13], take the entire terrain profile into account. In the presence of buildings, Lee proposed using an effective base station antenna height for those cases when LOS conditions exist to the rooftops in the vicinity of the mobile [14]. Later, diffraction over terrain obstacles was treated in connection with the influence of buildings [15]. In this section we discuss the basis for the formulations that combined effects of terrain and buildings and ray-based solutions for buildings on rolling terrain that can be used as the basis for deterministic predictions.

7.2a Paths with LOS to the rooftops near the subscriber

Transmission paths from an elevated base station are depicted in Figure 7–7 for three classes of subscriber locations in a metropolitan area built on rolling terrain. Again because the Fresnel zone is narrow for high-frequency radio waves, variations of the terrain out of the vertical plane between the base station and subscriber may be neglected. At locations such as A in Figure 7–7, where the rooftops are within sight of the base station, the path loss can be found using the approach described in Chapter 6. The path loss consists of the free-space loss, the reduction of the field reaching the last rooftop before the mobile due to the previous buildings, and the loss associated with the diffraction of the rooftop fields down to street level. Using the expressions and notation of (6–2) and (6–9) for the path gain associated with free space and diffraction down to the mobile, the path loss in decibels is

$$L = -10\log\left(\frac{\lambda}{4\pi R}\right)^2 - 10\log\left[\frac{\lambda\rho_1}{2\pi^2(H_B - h_m)^2}\right] - 10\log[Q(g_p)]^2 \qquad (7\text{–}1)$$

Here we have used the plane wave settled field $Q(g_p)$ for the reduction of the rooftop fields due to propagation over the previous rows, with g_p found from the local angle α_A shown in Figure 7–7. In the absence of the buildings, effective antenna height above the tangent to the terrain in the vicinity of the subscriber may be used in conjunction with the two-ray model of Chapter 4 to provide a simple approximation for the path loss. However, depending on the curvature of the terrain, there may be additional ground reflected rays from the base station to the subscriber [16].

The dimensionless parameter g_p may be expressed in terms of the effective height $h_{\rm eff}$ of the base station antenna above the tangent plane to the terrain in the vicinity of the mobile, as shown in Figure 7–7. This expression takes the form

$$g_p = \sin\alpha_A\sqrt{\frac{d}{\lambda}} \approx \frac{h_{\rm eff}-H_B}{R}\sqrt{\frac{d}{\lambda}} \tag{7–2}$$

The reduction term $Q(g_p)$ can be found from (6–32) or approximately from (6–34). When the simple approximation (6–34) for $Q(g_p)$ is valid, the path loss becomes

$$L = -10\log\left(\frac{\lambda}{4\pi R}\right)^2 - 10\log\left[\frac{\lambda\rho_1}{2\pi^2(H_B-h_m)^2}\right] - 10\log\left[2.347\left(\frac{h_{\rm eff}-H_B}{R}\sqrt{\frac{d}{\lambda}}\right)^{0.9}\right]^2 \tag{7–3}$$

Using the effective antenna height in the expression for base station height gain was first proposed by Lee [14], who employed the concept in conjunction with measurement-based slope-intercept models of path loss.

If the terrain slopes upward away from the base station, as shown at point A in Figure 7–7, the effect on the path loss in (7–1) or (7–3) can be significant. For example, a fairly steep grade of 10% corresponds to an angle arctan(0.1) = 5.7°. If the propagation path from the base station to the rooftops at point A is essentially horizontal, the angle of incidence $\alpha_A = 5.7°$. For $d = 50$ m and $\lambda = 1/3$ m, at this angle $g_p = 1.2$, which is outside the range of validity of (7–3). From Figure 6–7, it is seen that $Q \approx 1$, so that the fields reach the rooftops near the mobile under free-space propagation conditions. Note that (7–1) and (7–3) apply as well to terrain that slopes downward away from the base station, provided that the rooftops are within LOS of the base station antenna.

7.2b Paths with diffraction over bare wedge-shaped hills

For subscribers at locations such as B in Figure 7–7, the path loss in (7–2) is modified to account for diffraction over the intervening hill, or hills if more than one is present. If the diffraction angle θ over the hill is small enough, or the hill has a sharp peak, it can be replaced by a diffracting edge [5,6], or by a diffracting wedge [7,9], as shown in Figure 7–8. In this case the free-space loss in (7–2) must be replaced by the edge diffraction loss found from (5–62). For small diffraction angles it may be necessary to use the UTD diffraction coefficient D_T, in which case the path loss is given by

$$L = -10\log\left(\frac{\lambda|D_T|}{4\pi\sqrt{R_1R_2R}}\right)^2 - 10\log\left[\frac{\lambda\rho_1}{2\pi^2(H_B-h_m)^2}\right] - 10\log[Q(g_p)]^2 \tag{7–4}$$

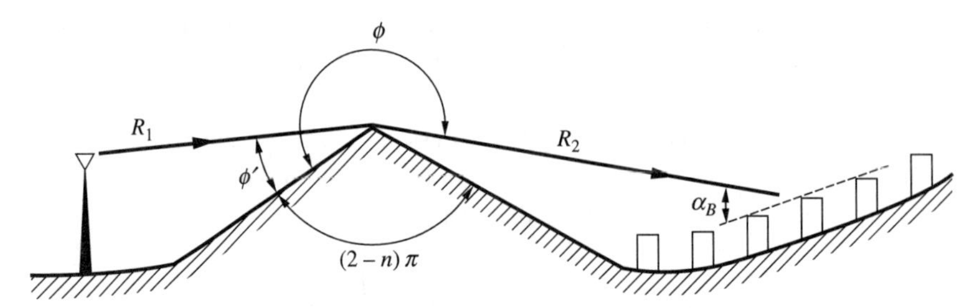

Figure 7-8 Wedge approximation for diffraction over a hill.

The distances R_1, R_2 and $R = R_1 + R_2$ in (7–4), are shown in Figure 7–8. These distances can be replaced by their horizontal components when they are large compared to the vertical distances. Also, g_p is found from (7–2) in terms of the local angle α_B shown in Figures 7–7 and 7–8.

Very near the shadow boundary, the diffraction coefficient D_T for a knife edge may be used in (7–4). However, Luebbers [7] has shown that away from the shadow boundary of a hill, better agreement with measurements is achieved using the diffraction coefficients for a wedge whose surfaces have the reflection coefficients for earth. The interior angle of the wedge in Figure 7–8 is $(2 - n)\pi$, which for rolling terrain is only slightly smaller than π. For example, if both sides of the hill have a grade of 10%, corresponding to a pitch of 5.71° the interior angle is 168.58°, in which case $2 - n = 0.937$ and $n = 1.063$. In terms of the angles ϕ and ϕ' shown in Figure 7–8, the diffraction coefficient for the wedge [7] is

$$D_T(\phi, \phi') = D_1^w + D_2^w + \Gamma_n D_3^w + \Gamma_0 D_4^w \tag{7–5}$$

Depending on the polarization, Γ_0 is the plane wave reflection coefficient at the ground for the component of the **E** or **H** that is perpendicular to the plane of incidence. In computing Γ_0, the angle of incidence is $\pi/2 - \phi'$, as in the case of the incident ray in Figure 7–8. The reflection coefficient Γ_n is similarly defined using the angle $\phi - (n - 1/2)\pi$ between the diffracted ray and the normal to the second surface of the wedge. The terms D_i^w in (7–5) are defined by

$$D_{1,2}^w = \frac{-1}{2n\sqrt{2\pi k}} \cot\left[\frac{\pi \pm (\phi - \phi')}{2n}\right] F[kLa^{\pm}(\phi - \phi')] \tag{7–6}$$

$$D_{3,4}^w = \frac{-1}{2n\sqrt{2\pi k}} \cot\left[\frac{\pi \pm (\phi + \phi')}{2n}\right] F[kLa^{\pm}(\phi + \phi')]$$

In (7–6) the factors F[·] are transition functions, as defined in (5–41) and (5–45), and are approximately unity outside their transition regions. Inside their transition regions, the arguments are evaluated using $L = R_1R_2/(R_1 + R_2)$ and

$$a^{\pm}(\beta) = 2\cos^2(n\pi N^{\pm} - \beta/2) \tag{7–7}$$

where $N^{\pm}$ are the integers closest to $(\beta \pm \pi)/(2\pi n)$. In the case of forward diffraction over low hills n is close to 1, ϕ - ϕ' is close to π, so that $N^+ = 1$ and $N^- = 0$ in D_1^w and D_2^w, respectively. Similarly, $\phi + \phi'$ is close to π so that again $N^+ = 1$ and $N^- = 0$ in D_3^w and D_4^w, respectively.

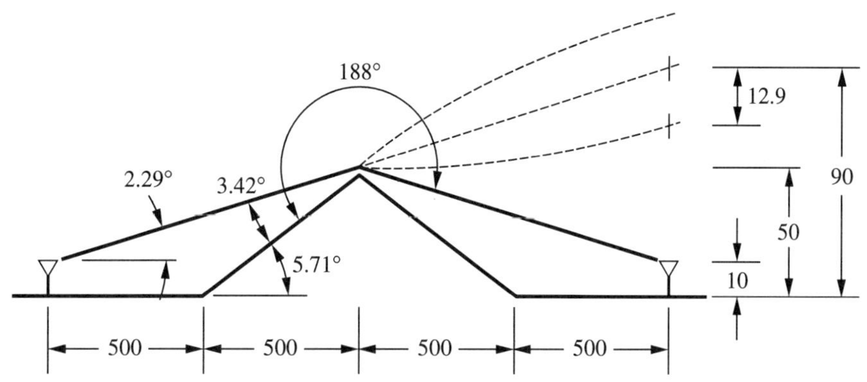

Figure 7-9 Symmetric diffraction path over a wedge-shaped hill (distances are in meters).

As an example of the evaluation of D_T from (7–4), consider the symmetric path shown in Figure 7–9, where the transmitting and receiving antennas are 10 m above the flat portion of the earth and 1000 m from the peak of a hill that is 50 m high. As seen from either antenna, the elevation angle to the peak is arctan(40/1000) = 2.29°. From this angle it is found that the shadow boundary of the incident wave passes over the receiver at an elevation of 90 m. The slope angle of the hill is arctan(50/500) = 5.71°, so that $n = 1.063$, as discussed prior to (7–5). Thus the angle ϕ' between the incident ray and the face of the hill is 3.42°, and the angle of incidence is 86.58°. For vertical electric field, and assuming that $\varepsilon_r = 15$, from Snell's law the angle of transmission θ_T into the ground is 14.93°, and the reflection coefficient at the ground surface is found from (3–40) to be $\Gamma_H = -0.614$. Since the path is symmetric, this value is taken for both Γ_0 and Γ_n in (7–5). Finally, the diffraction angle ϕ is the sum of hill angle of 5.71°, the elevation angle 2.29°of the diffracted ray, and 180°, for a total of 188.00°.

The half-width of the transition region for a receiver located about the shadow boundary is the same as the width of the first Fresnel zone when the receiver is located on the shadow boundary. From (5–13) with $n = 1$, the width is given by

$$w = \sqrt{\lambda \frac{R_1 R_2}{R_1 + R_2}} \tag{7–8}$$

For 900-MHz signals and for distances $R_1 = R_2 = 1000$ m, from (7–8) it is found that $w = 12.9$ m, which places the transition region far above the receiver. Thus we may set the transition functions to unity in (7–6). Using the parameter values found above, the diffraction coefficient in (7–5) is found to be

$$D_T = \frac{-1}{\sqrt{2\pi k}}[-3.36 - 10.03 - 0.614(-4.61 - 5.55)] = \frac{7.14}{\sqrt{2\pi k}} \tag{7–9}$$

For comparison, the Felsen diffraction coefficient for an absorbing screen, with $\theta = \pi + \phi' - \phi$ or -0.0997 rad, is $D = 9.87/\sqrt{2\pi k}$.

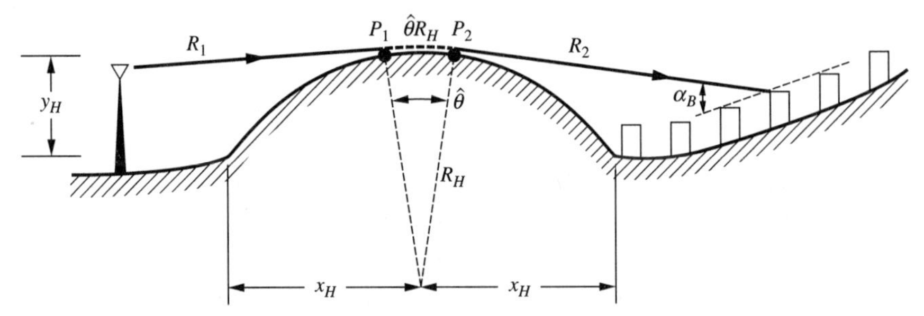

Figure 7-10 Cylindrical approximation for diffraction over a hill.

When no houses are present, as in Figure 7–9, the path loss between antennas is given by the first term in (7–4). Using D_T of (7–9), the path loss between the antennas in Figure 7–9 is found to be $L = 125.8$ dB, which is 28.3 dB greater than the free-space path loss $L_o = 97.5$ dB between 900-MHz antennas separated by 2 km. Note that in Figure 7–9 there will be three additional contributions to the received signal, two coming from fields reflected on the flat surfaces on either side of the hill and one contribution that involves reflections on both sides of the hill. Because the reflection angles are smaller and diffraction angles are greater, these contributions will be somewhat smaller than that of the path in Figure 7–9. However, for transmitter and/or receiver at a greater distance from the hill, these fields will be nearly equal in amplitude and will destructively interfere with that of the path that goes directly over the hill.

7.2c Paths with diffraction over bare cylindrical hills

When the hill is smooth and not covered by trees or houses, and if diffraction takes place through a larger angle $\hat{\theta}$, the diffraction process is described more accurately in terms of creeping rays [17,18]. At a smoothly curved, bare hill, creeping rays are launched by the ray that grazes the surface, as at point P_1 in Figure 7–10. The creeping ray fields follow the curvature of the surface but shed energy in the form of rays that leave the surface along tangent lines, as at point P_2 in Figure 7–10. Note that $\hat{\theta}$ is the negative of the diffraction angle θ introduced in Chapter 5. Because of this shedding of energy, the amplitude of the creeping ray fields decreases exponentially with angle $\hat{\theta}$. For $\hat{\theta} \geq (\lambda/\pi R_H)^{1/3}$, the asymptotic expression [17,18] for the creeping ray fields at a conducting cylinder can be approximated by the first term of their series expansion. The field component E_z or H_z perpendicular to the plane of incidence of the creeping ray in this case is given by

$$\left.\begin{matrix} E_z \\ H_z \end{matrix}\right\} = \left\{\begin{matrix} E^{\text{inc}} \\ H^{\text{inc}} \end{matrix}\right\} e^{-jk\hat{\theta}R_H} \frac{e^{-jkR_2}}{\sqrt{R_2}} \frac{1}{\sqrt{k}} D_1 e^{-\psi_1\hat{\theta}} \tag{7–10}$$

where E^{inc} or H^{inc} is the corresponding component of the field incident on the tangent point P_1 in Figure 7–10, and $\hat{\theta}$ is in radians.

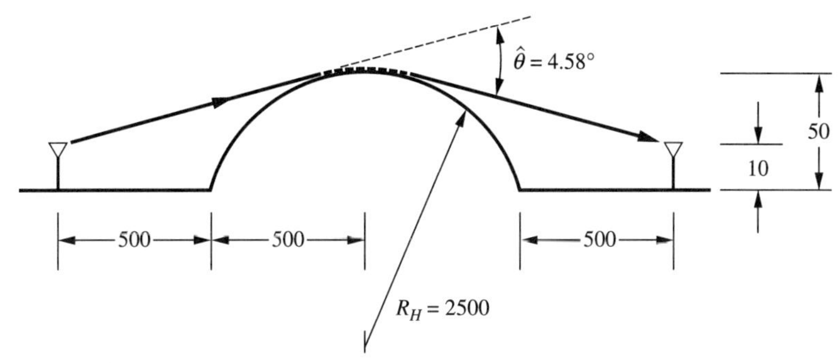

Figure 7-11 Symmetric diffraction path over a cylindrical hill (distances are in meters).

In (7–10) the coefficients D_1 and ψ_1 depend on the polarization and the ratio of cylinder radius R_H to wavelength λ. Using the superscripts e and h to denote the TE and TM polarizations, the coefficients for a conducting cylinder are given by [17,18]

$$\left.\begin{matrix}\psi_1^e \\ \psi_1^h\end{matrix}\right\} = \left(\frac{\pi R_H}{\lambda}\right)^{1/3} e^{j\pi/6} \begin{cases} 2.338 \\ 1.019 \end{cases}$$
$$\left.\begin{matrix}D_1^e \\ D_1^h\end{matrix}\right\} = \frac{1}{\sqrt{2\pi}}\left(\frac{\pi R_H}{\lambda}\right)^{1/3} e^{j\pi/6} \begin{cases} 2.034 \\ 3.421 \end{cases} \tag{7–11}$$

The difference in the coefficients between the two polarizations is most significant in ψ_I, since its real part is an attenuation constant and strongly affects the field amplitude as the wave goes around the cylinder. Terrain roughness and trees on the hill will significantly influence the values of D_1 and ψ_1, making the use of these formulas questionable [19]. In the next section we evaluate the changes in D_1 and ψ_1 due to buildings located on the hill.

For rolling terrain, the height of a hill y_H shown in Figure 7–10 is much less than its half-width x_H from the peak to the base of the hill. Fitting a circle to these points for $y_H \ll x_H$ gives the hill radius

$$R_H \approx \frac{x_H^2}{2y_H} \tag{7–12}$$

For example, if $y_H = 50$ m and $x_H = 500$ m, as in Figure 7–11, R_H is 2.5 km. For a hill of this radius and 900-MHz signals, expression (7–10) for the creeping ray field is valid when $\hat{\theta} \geq (\lambda/\pi R_H)^{1/3} = 0.0349$ rad (2°). If $\hat{\theta} = 0.0349$ rad, the length of the creeping ray path over the hill is $\hat{\theta} R_H \geq 87$ m. Because $\hat{\theta} R_H$ is small compared to the distance from either antenna to the top of the hill, the distances R_1 and R_2 in Figure 7–10 are close to the distance to the peak when the hill is replaced by a wedge, as in Figure 7–8. Thus the group of terms $D_1 e^{-\psi_1 \theta}/\sqrt{k}$ in

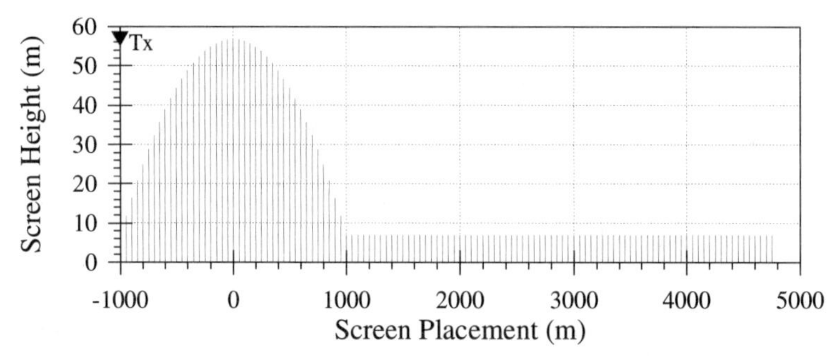

Figure 7-12 Screens representing rows of buildings of height H_B = 7 m located on a 50-m high cylindrical hill (R_H = 10 km) and the following plane [20](©1998 IEEE).

(7–10) takes the place of the diffraction coefficient D_T in (7–4), which allows a simple comparison between the path loss computed for cylindrical and wedge-shaped hills.

As a numerical example, consider the cylindrical hill shown in Figure 7–11, which has the same height and width as the wedge-shaped hill in Figure 7–9, with symmetrically located transmitter and receiver positions. The elevation angle from the transmitter or receiver to the point of ray tangency will be only slightly greater than arctan(40/1000) = 2.29°. Using this value for the elevation angle, $\theta = 4.58°$ or 0.0799 rad. For 900-MHz signals with vertical electric field (TM polarization), $(\pi R_H/\lambda)^{1/3} = 28.67$ so that $|D_1|/\sqrt{k} = 98.02/\sqrt{2\pi k}$ and $\mathrm{Re}\{\psi_1\} = 25.3$. The real part of the exponent in (7–10), $\mathrm{Re}\{\theta\hat{\psi}_1\}$, is then found to be 2.02, so that $\left|D_1 e^{-\psi_1\theta}\right|/\sqrt{k} = 13.0/\sqrt{2\pi k}$, which is 5.2 dB greater than the value in (7–9) for diffraction by a wedge. However, the creeping ray diffraction loss in decibels is proportional to θ, whereas the knife-edge loss is proportional to $\log\theta$. Thus doubling the angle θ doubles the creeping ray loss in decibels, whereas the knife-edge diffraction loss increases by only 6 dB. At a conducting cylinder, the creeping ray fields will be significantly weaker if the electric field is horizontal (TE polarization).

7.2d Diffraction of cylindrical waves over hills with buildings

When buildings are located on the hills, their effect on the diffraction process can be modeled numerically using the formulation discussed in Section 6–5. This formulation is essential in the case of subscribers located among buildings on the back side of a hill, such as at C in Figure 7–7, since each successive rooftop is in the shadow region of the previous rooftop. Various examples of wedge-shaped hills and smoothly varying hills are discussed in references [20] and [21]. Figure 7–12 shows the absorbing screens used to carry out one such simulation for 7-m-high buildings that are located on a cylindrical hill of radius R_H = 10 km [20]. The hill rises above the plane to a height of y_H = 50 m, and its base is at a distance x_H = 1000 m from the peak. For the simulation, the base station was assumed to be located at the point $x = -1000$ m from the hilltop and at a height above the plane h_{BS} = 57 m, which places it at the same height as the highest rooftops. As discussed in Chapter 6, the numerical calculations assume that a cylindrical

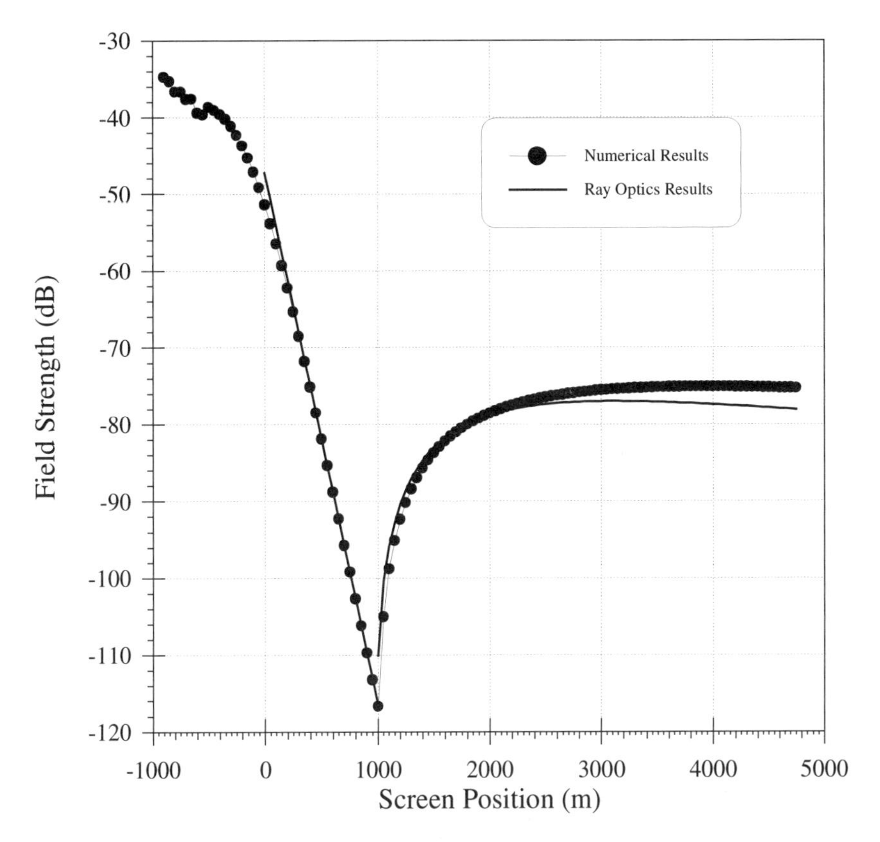

Figure 7-13 Field amplitude, shown as dots, at the roof tops of Figure 7–12 for a cylindrical wave radiated by a line source at position Tx in Figure 7–12. The continuous curves near the dots are the creeping ray approximations for the field on the back side of the hill and on the flat plane [21](©1998 IEEE).

wave generated by a line source at the base station location is incident on the first screen. For these computations, the electric field of the cylindrical wave was taken to have amplitude $1/\sqrt{k\rho}$.

The strength of the vertical component of electric field reaching rooftops is plotted in Figure 7–13 for the buildings and hill of Figure 7–12. On the back side of the hill, out to a distance of 1000 m, the field in decibels decreases linearly with distance, corresponding to an exponential decrease when the field is expressed in volts or amperes per meter. This may be understood in terms of the creeping wave representation of the fields traveling over the houses. Past the foot of the hill ($x > 1000$ m), the rooftop field increases initially to a maximum and then decreases slowly. With increasing distance from the foot of the hill, the ray reaching the rooftop is tangent to the hill at a point farther up the hill. Thus the diffraction angle $\hat{\theta}$ is smaller, leading to a smaller diffraction loss, which is initially more significant than the dependence on distance or

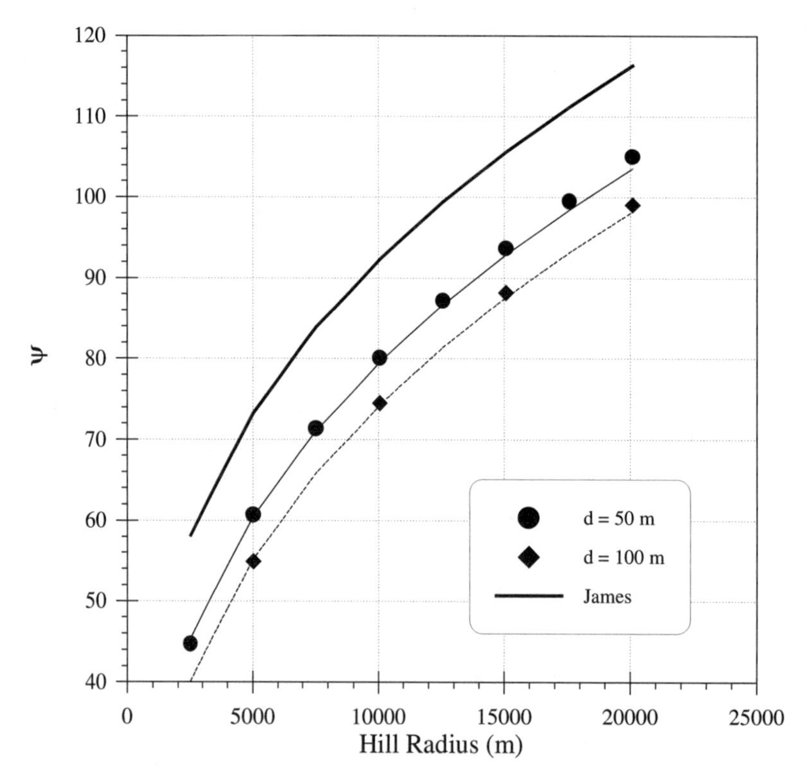

Figure 7-14 Exponential attenuation coefficient ψ at 900-MHz obtained by fitting the creeping ray approximation to the numerically computed fields on the back side of a hill for various hill radii R_H and row spacings d. The dark continuous curve is the theoretical result for diffraction of a TE polarized wave by a conducting cylinder [20](©1998 IEEE).

angle of incidence at the rooftops. Note that the results numerically generated from the physical optics approximation are the same for both TE and TM polarizations, whereas the conducting cylinder results are significantly different.

On the back side of the hill the creeping ray representation of the rooftop field amplitude for an incident cylindrical wave has the form

$$E(x_n, H_n) = \frac{1}{\sqrt{kR_1}} D_H e^{-\psi\hat{\theta}} \tag{7–13}$$

In (7–13) R_1 is the distance from the base station antenna to the tangent point of the ray with the hill, D_H is the diffraction coefficient, and ψ the attenuation coefficient. By choosing ψ and D_H it is possible to fit (7–13) to the computed fields on the back side of the hill. The resulting fit is shown in Figure 7–13. Computations such as those in Figure 7–13 were made for various values of row separation d, hill radius R_H, and at both 900 MHz and 1.8 GHz. From these simulations, the attenuation factor and diffraction coefficients were extracted by fitting (7–13) to the numeri-

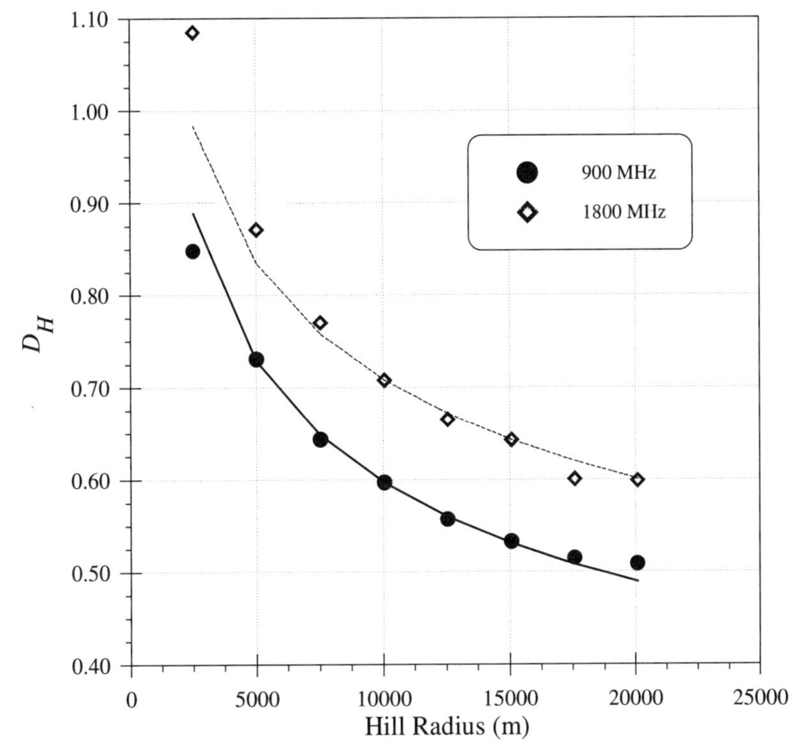

Figure 7-15 Diffraction coefficient D_H obtained by fitting the creeping ray approximation to the numerically computed fields on the back sides of various size hills for row spacing d = 50 m and frequencies 900 MHz and 1.8 GHz [20](©1998 IEEE).

cally computed fields at the rooftops [20,21]. The attenuation factors ψ found for different simulation conditions is indicated by the dots and diamonds in Figure 7–14 for 900-MHz signals. The solid dark curve labeled "James" in Figure 7–14 is $RE\{\psi_1^e\}$ found from (7–11) for the TE polarization at a smooth conducting cylinder.

A simple formula that has been fitted to the simulation points is given by [20]

$$\psi = 2.02\left(\frac{\pi R_H}{\lambda}\right)^{1/3} - 1.04\sqrt{\frac{d}{\lambda}} \tag{7–14}$$

where the first term is $RE\{\psi_1^e\}$ for TE polarization. The values obtained from (7–14) for $d = 50$ and 100 m are also shown in Figure 7–14 and are seen to closely fit the numerically generated values. Expression (7–14) was also found to apply to the simulations for 1.8 GHz. It is seen from (7–14) that the fit reduces to the smooth cylinder result for TE polarization when $d/\lambda = 0$. The coefficients D_H found by fitting (7–13) to the numerically computed rooftop fields are shown in Figure 7–15.

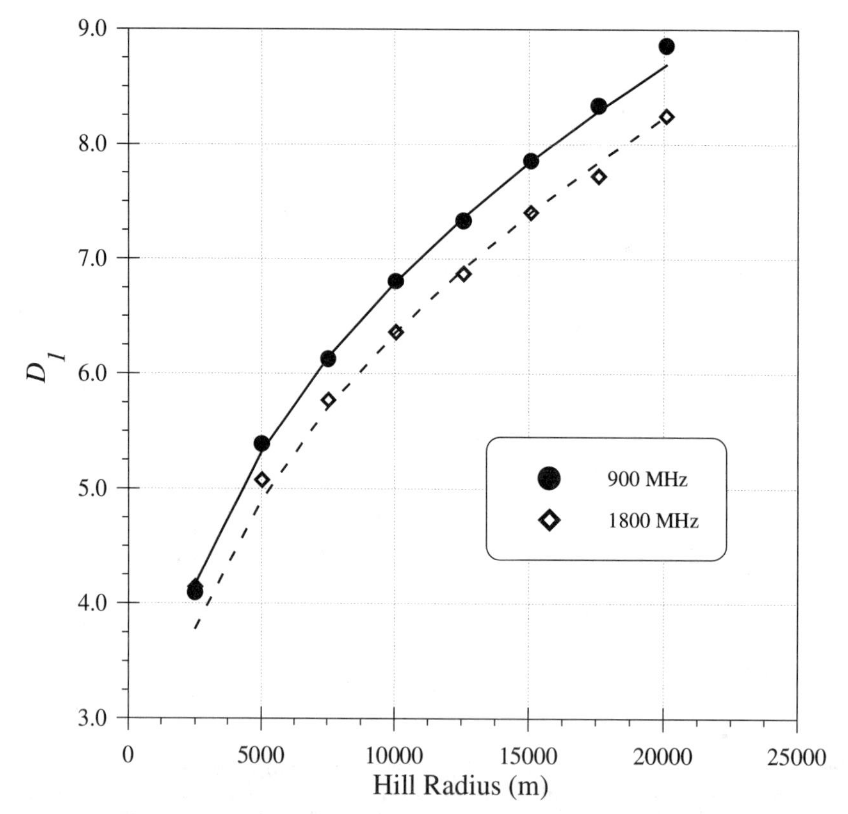

Figure 7-16 Diffraction coefficient D_1 obtained by fitting the creeping ray approximation to the numerically computed fields on the plane after various-size hills for row spacing $d = 50$ m and frequencies 900-MHz and 1.8 GHz [20](©1998 IEEE).

At rooftops on the flat terrain past the hill in Figure 7–12, the creeping ray representation of the fields due to an incident cylindrical wave takes the form

$$|E(x_n, h_n)| = \frac{1}{\sqrt{kR_1}} \frac{1}{\sqrt{R_2}} \frac{|D_1|}{\sqrt{k}} e^{-\psi\hat{\theta}} Q(g_p) \tag{7–15}$$

Here the factor $1/\sqrt{kR_1}$ gives the field amplitude of the incident ray that is tangent to the equivalent cylinder at the rooftops on the hill. The factor $1/\sqrt{R_2}$ is due to the spreading of the rays radiated from the vicinity of the launch point where the departing ray is tangent to the equivalent cylinder. The factor $|D_1|/\sqrt{k}$ represents the coupling from the incident field to the creeping ray fields and back out to the radiated ray, and $e^{-\psi\hat{\theta}}$ represents the attenuation of the creeping ray between the points of incidence and radiation. Finally, $Q(g_p)$ gives the reduction of the rooftop fields due to diffraction past the previous row of buildings on the flat terrain.

The attenuation coefficient ψ in (7–15) is that of Figure 7–13 or expression (7–14). Using this value of ψ, the diffraction coefficient $|D_1|$ in (7–15) is chosen so that expression (7–15) is equal to the numerically generated field, such as in Figure 7–13, at a point corresponding to an angle $\hat{\theta} = 1.7°$. With this choice, (7–15) will match the numerically generated field to within a few decibels over the entire range from the base of the hill out to 5 km, as shown in Figure 7–13 [21]. The diffraction coefficient obtained in this way is plotted in Figure 7–16 as dots and diamonds for 900 and 1800 MHz. The values of $|D_1|$ in Figure 7–16 for a hill with buildings can be compared to those given by (7–11) for a conducting cylinder. If the hill radius is 2.5 km, for 900-MHz signals it is found from (7–11) that $|D_1|$ is 23.3 for TE polarization and 39.1 for TM polarization. Both values are far from the value about 4 seen in Figure 7–16 for a hill of radius 2.5 km with buildings. Trees and terrain roughness are expected to act like houses to greatly reduce the value of $|D_1|$ and somewhat reduce the value of ψ. Thus the creeping ray coefficients found for a conducting cylinder are not reliable in predicting the diffraction over real hills. For example, consider the propagation path of Figure 7–11, assuming the houses to be located on the hill and the two antennas to be 10 m above the houses on either side of the hill. For a hill radius of 2.5 km and row spacing $d = 50$ m, $\psi = 44$ at 900 MHz. For this hill $\hat{\theta} = 4.58°$ or 0.0799 rad and $|D_1| = 4$, so that $|D_1|e^{-\psi\hat{\theta}}/\sqrt{k} = 0.298/\sqrt{2\pi k}$. This value is 32.8 dB smaller than found in the example of Section 7–2c for diffraction of a vertically polarized wave over a cylindrical conducting hill of the same dimensions and is 27.6 dB smaller than found in Section 7–2b for diffraction over the wedge-shaped hill in Figure 7–9.

As an example of the field variation for multiple hills, and as a demonstration of the use of the creeping ray representation (7–15) of the rooftop field, numerical simulations were carried out for rows of buildings separated by $d = 50$ m on two hills of sinusoidal shape. The rooftops are at the heights shown in Figure 7–17a and excitation is by a 900-MHz line source at the position indicated by Tx. For this set of roof heights, the rooftop fields due to the cylindrical wave from the line source are as shown in Figure 7–17b [21]. It is seen that when expressed in decibels, the fields on the back sides of the hills decrease nearly linearly between the peak of each hill ($x_n = 0$ or 2900 m) and the following inflection point ($x_n = 750$ or 3650 m) on the downslope. As x_n increases beyond the inflection point, up to the peak of the following hill, the rays illuminating the rooftops come from tangent points that move backward up the hill, and hence experience less attenuation.

A comparison between the rooftop fields predicted by the numerical simulation and the creeping ray expression (7–15) is also shown in Figure 7–17b for $900 \leq x_n \leq 2850$ m, which is the range from slightly past the inflection point of the first hill in Figure 7–17 to the peak of the second hill [21]. In applying (7–15), the peak of the first hill is replaced by a cylinder to determine the creeping ray coefficients, and the local terrain slope is used in finding the incidence angle α, so as to compute the reduction factor $Q(g_p)$. Near the inflection point α is very small, so that the plane wave reduction factor is overly pessimistic, since it applies to propagation past many more rows of buildings. Near the top of the second hill the predictions are again pessimistic for two reasons, one of which is that α again has a low value, so that the plane wave reduction

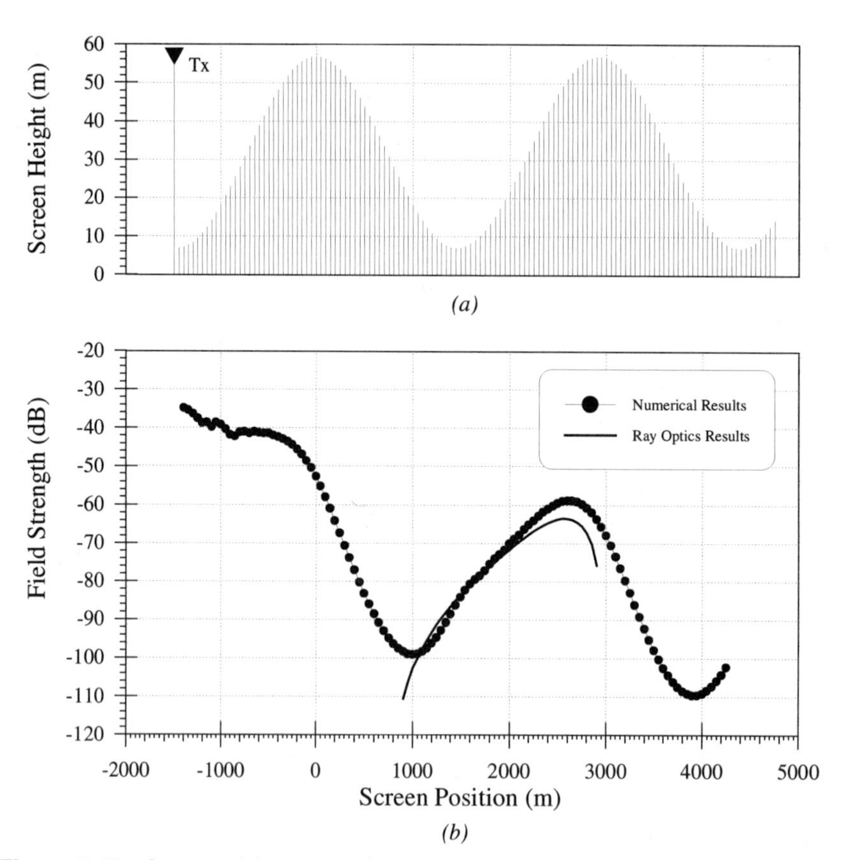

Figure 7-17 Screens (a) representing rows of buildings of height $H_B = 7$ m located on 50-m-high sinusoidal hills, and (b) the rooftop fields due to a cylindrical wave radiated by a line source located at Tx. The continuous curve is the creeping ray approximation from just beyond the inflection point on the back side of the first hill to the top of the second hill [21](©1998 IEEE).

factor is not applicable. The other reason is that the rooftops are approaching the transition region about the shadow boundary of the first hill, so that expression (7–10) no longer describes the creeping ray fields. However, between the two extremes, the simple creeping ray formulation (7–15) is seen to give a good approximation.

7.2e Path loss formulas for building-covered hills

The path loss from an elevated base station to subscribers located at the various points A, B, and C in Figure 7–7 is summarized here for a single intervening hill. First, at points such as A where there is a LOS path from the base station to the rooftops near the mobile:

$$L = -10\log\left(\frac{\lambda}{4\pi R}\right)^2 - 10\log\left[\frac{\lambda\rho_1}{2\pi^2(H_B - h_m)^2}\right] - 10\log[Q(g_p)]^2 \qquad (7\text{–}16)$$

where g_p is found for incidence at the angle α_A between the ray from the base station and the local tangent plane to the terrain.

When the path from the base station to the rooftops near the mobile is blocked by a hill, the blockage increases the loss. At points such as B in Figure 7–7 that are beyond the inflection point on the back side of the hill, the path loss is

$$L = -10\log\left(\frac{\lambda|\hat{D}|}{4\pi\sqrt{R_1 R_2 R}}\right)^2 - 10\log\left[\frac{\lambda\rho_1}{2\pi^2(H_B - h_m)^2}\right] - 10\log[Q(g_p)]^2 \qquad (7\text{–}17)$$

Here $\hat{D}$ represents one of the following:

1. D_T computed from (7–5) – (7–7) for a wedge-shaped hill or one of the diffraction coefficients discussed in Chapter 5
2. $(D_1/\sqrt{k})\exp(-\psi\hat{\theta})$, where D_1 is taken from Figure 7–16, and ψ from Figure 7–14 or expression (7–14)

In rolling terrain, where the horizontal distances are large compared to the vertical dimensions, little error will result in (7–17) if the distances R_1 and R_2 are taken as the horizontal distance from the base station to the peak of the hill and from the peak to the subscriber, and R is taken as the horizontal separation between the base station and subscriber. The parameter g_p is found for incidence at the angle α_B to the local tangent plane to the terrain at the subscriber.

When there are buildings on a cylindrical hill, the path loss to subscribers on the back side of the hill and above the inflection point is given by

$$L = -10\log\left(\frac{\lambda D_H e^{-\psi\hat{\theta}}}{4\pi\sqrt{R_1 R}}\right)^2 - 10\log\left[\frac{\lambda\rho_1}{2\pi^2(H_B - h_m)^2}\right] \qquad (7\text{–}18)$$

For rolling terrain, $R = R_1 + R_H\hat{\theta}$ in (7–18) may be approximated by the horizontal distance between the base station and the mobile, and R_1 by the horizontal distance between the base station and the top of the hill. Also, D_H is taken from Figure 7–15 and ψ from Figure 7–14 or expression (7–14).

7.3 Modeling the effects of trees

The leaves and branches of trees offer significant attenuation to UHF and microwave signals. Measurements of transmission loss at 869 MHz for low elevation angles through the canopies of large isolated trees [22] found attenuations of 10 dB and more. For these measurements an average specific attenuation α_{dB} was found to be 1.1 dB/m. Attenuation depends on the frequency and on the type of tree through the size, shape, and angular distribution of the leaves and branches. As an example, the specific attenuation for distance up to 20 m extracted from measurements at 11.2 GHz in an apple orchard [23] are α_{dB} = 2.0 dB/m when the trees are in leaf and 1.7 dB/m when the trees are without leaves.

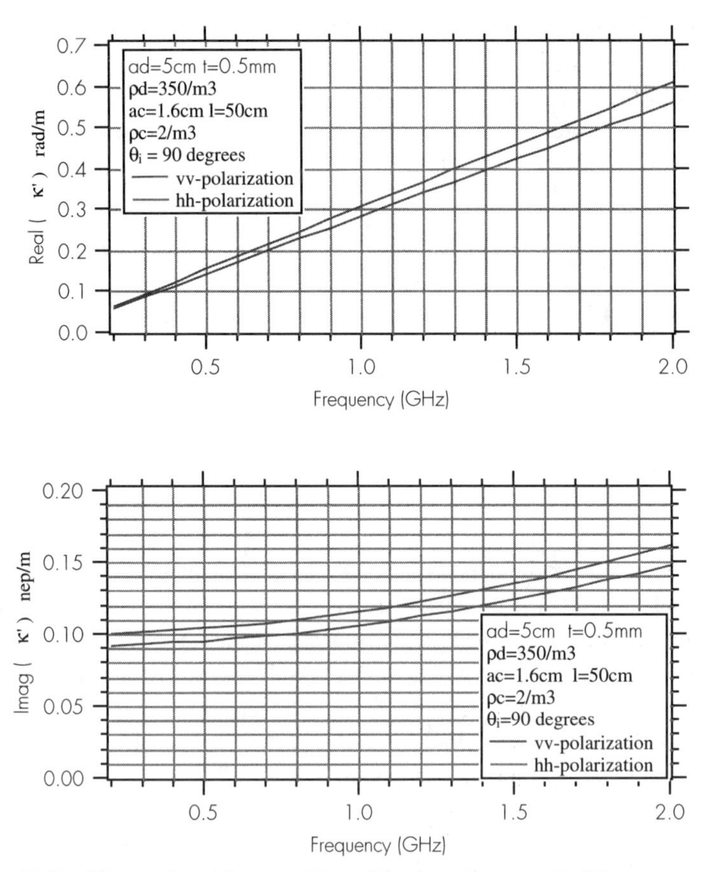

Figure 7-18 The real and the negative of the imaginary part of the wavenumber difference $\kappa = k_F - k$ for a particular set of parameters describing the size and orientation distribution of the leaves and branches in trees [27].

Treating the leaves and branches as a random medium, the attenuation may be evaluated theoretically [23–25]. Inside the random medium, the total fields consist of a coherent part and a diffuse part. The coherent fields are associated with the forward scatter past the individual leaves and branches, while the diffuse fields are those resulting from scattering away from the forward direction. As the radio waves propagate into the random medium, the coherent fields are initially dominant. However, they decay relatively rapidly as a result of absorption in the leaves and branches, and through scattering into the diffuse fields. After several nepers of decay of the coherent fields, the diffuse fields become dominant and continue to decay, but at a much lower rate.

The propagation constant for the coherent fields through the leaves and branches may be written as $k_F = k + \kappa$ through the forest, where k is the wavenumber in air and the deviation $\kappa = \kappa' - j\kappa''$ has a positive real part and a negative imaginary part. The frequency dependencies of $\kappa' = \mathrm{Re}\{\kappa\}$ and $\kappa'' = -\mathrm{Im}\{\kappa\}$ are plotted in Figure 7–18 for propagation parallel to the ground and for a particular set of parameters of the random medium [26,27]. Since the branches and trunks are distributed only over a limited range of angles about the vertical, the value of κ is different for vertically and horizontally polarized waves and will depend on the direction of propagation with respect to the vertical. Multiplying κ'' in Figure 7–18b by 8.69 gives the specific attenuation α_{dB}, which for 850 MHz gives values close to the measured average of 1.1 dB/m. For individual trees, or distances through a forest that are less than several skin depths [skin depth = $1/\kappa''$ = $8.69/\alpha_{dB}$], the coherent fields dominate and we may define an effective dielectric constant and polarizability of the tree canopy. Because the canopy is anisotropic, the relative dielectric constant and polarizability are properly represented by dyadics. Since we are concerned with propagation in directions that are close to horizontal, for simplicity we may think in terms of scalar quantities, which will be different for TE and TM polarization. The scalar dielectric constant ε_r and polarizability χ are defined by the relation

$$\varepsilon_r \equiv 1 + \chi = \frac{(k+\kappa)^2}{k^2} \tag{7–19}$$

Taking the values of κ from Figure 7–18 for vertical polarization at 1 GHz, it is found that real and imaginary parts of the polarizability $\chi = \chi' - j\chi''$ have values $\chi' = 0.027$ and $\chi'' = 0.010$. At 200 MHz, χ' is unchanged, but $\chi'' = 0.043$.

In addition to the problem of characterizing individual types of trees, modeling the effects of trees is complicated by the many ways they are grouped in and around metropolitan areas. At one extreme is a natural dense forest in which the trees form a continuous canopy whose top is nearly flat. In this case commercial wireless communications will primarily be to subscribers located on roads cut through the forest, or in clearings used for parking lots and buildings. In distant suburbs and urban parks, the density of trees may be less than in a forest, so the canopy will be broken. Close-in suburbs may have trees planted in rows along the streets, and towards the city center there may be isolated trees or small groups of trees. In this chapter we examine the effect of trees on path loss from a base station antenna to subscribers located in (1) a forest area, (2) a clearing in the forest, and (3) a suburban area with trees in rows along the streets.

7.3a Propagation to subscribers in forested areas

Because the fields passing through the forest are highly attenuated, the dominant contribution to the signal received by a subscriber in the forest must come from fields that propagate via other paths that are predominantly in the air. Evaluating measurements of the signal strength for mobile-to-mobile communications made in forests at frequencies in the band 2 to 200 MHz, Tamir [28] recognized that for large separations between transmitter and receiver the main segment of the propagation path lay in the air region above the forest canopy. To model the propagation Tamir represented the forest by a lossy dielectric layer with complex dielectric constant. With this representation, he was able to show that at long distances mobile-to-mobile propagation takes place via a lateral wave in the air propagating parallel to the forest canopy, as indicated in Figure 7–19. The lateral wave is excited by a ray reaching the treetops from the

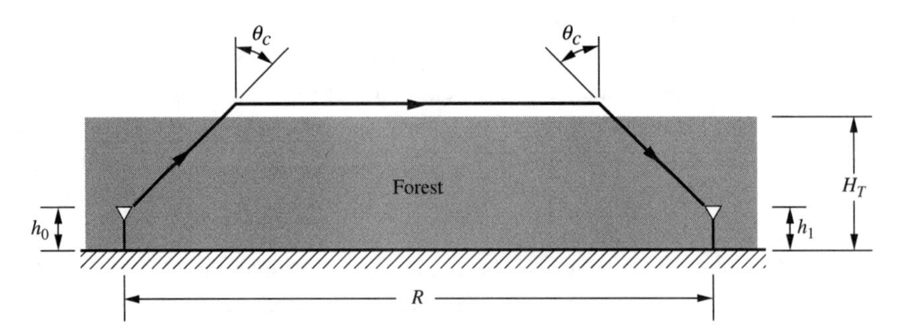

Figure 7-19 Mobile-to-mobile propagation via the lateral wave in a forest.

transmitter at the critical angle $\theta_c = \arcsin(1/\sqrt{\varepsilon_r})$, and it radiates back into the forest at the critical angle. Derivation of the lateral wave fields requires mathematical techniques not covered in this book, so that we merely give the result. Neglecting ground reflection, and for horizontal separations R that are large compared to the vertical dimensions, the path gain PG is found to be [27]

$$\mathrm{PG} = \left(\frac{1}{|\chi|}\right)^2\left(\frac{\lambda}{2\pi R}\right)^4 \exp(2Sk\,\mathrm{Im}\{\sqrt{\chi}\}) \tag{7–20}$$

(recall that the path loss L in decibels is $-10\log\mathrm{PG}$). Here

$$S = (H_T - h_0) + (H_T - h_1) \tag{7–21}$$

is the sum of the vertical distance from each mobile to the top of the forest canopy, whose height is H_T. Note that $k\,\mathrm{Im}\{\sqrt{\chi}\} < 0$ and is equal to $\mathrm{Im}\{k\sqrt{\varepsilon_r - 1}\}$, which is the imaginary part of the vertical component of the wavenumber in the forest canopy when the transverse wavenumber is k. It is seen from (7–20) that the received signal decreases as $1/R^4$, corresponding to a slope index $n = 4$. For example, if $\chi = 0.027 - j0.043$ at 200 MHz, then $|\chi| = 0.051$ and $k\,\mathrm{Im}\{\sqrt{\chi}\} = -0.11\ \mathrm{m}^{-1}$. A symmetric link of length $R = 1$ km, with $H_T - h_0$ and $H_T - h_1$ both equal to 10 m, has a path loss of 134.6 dB.

Selecting values of χ' and χ'' to give a good fit between predictions given by the model and measurements, Tamir [28] and others [15] have found values of χ' between about 0.04 and 0.1, which are somewhat larger than the computed value of 0.027. Their values of χ'' at 200 MHz are between about 0.009 and 0.014, which are smaller than the computed value of 0.043. The Tamir model has been applied at higher frequencies [15], but some caution should be observed in this extension when the critical angle ray through the trees is longer than the distance in which the mean field is greater than the diffuse field. Note that the relative dielectric constant of the forest is only slightly greater than unity, so that the critical angle θ_c is close to 90° and the length of the critical angle ray through the trees is much longer than the height of the trees H_T. Using the computed values of χ' and χ'', the critical angle is $\theta_c = \arcsin(1/\sqrt{1.027}) = 80.7°$ and the length of the critical ray is $(H_T - h_m)/\cos\theta_c = 6.2(H_T - h_1)$. In an old forest, the

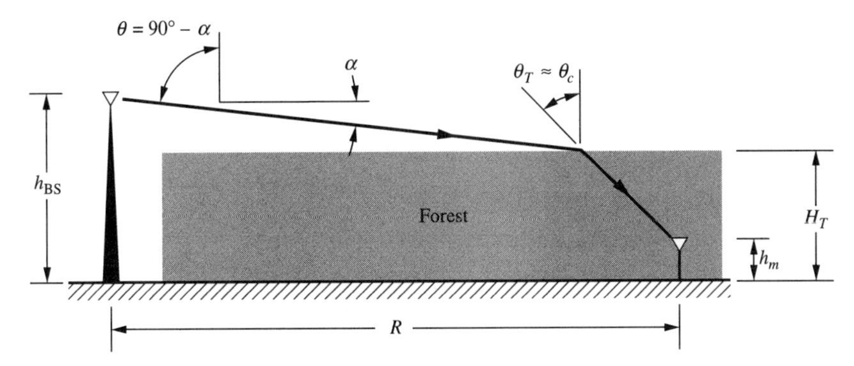

Figure 7-20 Refraction path for propagation to a subscriber located inside a forest.

trees may be well over 10 m high, and the length of the critical angle ray is therefore more than 60 m. At 1 GHz, the skin depth is about 10 m, so that the length of the critical angle ray is in the range where the diffuse field dominates.

When the base station is elevated above the trees ($h_{BS} > H_T$) as in Figure 7–20, and R is large, the path gain can be found by accounting for the free-space path gain, transmission into the forest canopy, and the attenuation from the top of the canopy down to the subscriber. Thus the path gain is given by

$$\mathrm{PG} = \left(\frac{\lambda}{4\pi R}\right)^2 |T|^2 \exp[2(H_T - h_m)k\,\mathrm{Im}\{\sqrt{\varepsilon_r - \sin^2\theta}\}] \tag{7–22}$$

where θ is the angle of incidence, as measured from the vertical.

Further approximations to (7–22) are valid when R is large and highlight the spatial parameter dependence of PG. Using Snell's law for the angle of transmission θ_T into the forest canopy, when the glancing angle α in Figure 7–20 is small, the following conditions hold:

$$\begin{aligned} \cos\theta &= \sin\alpha \ll 1 \\ \cos\theta_T &= \sqrt{1 - \frac{1}{\varepsilon_r}\sin^2\theta} = \sqrt{1 - \frac{1}{\varepsilon_r}\cos^2\alpha} \approx \sqrt{1 - \frac{1}{\varepsilon_r}} \end{aligned} \tag{7–23}$$

Then using (7–23) in conjunction with (3–34) and (3–41), it is seen that the transmission coefficients of the TE and TM polarizations are

$$\begin{aligned} T_E &= \frac{2\cos\theta}{\cos\theta + \sqrt{\varepsilon_r}\cos\theta_T} = \frac{2\sin\alpha}{\sin\alpha + \sqrt{\varepsilon_r - 1}} \\ T_H &= \frac{2\sqrt{\varepsilon_r}\cos\theta}{\sqrt{\varepsilon_r}\cos\theta + \cos\theta_T} = \frac{2\varepsilon_r\sin\alpha}{\varepsilon_r\sin\alpha + \sqrt{\varepsilon_r - 1}} \end{aligned} \tag{7–24}$$

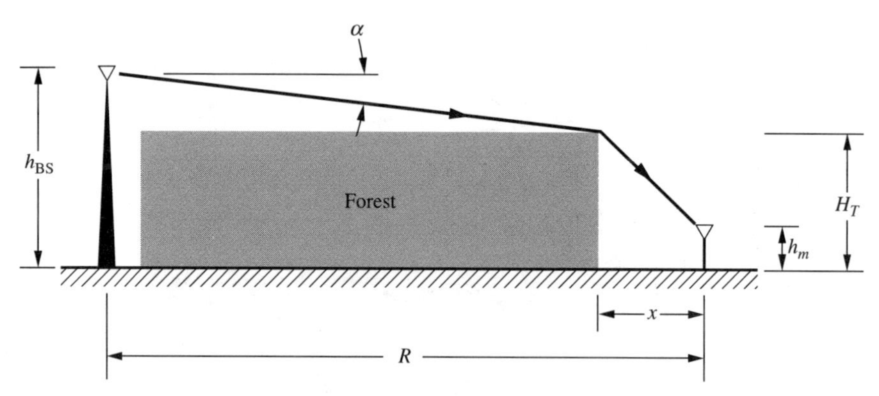

Figure 7-21 Diffraction path for propagation over a forest to a subscriber located in a clearing.

Since $|\varepsilon_r - 1| > 0.01$, $\sqrt{\varepsilon_r - 1} > 0.1$, so that the terms $\sin\alpha$ in the denominator of (7–24) can be neglected. Also, since $0.1 > |\varepsilon_r - 1|$, $\varepsilon_r \sin\alpha \approx \sin\alpha$ in the numerator. With these approximations, both transmission coefficients are the same and can be written

$$T = \frac{2\sin\alpha}{\sqrt{\varepsilon_r - 1}} \approx \frac{2(h_{BS} - H_T)}{R\sqrt{\chi}} \tag{7–25}$$

With the help of (7–25), and setting $\sin^2\theta \approx 1$ in the exponent of (7–22), the path gain can be written

$$PG = \left(\frac{\lambda}{4\pi}\right)^2 \frac{4}{R^4} \frac{(h_{BS} - H_T)^2}{|\chi|} \exp[2(H_T - h_m)k\,\mathrm{Im}\{\sqrt{\chi}\}] \tag{7–26}$$

As in the case of mobile-to-mobile transmission, the received signal has range dependence $1/R^4$. Using the same values for $\chi' = 0.027$ and $\chi'' = 0.043$ indicated in the example after (7–21), the path loss at 200 MHz between a base station located 10 m above the top of the forest canopy and a mobile located 10 m below is 109.1 dB at a distance $R = 1$ km. For mobile-to-mobile communication, the path loss found after (7–21) is 30.7 dB greater than this value, of which 9.5 dB is due to the attenuation through the trees at the base station end of the link. Whereas (7–20) remains valid as a base station antenna inside the forest is raised to the top of the forest canopy, (7–26) is valid only when the base station is a few wavelengths above the trees [29].

7.3b Path loss to subscribers in forest clearings

Modeling the path loss to subscribers located in a cleared section of the forest using the path indicated in Figure 7–21 treats the edge of the forest as a source of diffracted fields. This approach is suggested by the measurements made by LaGrone, who observed that the edge of a forest acts as a diffracting knife edge for signals with a horizontally polarized electric field for frequencies ranging from 82 MHz to 2.95 GHz [30]. For subscribers close to the treeline, and for the lowest frequency, the height of the equivalent knife was somewhat lower than the height of the trees, indicating penetration through the trees. However, at UHF frequencies and away from the treeline, LaGrone found that the height of the diffracting edge approaches the height of the trees. On narrow road cuts through forests, seasonal variations of several decibels have been observed at 457 and 914 MHz [31], indicating that the equivalent height depends on the specific attenuation. For this analysis we take the height of the edge to be that of the trees.

The field just above the top of the forest canopy is the incident field plus the reflected field, or $(1 + \Gamma) = T$ times the incident field. This field is diffracted down to ground level. If R is large compared to the distance x from the treeline to the subscriber, and neglecting ground reflection, the path gain from an elevated base station to the subscriber is

$$PG = \left(\frac{\lambda}{4\pi R}\right)^2 |T|^2 \frac{|D|^2}{\rho} \tag{7–27}$$

where $\rho = \sqrt{x^2 + (H_T - h_m)^2}$ is the distance from the edge of the treetops to the subscriber. Note that the path gain in (7–27) is like that for propagation over rows of buildings, with T for the forest playing the role of Q for buildings. Any of the knife-edge diffraction coefficients discussed in Chapter 5 may be used for D in (7–27). If the diffraction angle is not large, all diffraction coefficients will give similar results. When x is large, a ground reflected wave must be included since it will partially cancel the direct ray from the diffracting edge. If Γ_G is the ground reflection coefficient and $\rho_G = \sqrt{x^2 + (H_T + h_m)^2}$ is the length of the ground reflected ray, ground reflection may be accounted for by multiplying (7–27) the factor

$$\left|1 + \frac{\rho}{\rho_G}\Gamma_G e^{-jk(\rho_G - \rho)}\right|^2 \approx 4\sin^2\frac{kH_T h_m}{x} \tag{7–28}$$

The approximate form in (7–28) holds when x is large enough so that ρ_G is close to ρ, and $\Gamma_G \approx -1$, as discussed in Chapter 4. The excess loss in (7–27) can also be obtained from the diffraction formulation in (7–5). Recognizing that the glancing angle $\alpha = \phi'$ is very small, from (7–6) $D_1 \approx D_3$ and $D_2 \approx D_4$ so that (7–5) becomes $D_T = (1 + \Gamma)(D_1 + D_2)$. Letting $D = D_1 + D_2$ and since $T = 1 + \Gamma$, one obtains the same diffraction loss as indicated in (7–27).

Using approximation (7–25) for T and the first term in the Felsen diffraction coefficient for D, the path gain is

$$PG = \frac{\lambda^3}{(2\pi)^4 R^4}\frac{1}{\rho\theta^2|\chi|}\frac{(h_{BS} - H_T)^2}{} \approx \frac{\lambda^3}{(2\pi)^4}\frac{1}{R^4}\frac{\rho}{|\chi|}\frac{(h_{BS} - H_T)^2}{(H_T - h_m)^2} \tag{7–29}$$

where the last approximation is obtained by replacing θ by $\sin\theta$. As in previous cases of propagation over forests, the received signal is seen from (7–29) to have the range dependence $1/R^4$. To compare the 200-MHz path loss with that of a mobile under trees, as given by (7–26), let $R = 1$ km and assume that $h_{BS} - H_T$ and $H_T - h_m$ are both 10 m, while $x = 50$ m. With these assumptions and using the same value for χ, the path loss obtained from the final approximation in (7–29) is $L = 116.1$ dB. This is 7.5 dB more than found for a mobile under the trees, indicating that

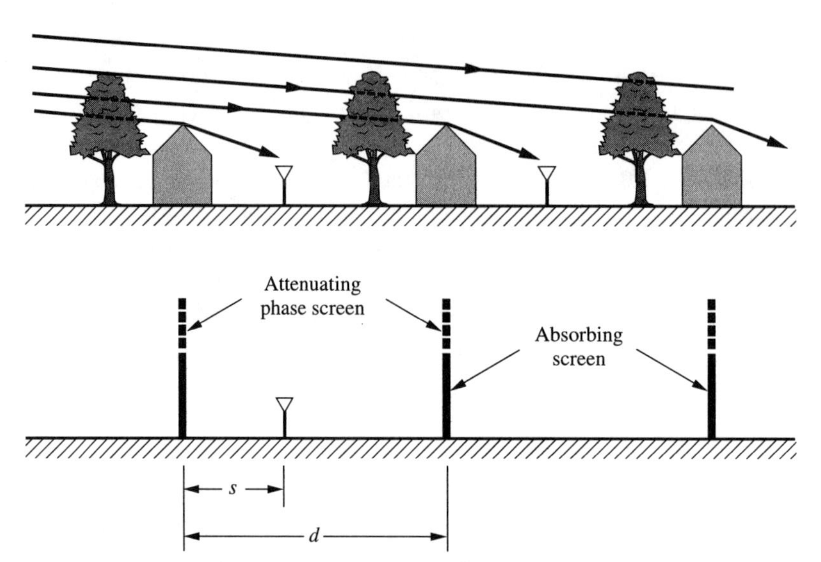

Figure 7-22 Rows of trees planted next to the row of buildings and their representation in terms of screens having both phase shift and attenuation.

for the choice of parameters the diffraction loss is slightly greater than the attenuation in the forest canopy. At higher frequencies, the exponential attenuation in the forest canopy will cause the path loss for a mobile under trees to be greater than for a mobile in a clearing.

7.3c Rows of trees in residential areas

In residential and suburban areas, trees are often planted in conjunction with houses. Mature trees will often stand well above two-story houses, and because of their attenuation they may have a comparable effect on the propagation. To gain insight into the added effects of trees located in conjunction with houses, consider the idealization of continuous rows of trees planted next to the buildings, as shown in side view in Figure 7–22. This two-dimensional geometry is amenable to the numerical integration techniques discussed in Section 6–5. The rows of buildings are replaced by absorbing screens, as in the past, while the rows of trees are replaced by partially attenuating screens that also account for the additional phase shift due to the leaves and branches. For simplicity, the phase and attenuation screen is located directly above the absorbing screen.

The phase shift and attenuation of the screens representing a row of trees is described mathematically by the factor

$$\exp[-j(k_T - k)w(y)] = \exp[-j\kappa w(y)] \tag{7–30}$$

where $w(y)$ is the width of the tree at a height y above ground, as indicated in Figure 7–23. Here k_T is the complex wavenumber in the tree, and its deviation from the free-space wavenumber is κ

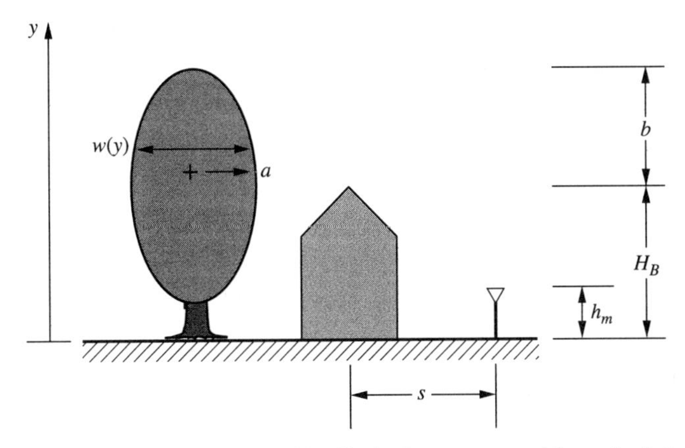

Figure 7-23 Dimensions of trees with elliptical crowns used for calculations.

$= k_T - k$, whose real and imaginary parts are plotted in Figure 7–18. To incorporate this factor into the numerical calculations of Section 6–5, ρ[p,m] defined in (6–66) is modified as follows:

$$\rho[p, m] = \sqrt{d_n^2 + (p-m)^2(\Delta y)^2} + \frac{\kappa}{k} w(m\Delta y) \qquad (7\text{–}31)$$

Note that ρ[p,m] is now complex. Above the trees $w(y)$ is set to zero.

Computations have been made for trees having elliptical shape with their centers located at the height of the buildings. If $2a$ is the minor axis and $2b$ the major axis, the width of the tree is given by

$$w(y) = 2a\sqrt{1 - \frac{(y-H_B)^2}{b^2}} \qquad (7\text{–}32)$$

for $H_B \le y \le H_B + b$. Results obtained from the numerical computations for an incident 900-MHz plane wave are shown in Figures 7–24 and 7–25, assuming that the rows are spaced $d = 50$ m apart, the two-story houses have height $H_T = 8$ m, the trees are 12 m high (so that $b = 4$ m), and the maximum width of the trees is 4 or 8 m ($a = 2$ or 4 m).

Figure 7–24 shows the height gain of the field at a distance $s = 20$ m past the twentieth row of buildings and trees for plane wave incidence at an angle $\alpha = 0.5°$. Trees of either width are seen to have an effect on the field strength similar to that when there are houses but no trees. Aside from the deep interference minima that would be highly modified by other multipath contributions, the signal at a street-level subscriber is around 5 dB lower than would occur in the absence of trees. Figure 7–25 is a plot of the field incident at the height H_B of the buildings in the $N + 1$ row for plane waves incident at the angles θ to the vertical of 90° and 89.5°. The plots for 89.5° show the settling behavior found previously for buildings alone, while those for 90° show that the fields continue to decrease with N. However, as N increases, the difference between the curves with trees and without trees in both cases approaches a constant value of 4 to 5 dB for the

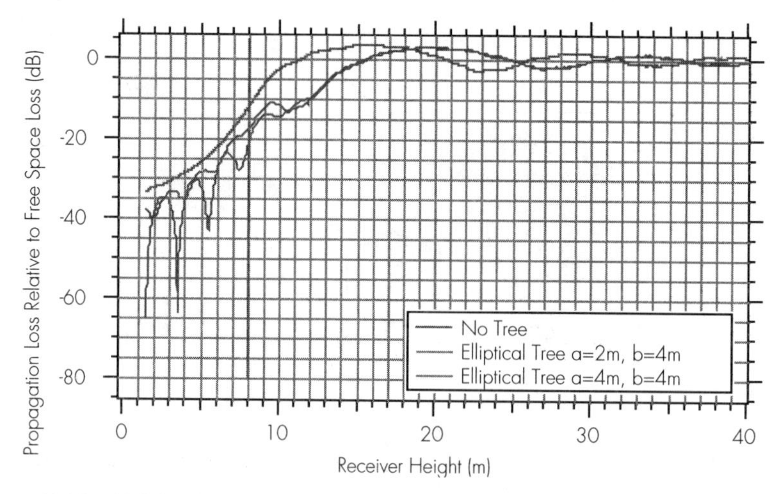

Figure 7-24 Height dependence of the field amplitude at a distance $s = 20$ m beyond the twentieth screen for a 900-MHz plane wave incident at $\alpha = 0.5°$. Results are plotted assuming that $d = 50$ m for two sizes of the trees and in the absence of the trees [26](©1998 IEEE).

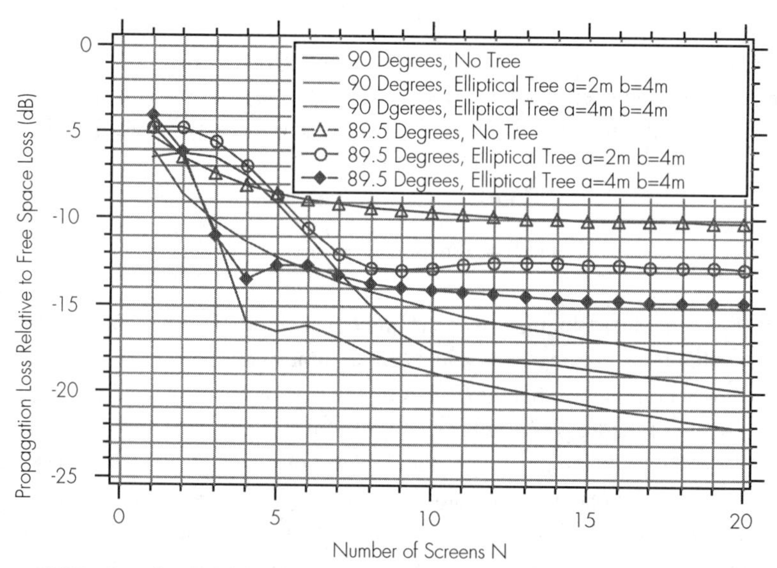

Figure 7-25 Amplitude of the field incident on the $N + 1$ screen at the rooftop height $H_B = 8$ m for a 900-MHz plane wave incident at $\alpha = 0$, 0.5°. Assuming that $d = 50$ m, results are plotted for two sizes of the trees and in the absence of the trees [26](©1998 IEEE).

wide trees. Because the effect of the branches is roughly half that of the leaves and branches together [25], in winter when the leaves are missing the path loss will be reduced by 2 or 3 dB.

7.4 Summary

In this chapter we have examined how the diffraction model for low building environments can be used to quantify the influence of randomness in the building environment, terrain variation, and the presence of trees, and thus introduce greater realism into prediction models. In studying randomness, it was found that the lognormal distribution observed for the shadow fading at street level can be the result of a cascade of two random factors acting on the signal. Building height variations by themselves do not produce a lognormal distribution, but can when combined with randomness that effects how the rooftop fields diffract down to street level. This result is independent of the distribution function of building heights. Conversely, for antennas located on rooftops, the distribution of amplitudes is not expected to be lognormal. The standard deviation of the shadow fading increases with the irregularity of building height and is weakly dependent on frequency.

For metropolitan areas on rolling terrain, the local ground slope can easily be taken into account when rooftops near the mobile are within LOS of the base station antenna. When intervening hills obscure the LOS path, their diffraction effects must be accounted for. Well beyond a hill, its diffraction may be computed using the wedge model, or even a simple knife edge. Caution must be used in treating the hill as a diffracting cylinder, since houses, trees, or small-scale roughness will alter the coefficients of the creeping wave, as demonstrated for rows of buildings on a cylindrical hill. On the back side of the hill and near the base of the hill, the shape of the hill and the buildings on it will have a significant impact on the rooftop fields. From simulations for rows of buildings on cylindrical hills, it is found that the rooftop field can be represented by a creeping ray, and the necessary coefficients are obtained numerically. The creeping ray model gives a simple expression for the rooftop fields on the back side of the hill and for locations near and beyond the base of the hill.

Attenuation of UHF radio waves through trees is significant. In forests, the trees act much as buildings by attenuating the direct waves, so that the primary contribution to the received signal comes from waves propagating over them. As a result, the range index of the path loss in the presence of a forest is found to be $n = 4$, as was found for propagation over flat earth, and close to the range index found when buildings dominate. Replacing the forest canopy by a dielectric layer having complex dielectric constant allows simple evaluation of the signal received inside the forest. Because of the attenuation, simple diffraction models can be used to find the path loss to subscribers located in clearings. When trees are planted in conjunction with the buildings, as along tree-lined suburban streets, the combined effect of the trees and buildings gives a path loss that is only a few decibels greater than that of the buildings alone.

Problems

7.1 Estimate the range of variation in decibels of the path loss due to differences in building shape and location. To do so, find the differences in the diffraction loss down to the mobile, which is given b y $-10 \log PG_2$, for the two extreme cases shown in Figures P7–1a and b. Assume that $d = 50$ m, $w = 10$, m, $h_m = 1.8$ m, $f = 900$ MHz, and that all buildings have the average height $H_B = 11$ m. Use Felsen's diffraction coefficient and take $\alpha \approx 0$.

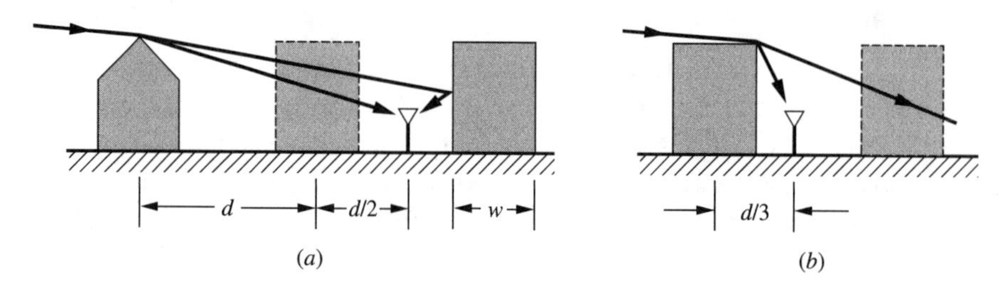

Figure P7–1

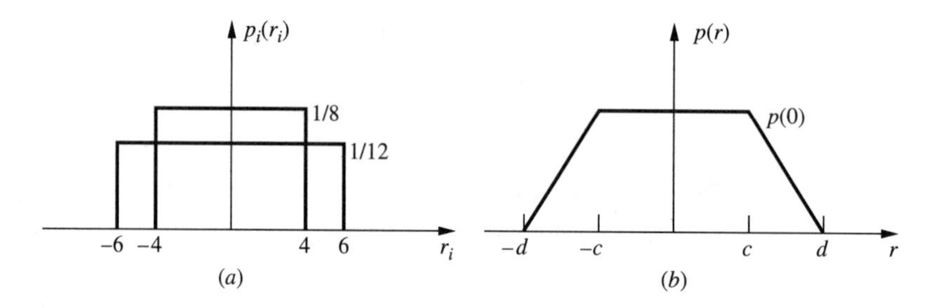

Figure P7–2

7.2 Let the random variable r be the sum of two random variables r_1 and r_2, both with zero means, whose PDFs are $p_1(r_1)$ and $p_2(r_2)$. The PDF of r is the conditional probability that for any value of r_1, the variable r_2 has the value $r - r_1$. Thus

$$p(r) = \int_{-\infty}^{\infty} p_1(r_1)p_2(r - r_1)\, dr_1 \tag{7–33}$$

Suppose that r_1 represents the deviation of the street level signal in decibels due to random building heights and, as seen from Figure 7–4, is uniformly distributed over a range of ±6 dB about its mean. Assume that r_2 represents the deviation of the signal due to differences in building shape and location and that it is uniformly distributed over a range of ±4 dB. These distributions are shown in Figure P7–2a.

(a) Evaluate the integral for $p(r)$, showing that it is trapezoidal, as depicted in Figure P7–2b, and find the values of c, d, and $p(0)$.

(b) Compute the standard deviations of $p_1(r_1)$, $p_2(r_2)$, and $p(r)$.

(c) Plot $p(r)$ on graph paper, and for comparison plot a Gaussian distribution having the same standard deviation.

7.3 When no buildings are present, Figure 7–9 needs to be modified to include three additional rays that are reflected at the flat ground near the transmitter and near the receiver. These rays can be found by considering the images of the transmitter and receiver in the flat ground. If the antennas are close to the ground, the diffraction angles for all four rays will be nearly the same and hence have nearly the same diffraction loss. In this case the additional rays can be accounted for using a correction factor at each end of the path that is given by (7–28), where x is the horizontal distance from the transmitter or receiver to the peak of the hill and H_T is replaced by the height of the hill y_H above the plane. For the dimensions of Figure 7–9:

(a) Compute and compare the diffraction coefficients for the ray going from one antenna to the other, and for the ray going from the image of one antenna to the image of the other. Does this result justify using the correction factor (7–28)?

(b) Compute the exact and approximate forms of the correction factor in (7–28) for each end of the link.

(c) Find the path loss accounting for all four rays.

7.4 In Figure 7–9, assume that the height of the hill is $y_H = 20$ m and that the antenna on the left is at a height of 20 m, but otherwise the geometry is unchanged. Compute the 900-MHz path loss neglecting ground reflection, but accounting for the computed value of $F(s)$.

7.5 A cylindrical hill rises $y_H = 30$ m above the surrounding plane and has width $2x_H = 2050$ m. The hill and surrounding plane are covered with houses of average height $H_B = 9$ m and row spacing $d = 50$ m. The base station antenna is located at the base of the hill on one side at a height $h_{BS} = 39$ m and is used in a 900-MHz wireless local loop system.

(a) Find the radius of the hill and the ground slope at the base of the hill.

(b) Compute and plot the path loss to antennas mounted at the rooftop height of the houses on the back side of the hill.

7.6 In Figure 7–11, let the antenna on the left be that of a base station at a height of 30 m. The hill and plane following are built up with houses of height 10 m and row spacing 50 m. The antenna on the right represents a subscriber between two rows of houses, and its height of $h_m = 1.5$ m places it below the surrounding rooftops. The dimensions of the hill and the horizontal distances of the antennas from the peak are as indicated in Figure 7–11.

(a) Find the path loss for 1800-MHz signals.

(b) For comparison, find the path loss if the hill and buildings on it are replaced by a single absorbing knife edge of height 60 m. Use Felsen's diffraction coefficient.

7.7 In Figure 7–20, the height of forest is $H_T = 12$ m, a base station operating at 500 MHz is above the forest at the height $h_{BS} = 30$ m, and the mobile height is $h_m = 2$ m.

(a) From Figure 7–18, find the real and imaginary parts of the polarizability χ for vertical polarization.

(b) Compute and plot the path loss to the mobile as a function of R in the interval $1 \le R \le 10$ km.

7.8 Referring to Figure 7–21, suppose that $h_{BS} = 25$ m, $H_T = 15$ m, $h_m = 1.6$ m, $f = 1.8$ GHz, and the distance from the base station to the edge of forest is $R - x = 2$ km. Compute and plot the path loss as a function of x in the range $10 \le x \le 200$ m.

References

1. C. Chrysanthou and H. L. Bertoni, Variability of Sector Averaged Signals for UHF Propagation in Cities, *IEEE Trans. Veh. Technol.*, vol. 39, pp. 352–358, 1990.
2. H. L. Bertoni and L. Maciel, Theoretical Prediction of Slow Fading Statistics in Urban Environments, *Proc. IEEE ICUPC Conference*, pp. 1–4, 1992.
3. P. E. Mogensen, P. Eggers, C. Jensen, and J. B. Andersen, Urban Area Radio Propagation Measurements at 955 and 1845 MHz for Small and Micro Cells, *Proc. IEEE GLOBECOM'91*, pp. 1297–1302, 1991.
4. K. Low, A Comparison of CW-Measurements Performed in Darmstadt with the COST-231-Walfisch-Ikegami Model, *Rep. COST 231 TD(91) 74*, Darmstadt, Germany, 1991.

5. Y. Okumura, E. Ohmori, T. Kawano, and K. Fukuda, Field Strength and Its Variability in VHF and UHF Land-Mobile Radio Service, *Rec. Electron. Commun. Lab.*, vol. 16, pp. 825–873, 1968.
6. K. Bullington, Radio Propagation for Vehicular Communications, *IEEE Trans. Veh. Technol.*, vol. VT-26, pp. 295–308, 1977.
7. R. J. Luebbers, Finite Conductivity Uniform GTD versus Knife Edge Diffraction in Prediction of Propagation Path Loss, *IEEE Trans. Antennas Propagat.*, vol. AP-32, pp. 70–76, 1984.
8. J. D. Parsons, The Mobile Radio Propagation Channel, Wiley, New York, Chap. 3, 1991.
9. G. Lampard and T. Vu-Dinh, The Effect of Terrain on Radio Propagation in Urban Microcells, *IEEE Trans. Veh. Technol.*, vol. 42, pp. 314–317, 1993.
10. D. E. Elaides, Alternative Derivation of the Cascaded Cylinder Diffraction Model, *IEE Proc.*, vol. 140, pp. 279–284, 1993.
11. M. F. Levy, Parabolic Equation Modeling of Propagation over Irregular Terrain, *Electron. Lett.*, vol. 26, pp. 1153–1155, 1990.
12. R. Janaswamy and J. B. Andersen, A Curvilinear Coordinate-Based Split-Step Parabolic Equation Method for Propagation Predictions over Terrain, *IEEE Trans. Antennas Propagat.*, vol. 46, pp. 1089–1097, 1998.
13. J. T. Hviid, J. B. Andersen, J. Toftgard, and J. Bojer, Terrain-Based Propagation Model for Rural Area—An Integral Equation Approach, *IEEE Trans. Antennas Propagat.*, vol. 43, pp. 41–46, 1995.
14. W. C. Y. Lee, Studies of Base-Station Antenna Height Effect on Mobile Radio, *IEEE Trans. Veh. Technol.*, vol. VT-29, pp. 252–260, 1980.
15. T. Kurner, D. J. Cichon, and W. Wiesbeck, The Influence of Land Usage on UHF Wave Propagation in the Receiver Near Range, *IEEE Trans. Veh. Technol.*, vol. 46, pp. 739–746, 1997.
16. K. A. Chamberlin and R. J. Luebbers, An Evaluation of Longly–Rice and GTD Propagation Models, *IEEE Trans. Antennas Propagat.*, vol. AP-30, pp. 1093–1098, 1982.
17. G. L. James, Geometrical Theory of Diffraction for Electromagnetic Waves, Peter Peregrinus, Stevenage, Herts, England, Chap. 6, 1976.
18. D. A. McNamara, C. W. I. Pistorius, and J. A. G. Malherbe, Introduction to the Uniform Geometrical Theory of Diffraction, Arctech House, Norwood, Mass., Chap. 8, 1990.
19. K. Hacking, R.F. Propagation over Rounded Hills, *IEE Proc.*, vol. 117, pp. 499–511, 1970.
20. L. Piazzi and H. L. Bertoni, Effect of Terrain on Path Loss in Urban Environments for Wireless Applications, *IEEE Trans. Antennas Propagat.*, vol. 46, pp. 1138–1147, 1998.
21. L. Piazzi and H. L. Bertoni, A Path Loss Formulation for Wireless Applications Considering Terrain Effects for Urban Environments, *Proc. IEEE-VTC'98 Conference*, pp. 159–163, 1998.
22. W. J. Vogel and J. Goldhirsh, Tree Attenuation at 869 MHz Derived from Remotely Piloted Aircraft Measurements, *IEEE Trans. Antennas Propagat.*, vol. AP-34, pp. 1460–1464, 1986.
23. M. O. Al-Nuaimi and A. M. Hammoudeh, Measurements and Predictions of Attenuation and Scatter of Microwave Signals by Trees, *IEE Proc. Microwave Antennas Propagat.*, vol. 141, pp. 70–76, 1994.
24. F. K. Schwering, E. J. Violette, and R. H. Espeland, Millimeter-Wave Propagation in Vegetation: Experiments and Theory, *IEEE Trans. Geosci. Remote Sensing*, vol. GRS-26, pp. 355–367, 1988.
25. G. M. Whitman, F. Schwering, A. A. Triolo, and N.Y. Cho, A Transport Theory of Pulse Propagation in a Strongly Forward Scattering Random Medium, *IEEE Trans. Antennas Propagat.*, vol. 44, pp. 118–128, 1996.
26. S. A. Torrico, H. L. Bertoni, and R. H. Lang, Modeling Tree Effects on Path Loss in a Residential Environment, *IEEE Trans. Antennas Propagat.*, vol. 46, pp. 872–880, 1998.
27. S. A. Torrico, *Theoretical Modeling of Foliage Effects on Path Loss for Residential Environments*, D.Sc. dissertation, George Washington University, Washington, D.C., 1998.
28. T. Tamir, On Radio-Wave Propagation in Forest Environments, *IEEE Trans. Antennas Propagat.*, vol. AP-15, pp. 806–817, 1967.

29. D. Dence and T. Tamir, Radio Loss of Lateral Waves in Forest Environments, *Radio Sci.*, vol. 4, pp. 307–318, 1969.
30. A. H. LaGrone, Propagation of VHF and UHF Electromagnetic Waves over a Grove of Trees in Full Leaf, *IEEE Trans. Antennas Propagat.*, vol. AP-25, pp. 866–869, 1977.
31. K. Low, UHF Measurements of Seasonal Field-Strength Variations in Forests, *IEEE Trans. Veh. Technol.*, vol. VT-37, pp. 121–124, 1988.

CHAPTER 8

Site-Specific Propagation Prediction

In Chapters 6 and 7, the building environment was described in terms of statistical parameters, such as the average row spacing and the building height distribution, as a basis for making propagation predictions. As an alternative approach, in this chapter we discuss site-specific predictions that make use of a database of the actual shapes of buildings in the region under study. In this work we classify such predictions into three categories: two for outdoor propagation around and over buildings and a third dealing with propagation between subscriber and access point located inside a building or tunnel. One category of outdoor predictions applies to low antennas in a tall-building environment, where the primary propagation paths lie around the sides of the buildings rather than over the tops. Prediction software for this case requires only a two-dimensional database of the building foot prints. For base station antennas mounted on roofs, or in an environment containing at least some low buildings, significant propagation paths have segments that go over the tops of buildings as well as around the sides. For this case it is necessary to have a database of the buildings in three dimensions. Propagation inside buildings and tunnels is treated here as a single class, even though it may involve propagation over a single floor, or propagation between floors.

Because of the size of the buildings compared to the wavelength at wireless frequencies, direct numerical solvers of Maxwell's equations, such as the finite element and finite difference methods, involve too many unknowns to be feasible at this time. Only ray optical methods can be used at these frequencies. These methods are referred to in the literature as the geometrical theory of diffraction (GTD) or the uniform theory of diffraction (UTD) when the transition functions discussed in Chapter 5 are included in the diffraction coefficients. Various ray-based computer codes have been developed for rendering images at optical frequencies and to determine the radar cross section of metallic objects, such as aircraft and the radiation characteristics of radio antennas mounted on them. The codes for these applications are not well suited to the problem of propagation prediction among buildings. At optical frequencies, diffuse scattering of light is the primarily physical process occurring at walls, and the codes need only account for one or two interactions of a light ray with walls. However, the inclusion of color effects and small architectural details is important. Predicting radar scattering by metallic targets involves sophisticated diffraction effects, such as creeping rays on surfaces with compound and variable curvature, but again requires few multiple interactions along a ray path. Also, radar targets are composed of relatively few geometric elements.

At UHF and microwave frequencies, rays incident on building walls reflect strongly in the specular direction, although there is some evidence that diffuse scattering may also be important. Unlike other ray applications, it is important to account for multiple interactions along a ray path to accurately predict the received signal. These interactions take the form of reflections from building walls and the ground, alone or in sequence with diffractions at building edges and/or terrain blockage. The number of such reflection and diffraction events along a ray path may be six to eight, or more, and the number of building walls in the database may number from hundreds to thousands. However, a simplifying feature is that building walls are almost always vertical and electrically flat, especially at UHF frequencies. To accommodate the special features of UHF propagation among buildings, various groups have written computer codes that work with either two-dimensional or three-dimensional building databases to make predictions for wireless systems.

The most obvious application of the ray methods is to predict the received signal level for coverage and interference evaluation. Such predictions may be of special importance when it is difficult or impractical to make tests at a site prior to system installation, or to make adjustments after it is installed. Although the ray methods might, in principle, be able to predict the multipath interference pattern for narrowband signals, which is experienced as fast fading, in practice they can only give its statistical properties, not the actual pattern in space. Several factors limit the accuracy with which the phase and amplitude of the arriving rays can be computed, and hence preclude computation of the fading pattern. First, it is difficult to create a building database with position accuracy better than 0.5 m, which is on the order or greater than the wavelength. Also, the databases do not include the architectural detail that can introduce phase shifts in the reflection coefficients. Furthermore, signals that are weakly scattered by small movable objects, such as people and automobiles, can shift the location of the interference minima. Finally, all codes truncate the number of rays accounted for because of computer memory and time limitations.

Although the exact fading pattern cannot be predicted, its statistical properties can. Of most interest is the small-area average received power, which is found by spatially averaging the field magnitude squared. Since this average is equal to the sum of the squares of the field magnitudes, as discussed in Chapter 2, it is usual to add ray powers rather than add the ray fields. However, because the ray codes can output the direction of the rays at both ends of the link, as well as ray path length and field amplitude, they can also be used to find time delay and angle of arrival and other statistical information. Thus ray codes can also serve as a simulation tool for finding statistical channel parameters, such as root-mean-square (RMS) delay spread or angle of arrival

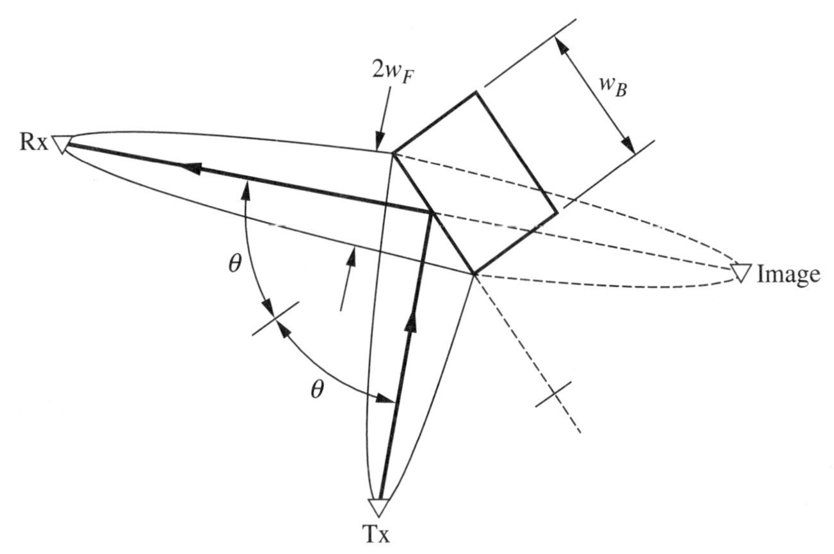

Figure 8-1 Specular rays can be used to describe reflections from buildings provided that the Fresnel zone is narrower than the building. This requirement limits the range over which ray predictions can be used.

spread and slow fading correlation, in different building environments, or a part of an overall system simulation tool.

There is a limitation on the size of the region over which ray methods can be used to predict the received signal. This limitation can be understood in terms of the Fresnel zone, as shown in Figure 8–1 for a simple two-dimensional example. Here a ray from a transmitting antenna Tx is reflected by a building to a receiving antenna Rx. The ray, and the Fresnel zone about the ray, can be constructed from the image of Tx in the reflecting face of the building. If the width of the reflecting face of the building w_B is greater than that of the Fresnel zone, the fields of the reflected ray can be found using the simple rules for reflection at a plane surface. This condition is $w_B \cos\theta \geq 2w_F$, where θ is the angle of incidence with respect to the normal. If S is the total unfolded length of the reflected ray, then from (5–14), the maximum value of $2w_F$ is $\sqrt{\lambda S}$. Thus the foregoing requirement on the building width is equivalent to

$$S \leq \frac{1}{\lambda}(w_B \cos\theta)^2 \tag{8–1}$$

If $\theta = 0$ and $w_B = 20$ m, (8–1) limits the ray length to 1.2 km at 900 MHz or 2.4 km at 1.8 GHz. For terrestrial base stations, condition (8–1) limits the distance from the base station over which ray methods may be used to about 1 km at 900 MHz and 2 km at 1.8 GHz. This limitation does not hold if the base station is an aircraft or satellite, since the widest part of the Fresnel zone then lies up in the air, or in space, where there are no buildings.

If the Fresnel zone is wider than the building, the building acts as a scatterer of the incident ray field rather than as a reflector [1,2]. For tall buildings, for which the scattering is essentially two-dimensional, the scattering cross section can be found by coherently adding the fields of the rays diffracted by the edges to the fields of the reflected rays. Besides being more complicated, this method fails when the Fresnel zone extends above the tops of low buildings, in which case the scattering is three-dimensional. To get an accurate representation of the scattered field in three dimensions, one must coherently add the fields diffracted by the vertical and horizontal edges, the fields diffracted by the corners, and the reflected fields. Accounting for the coherent addition of these contributions is possible for isolated surfaces but becomes very complex when there are multiple interactions between a number of surfaces. It appears that no one has attempted to use such an approach for building environments.

The primary problem in applying ray methods is to find the ray paths. The nature of ray codes is such that they start the rays at a single transmitter point, typically taken as the base station, and find the rays to many receiver locations in one process. Note that the ray paths are reciprocal, so that the results apply to transmission by both the base station and the subscriber unit. Two well-known methods for finding the ray paths are the image method and the pincushion method, which are discussed in greater detail in the following section on two-dimensional predictions. In even a slightly complex geometry, there can be an infinite number of rays going from a transmitting antenna to a receiving antenna. The field amplitude decreases as rays undergo more reflection or diffraction events and can eventually be discarded. Thus all ray codes have criteria that limit the number of rays arriving at a receiving antenna. Although the physical principles are the same for all codes, the algorithms used in the implementation can be very different, which affects the number and types of rays that are retained, (hence the accuracy of the predictions) and the running time.

8.1 Outdoor predictions using a two-dimensional building database

Examples of reflected and diffracted rays in two-dimensions are shown in Figure 8–2. The receiver Rx1 is illuminated by a ray undergoing two reflections and by rays diffracted at an edge, one of which is reflected after the diffraction. Rays can reach Rx2 only after a single diffraction at various corners, while diffraction at two corners is required to reach Rx3. The path loss upon diffraction through a large angle, as in turning a corner, is much greater than at a reflection, so that the twice reflected ray reaching Rx1 is likely to give the dominant contribution. However, in some cases there is no reflected ray, such as at Rx2 in Figure 8–2, or only those that have undergone many reflections, so that diffracted rays give the dominant contribution. For some locations, such as Rx3, rays must undergo diffraction at two corners. Rays that undergo diffraction at more than two corners are typically neglected because most areas can be reached by rays that are reflected and/or diffracted at two corners, and because of the significant addition to the computation time that would be required.

Each ray seen in top view in Figure 8–2 may represent two or more rays that have different components of displacement in the vertical direction. In the absence of terrain blockage, one such ray has a vertical component of travel that goes directly from Tx to Rx. If the ground is flat, a second ray exists that is reflected in the ground and appears to come from the image of Tx in

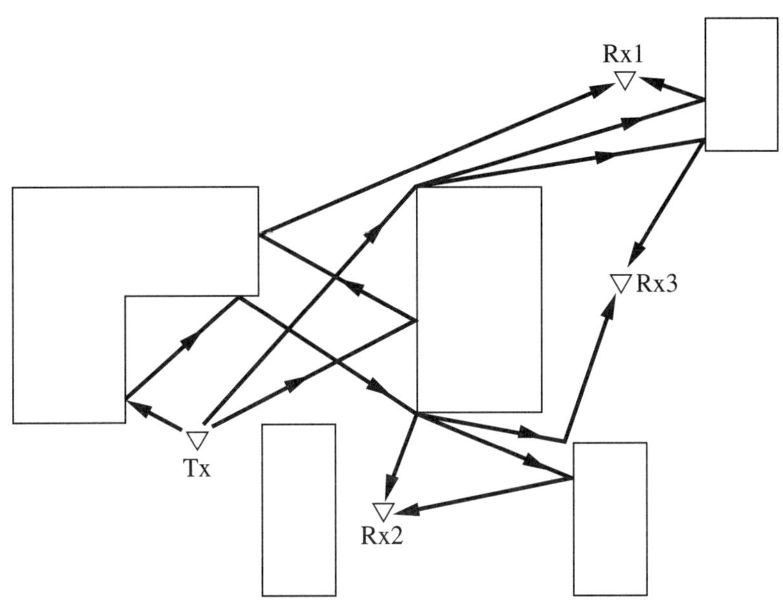

Figure 8-2 Rays from a base station can reach subscriber locations by a combination of multiple reflection and diffraction events, and via many such paths.

the ground plane, as discussed for LOS paths in Chapter 4. When viewed from above, the direct and ground reflected rays appear as one. If the terrain is not flat, there may be more than one ground reflected ray connecting Tx to Rx.

Several groups have written computer codes that work with a two-dimensional database of buildings to find the multiply reflected and diffracted rays [3–8]. These codes start by finding the rays that arrive at the subscriber locations and at each of the building corners directly or via reflections at the building walls. Each building corner is treated as a secondary source, so that the code must also find the direct and multiply reflected rays connecting each corner to the receiver locations. To find the rays that undergo diffraction at two corners, it is also necessary to find the rays connecting each corner to all other corners. By their nature, ray codes are efficient at finding the paths from a single base station to many subscriber locations at the same time. This results from the large fraction of the computation devoted to finding ray intersections with walls, or corner illumination, and the small fraction of the computations that are directly associated with the individual subscriber locations.

8.1a Image and pincushion methods

The image method and the pincushion method, also known as the shooting and bouncing ray (SBR) method, are the two approaches that have been used to find the multiply reflected rays from a source (or secondary source) to receiver points. These two methods are indicated in Figure 8–3. The *image method* starts by constructing the image of the source in all the building sur-

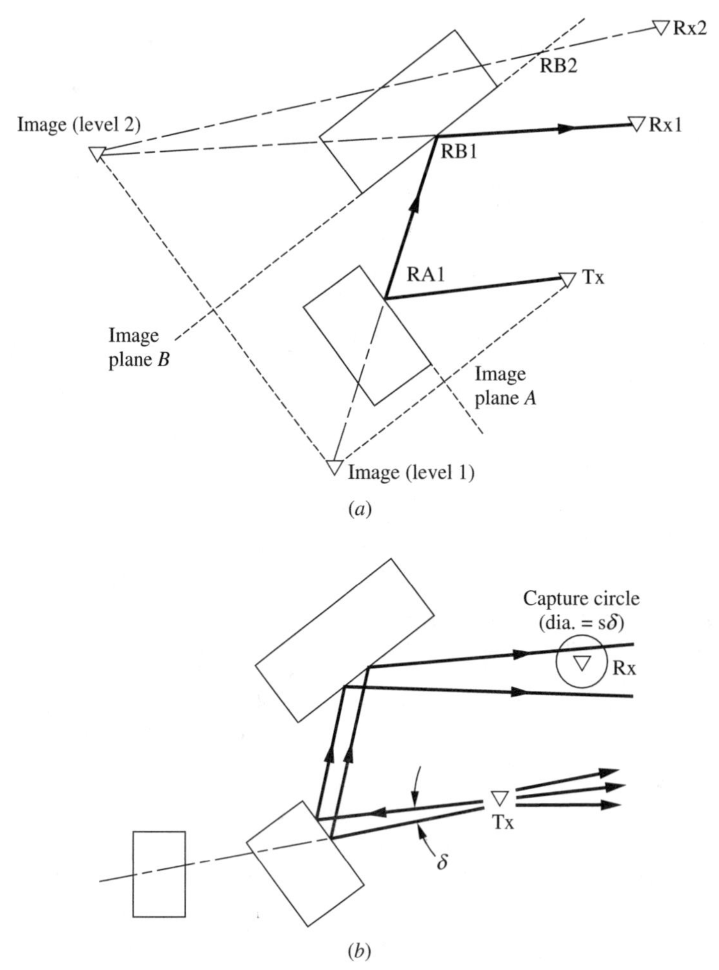

Figure 8-3 Comparison of the image method (a) and the pincushion method (b) for finding the ray paths in two dimensions.

faces that are visible to it. The image is itself imaged in all the surfaces visible to it, as seen for one such image in Figure 8–3a, and the process is repeated up to the number of multiple reflections that are to be accounted for. The lines between each image and the receiver points are then constructed. At this step in the process, it is necessary to determine if the lines cross the image plane at the location of the building wall, not just its analytic extension. For example, in Figure 8–3a the ray from the image (level 2) to Rx1 crosses the image plane B at point RB1, where a building surface exists, and therefore acts as a point of reflection. However, the ray from image (level 2) to Rx2 crosses the image plane at point RB2, where there is no wall, and hence is not a valid reflection point.

The foregoing process must be continued back to the next-lower level. Thus in Figure 8–3a the line from the image (level 1) to reflection point RB1 crosses plane A at point RA1 on the wall of the building and is thus a valid reflection point. At any step in the process the reflection point may not lie on the building surface and the path is not a legitimate ray. When the database of buildings is large and several reflections are allowed, the process of checking to see that all the reflection points lie on building surfaces is the most time-consuming aspect of the program. After identifying the legitimate multiple reflections, for each ray the angles of incidence at all surfaces and the unfolded path length S of the ray are found to compute the ray field.

The *pincushion method* depicted in Figure 8–3b starts rays from the source (or secondary source) with angular separation δ. The intersection of a ray with all the walls in the database is computed, and the one closest to the source is chosen as the reflecting wall, since the other intersections are shadowed by the first wall. The resulting ray segment is then tested to see if it illuminates any of the receiver locations. Next, computing the angle of incidence at the reflecting wall gives the direction of the reflected ray, whose intersection point with a subsequent wall is then found, thereby defining the reflected ray segment. The reflected ray segment is tested to see if it illuminates any of the receiver locations. This process is repeated up to the desired number of reflections and for each ray originating from the source. Finding the intersections of the rays with the walls is the most time-consuming aspect of pincushion codes.

An algorithm must be employed in the pincushion method to ensure that one ray, but only one, originating from the source and reflected from the same set of surfaces will illuminate the subscriber location. One approach is to give the receiver point a finite size that is dependent on the unfolded length of the ray, as indicated by the "capture circle" shown in Figure 8–3b. For this approach, the perpendicular projection of the subscriber location onto the ray segment is constructed and the total path length S along the ray to this point is found. The diameter of the capture circle is chosen to be δS, and the receiver location is taken to be illuminated if the ray segment crosses the circle. If a ray illuminates the receiver, neighboring rays that are separated by the angle $\pm\delta$ will fall outside the circle and will not be captured. The angular separation δ must be chosen to make the spacing between the rays at the edge of the computational area less than the building size. For example, if the distance between rays at $R = 1$ km is to be less than the building width $w_B = 10$ m, δ must be less than 0.01 rad, or about 0.6°.

8.1b Ray contributions to total power

To express the small-area averaged path gain, we separate the rays obtained from the image or pincushion methods into three classes: (1) rays that undergo only reflections at building walls (R); (2) rays that experience only one diffraction at building corners ($D1$), with or with out reflections before or after the diffraction; and (3) rays that are diffracted twice at building corners ($D2$), again with or without multiple reflections. The total path gain for isotropic antennas is written as the sum over the individual rays in each class as

$$\mathrm{PG} = \sum_i P_R^{(i)} + \sum_i P_{D1}^{(i)} + \sum_i P_{D2}^{(i)} \tag{8–2}$$

Because (8–2) calls for the addition of ray powers rather than ray fields, it will exhibit an error that is greatest on a shadow boundary, where one of the reflected rays and one of the diffracted rays add coherently. Next to the shadow boundary, and inside the illuminated region, the total field of these two rays is one-half that of the reflected ray field, giving a power that is one-fourth that of the reflected ray power. Instead, (8–2) gives the sum of the reflected ray power and the diffracted ray power, which itself is one-fourth that of the reflected ray. If these two are the only rays present, the power at one side of the shadow boundary is in error by a factor of 5. However, there are usually many other rays present, so that the fractional error is much smaller. This error can be eliminated by coherently adding the fields of the two rays near the shadow boundary.

For the rays that undergo two diffractions at building corners, let s_{i1}, s_{i2}, and s_{i3} be the two-dimensional unfolded path lengths, as seen from above, between the base station and the first diffracting corner, between the first and second diffracting corner, and between the second diffracting corner and the subscriber, respectively. For rays that undergo a single diffraction, let s_{i1} and s_{i2} be the unfolded path lengths from the base station to the diffracting corner and from the corner to the subscriber, respectively. Finally, for any of the classes, let S_i be the total unfolded ray length from the base station to the subscriber, as seen from above. With the foregoing definitions, and neglecting polarization conversion upon reflection or diffraction at walls and corners, the individual terms in (8–2) for vertical polarization are given by

$$P_R^{(i)} = P_0^{(i)} \prod_j |\Gamma_E(\theta_j)|^2 \tag{8–3}$$

$$P_{D1}^{(i)} = P_0^{(i)} \frac{S_i}{s_{i1}s_{i2}} |D_T(\theta_1)|^2 \prod_j |\Gamma_E(\theta_j)|^2$$

$$P_{D2}^{(i)} = P_0^{(i)} \frac{S_i}{s_{i1}s_{i2}s_{i3}} |D_T(\theta_1)|^2 |D_T(\theta_2)|^2 \prod_j |\Gamma_E(\theta_j)|^2$$

In (8–3) the products are taken over the reflections ($j = 1, 2,...$) that are undergone by the ray. The various terms in (8–3) are discussed below.

The reflection coefficient $\Gamma_E(\theta_j)$, where θ_j is the angle of incidence, is typically taken to be that of a dielectric half space, as given by (3–33), with dielectric constant ε_r chosen in the range 5 to 7. Values of ε_r in this range are suggested by direct measurement of the reflection properties of walls [9] and by fitting ray predictions to measurements [7,10]. The factor D_T is the UTD version of the diffraction coefficient at the corner, such as given in (5–47) for absorbing wedges, by (5–49) – (5–50) for right-angle conducting wedges, or by (7–5) – (7–7) for dielectric wedges of any angle. The directions that the rays make arriving and departing from the corners are used to compute these diffraction coefficients.

When viewed from the side, each ray found in the two-dimensional ray trace represents rays having vertical components of travel from Tx to Rx, either directly or after reflection in the ground. For large unfolded path lengths, the direct and ground reflected rays can interfere destructively, as discussed in Chapter 4 for propagation over a flat earth. To account for this

Figure 8-4 Footprints of the buildings over a 500- by 600-m area that covers the high-rise section of Rosslyn, Virginia. Various subscriber locations at height h_m = 2.5 m are indicated by x's and various base station sites by the Tx's [18].

effect, the term $P_0^{(i)}$ in (8–3) represents the path gain for propagation from the base station to the subscriber over the equivalent terrain that is found in the vertical plane when the ray path, and the terrain under it, are unfolded. For flat earth, $P_0^{(i)}$ can be found using the distances to the receiver from the transmitter and its image in the ground, which are

$$s_{id} = \sqrt{S_i^2 + (h_{\rm BS} - h_m)^2} \approx S_i + \frac{h_{\rm BS}^2 + h_m^2}{2S_i} - \frac{h_{\rm BS}h_m}{S_i} \tag{8–4}$$

$$s_{ig} = \sqrt{S_i^2 + (h_{\rm BS} + h_m)^2} \approx S_i + \frac{h_{\rm BS}^2 + h_m^2}{2S_i} + \frac{h_{\rm BS}h_m}{S_i}$$

Then

$$P_0^{(i)} = \left(\frac{\lambda}{4\pi}\right)^2 \left|\frac{e^{-jks_{id}}}{s_{id}} + \Gamma_H \frac{e^{-jks_{ig}}}{s_{ig}}\right| \approx \left(\frac{\lambda}{4\pi S_i}\right)^2 \left|e^{jkh_{\rm BS}h_m/S_i} + \Gamma_H e^{-jkh_{\rm BS}h_m/S_i}\right|^2 \tag{8–5}$$

where the approximate form in (8–5) is obtained using the approximate forms of (8–4) in the exponents, and using S_i for s_{id} and s_{ig} in the denominator.

8.1c Comparison of predictions with measurements

Predictions made using a two-dimensional pincushion code have been compared with measurements in the business section of Rosslyn, Virginia, where there are many tall buildings but also some as low as two stories. The footprints of the buildings covering an area of 600 x 500 m are shown in Figure 8–4, together with several base station locations and a sequence of subscriber locations, which are indicated by diamonds. Computations were made allowing for rays that

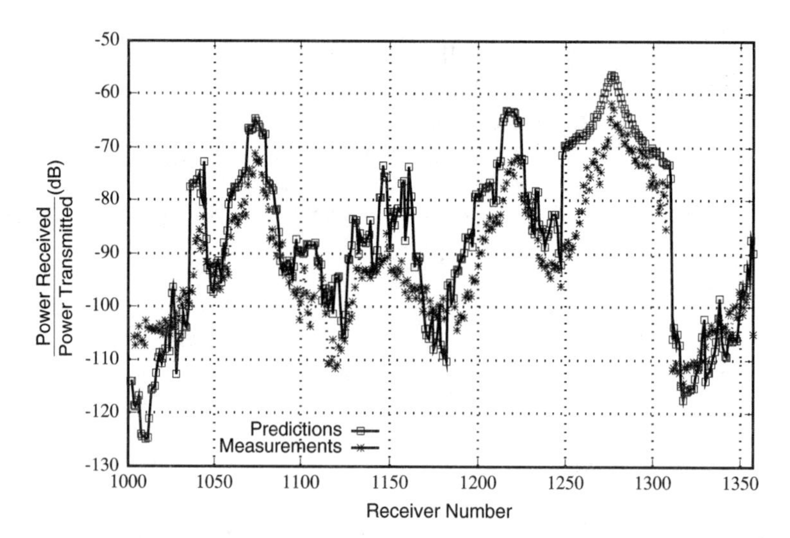

Figure 8-5 Comparison between 1.9-GHz measurements and predictions made using a two-dimensional building database for the footprints of Figure 8–4, and for a h_{BS} = 5 m base station antenna at site Tx4b.

experience two diffractions at building corners, and up to six reflections on each link between the base station, corners, and the subscriber locations. Plots are shown in Figure 8–5 of the predictions and measurements of the small-area average path gain for an omnidirectional base station antenna of height h_{BS} = 5 m located at site Tx4b in Figure 8–4, and for h_m = 2.5 m. The comparison in Figure 8–5 indicates the accuracy that can be obtained by site-specific predictions [7]. To the left of the base station there were several buildings that were only two stories high. (A perspective view of the site is shown in Figure 8–12, with the top of Figure 8–4 corresponding to the bottom of Figure 8–12.) Propagation over these buildings may explain why the signals measured at locations 1001 to 1020 are significantly higher than the predictions. Although the predictions follow the measurements, they show a greater variation than the measurements, which is typical of ray methods.

8.2 Two-dimensional predictions for a Manhattan street grid

Site-specific prediction methods can also be used to understand propagation in generic building environments. Large portions of Manhattan exemplify a generic environment of high-rise buildings organized along a rectangular street grid. Because of the high cost of land, the bases of buildings are built out to the lot line and form a nearly continuous facade around each city block. The resulting city canyons are shown in Figure 8–6 together with a base station located in the middle of a block. On the LOS street, the dominant contribution to the received signal is given by the direct ray from the base station to the subscriber. Rays reflected from the buildings at small glancing angles also contribute, as discussed in Chapter 4. On the perpendicular streets, rays arrive via reflection from the building walls, diffraction at the building corners, and by a

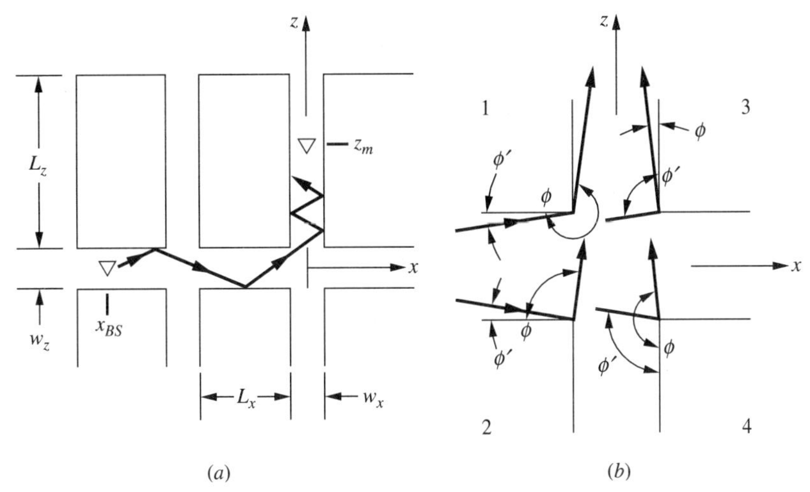

Figure 8-6 Turning a corner via multiple reflection (a) and diffraction (b) in a high-rise building environment with rectangular street grid, as in Manhattan.

combination of the two processes. Near the intersection, the reflected rays play a significant role, but at greater distances along the perpendicular street the diffracted rays become the dominant contribution [11], as discussed below.

8.2a Path loss in turning one corner

One of the rays that illuminate a perpendicular street via reflections only is depicted in Figure 8–6a. Other rays that undergo less reflections on the LOS street will experience more reflections to reach the same distance along the perpendicular street, and vice versa. For a dielectric constant of $\varepsilon_r = 6$, the plane wave reflection coefficient at normal incidence is $\Gamma_E(0) = -0.42$, corresponding to a 7.5-dB reflection loss. Thus for angles of incidence away from glancing, the loss due to several reflections is significant. Since the reflected ray must undergo many reflections at the building faces to reach points far away from the intersection along the perpendicular street, its amplitude will be smaller than that of diffracted rays that do not undergo reflections, such as the four rays depicted in Figure 8–6b, which turn the corner via diffraction at each of the building corners.

As an example of the number of reflections a ray can undergo, suppose that $|x_{BS}| = 100$ m in Figure 8–6a, with $L_z = 170$ m, $L_x = 80$ m, $w_z = 30$ m, and $w_x = 20$ m. Consider a ray that reflects $N_L = 2$ on the LOS street and just misses the building corner at $(-w_x/2, w_z/2)$, as shown in Figure 8–6a. On the LOS street the ray will travel a distance in the $\pm z$ direction that is $2.5w_z = 70$ m, while it travels a distance $|x_{BS}| - w_x/2 = 90$ m along x. Thus the angle ψ the ray makes with the x axis is $\psi = \arctan(70/90) = 37.9°$. After N_P reflections along the perpendicular street, the ray will cross the z axis at a point whose distance is $z_m = (N_P + 0.5)w_x \tan\psi$, which for this example is $z_m = (N_P + 0.5)15.56$. Choosing $N_P = 6$ reflections gives a distance of 101.1 m. Thus, when base station and subscriber are both about 100 m from the center of the intersection, one of the rays connecting them experiences a total of eight reflections, $N_L = 2$ reflections on the LOS street and $N_P = 6$ reflections on the perpendicular street. Other multiply reflected rays with different combinations of N_L and N_P will also connect the base station and mobile, but the total number of reflections will still be large unless the base station or subscriber are close to the intersection. For the parameters used in the example, the total path length of the ray is $S_i = 254.2$ m. The angle of incidence on walls lining the LOS street is $90 - \psi = 52.1°$, leading to a reflection coefficient $\Gamma_E = -0.581$ for $\varepsilon_r = 6$, whereas for the reflections on the perpendicular street, $\Gamma_E = -0.501$. For this ray it is seen from (8–3) that $P_R^{(i)} = (0.581)^4(0.501)^{12}P_0^{(i)}$, which is 45.5 dB lower than the path gain $P_0^{(i)}$ for two antennas separated by 254.2 m on an LOS path.

To evaluate the contributions from the diffraction paths in Figure 8–6b, the UTD diffraction coefficient at a 90° dielectric corner is used. As discussed in Chapter 7, the diffraction coefficient is

$$D_T(\phi, \phi') = D_1^w + D_2^w + \Gamma_n D_3^w + \Gamma_0 D_4^w \tag{8–6}$$

For a vertical electric field, Γ_0 is the TE plane wave reflection coefficient evaluated for the angle between the incident ray and the normal to that building surface from which the angle ϕ' is measured. The reflection coefficient Γ_n is similarly defined for the angle between the diffracted ray and the normal to the second surface of the building. While (8–6) is rigorously valid for a conductor, it is only a heuristic approximation for a dielectric. Depending on the surface that is chosen as the reference, different values of the angles of incidence for determining Γ_0 and Γ_n are obtained. A heuristic rule for choosing the reference surface is to pick the one for which ϕ' has the smallest value, as in Figure 8–6b. The limitations of the heuristic approach can be overcome by using the rigorous but considerably more complicated diffraction coefficient for corners with finite conductivity, which is given in reference [12].

For a 90° corner, $n = 3/2$, so that the terms D_i^w in (8–6) are defined by

$$\begin{aligned} D_{1,2}^w &= \frac{-1}{3\sqrt{2\pi k}}\cot\left[\frac{\pi \pm (\phi - \phi')}{3}\right]F[kLa^{\pm}(\phi - \phi')] \\ D_{3,4}^w &= \frac{-1}{3\sqrt{2\pi k}}\cot\left[\frac{\pi \pm (\phi + \phi')}{3}\right]F[kLa^{\pm}(\phi + \phi')] \end{aligned} \tag{8–7}$$

In (8–7) the factors $F[\cdot]$ are transition functions, as defined in (5–41) and (5–45), and are approximately unity outside their transition regions. Inside their transition regions, the arguments are evaluated using $L = s_{i1}s_{i2}/(s_{i1} + s_{i2})$, where s_{i1} and s_{i2} are the path lengths before and after the corner, as seen from above. Also in (8–7)

$$a^{\pm}(\beta) = 2\cos^2\left(\frac{3}{2}\pi N^{\pm} - \frac{1}{2}\beta\right) \tag{8–8}$$

where $N^{\pm}$ are the integers closest to $(\beta \pm \pi)/(3\pi)$.

Diffraction at each of the four corners of the intersection contributes to the received signal. Adding the powers of each contribution gives the small-area average [11]. As a simple example, assume that the base station and subscriber are far from the intersection, so that $\phi' \approx 0$ for corners 1 and 2 in Figure 8–6b, and $\phi' \approx \pi/2$ for corners 3 and 4. Similarly, $\phi \approx 3\pi/2$ for corner 1, $\phi \approx \pi/2$ for corner 2, $\phi \approx 0$ for corner 3, and $\phi \approx \pi$ for corner 4. With these approximations, the angle used in computing Γ_0 and Γ_n are close to $\pi/2$ for corners 1 and 2, so that $\Gamma_0 \approx \Gamma_n \approx -1$ there. For corners 3 and 4, the angle is nearly 0 for Γ_0 and $\pi/2$ for Γ_n, so that $\Gamma_0 \approx \Gamma_E(0)$ and $\Gamma_n \approx -1$. Also, the subscriber is away from the transition regions so that the transition functions are unity. Since $\phi' \approx 0$ at corners 1 and 2, it is seen from (8–7) that $D_1^w = D_3^w$ and $D_2^w = D_4^w$ there. Further, since the reflection coefficients are –1, from (8–6) it is seen that $D_T^{(1)} = D_T^{(2)} = 0$. Using the values of ϕ and ϕ' indicated above for corners 3 and 4 yields,

$$D_T^{(3)} = -\frac{1}{3\sqrt{2\pi k}}\left(\cot\frac{\pi}{6} + \cot\frac{\pi}{2} - \cot\frac{\pi}{2} + \Gamma_E(0)\cot\frac{\pi}{6}\right) \tag{8–9a}$$

$$= -\frac{1}{3\sqrt{2\pi k}}\{[1 + \Gamma_E(0)]\sqrt{3}\}$$

$$D_T^{(4)} = -\frac{1}{3\sqrt{2\pi k}}\left(\cot\frac{\pi}{2} + \cot\frac{\pi}{6} - \cot\frac{5\pi}{6} + \Gamma_E(0)\cot\frac{-\pi}{6}\right) \tag{8–9b}$$

$$= -\frac{1}{3\sqrt{2\pi k}}\{[2 - \Gamma_E(0)]\sqrt{3}\}$$

The distances from the source to corners 3 and 4 in Figure 8–6 have the same value $s_{i1} \approx |x_{BS}|$ as in (8–3). Because the distances from the corners to a subscriber are $s_{i2} \approx z_m \mp w_z/2$, where w_z is the width of the LOS street, $s_{i2} \approx z_m$ for $z_m \gg w_z/2$. Assuming no reflections, the path gain due to the diffracted rays, as given by (8–3), is

$$P = P_0^{(i)}\frac{|x_{BS}| + z_m}{|x_{BS}|z_m}\frac{1}{6\pi k}\{[1 + \Gamma_E(0)]^2 + [2 - \Gamma_E(0)]^2\} \tag{8–10}$$

Substituting the value $\Gamma_E(0) = -0.42$ obtained for $\varepsilon_r = 6$ and for a frequency of 900 MHz, expression (8–10) becomes

$$P = P_0^{(i)}\frac{|x_{BS}| + z_m}{|x_{BS}|z_m}(0.0174) \tag{8–11}$$

When converted into path loss in decibels, the factor in parentheses contributes –17.6 dB. If $|x_{BS}|$ and z_m are both 100 m, the fraction in (8–11) is 0.02 and contributes –17 dB to the path gain. Thus the received power is 34.6 dB lower than it would be over plane earth at a distance of 200 m. For this example, the path gain contribution from the diffracted rays of Figure 8–6b is about 10 dB stronger than the contribution found for the reflected ray of Figure 8–6a. At greater distances along the perpendicular street, the contributions from reflected rays will be even smaller compared to the contributions from the diffracted rays.

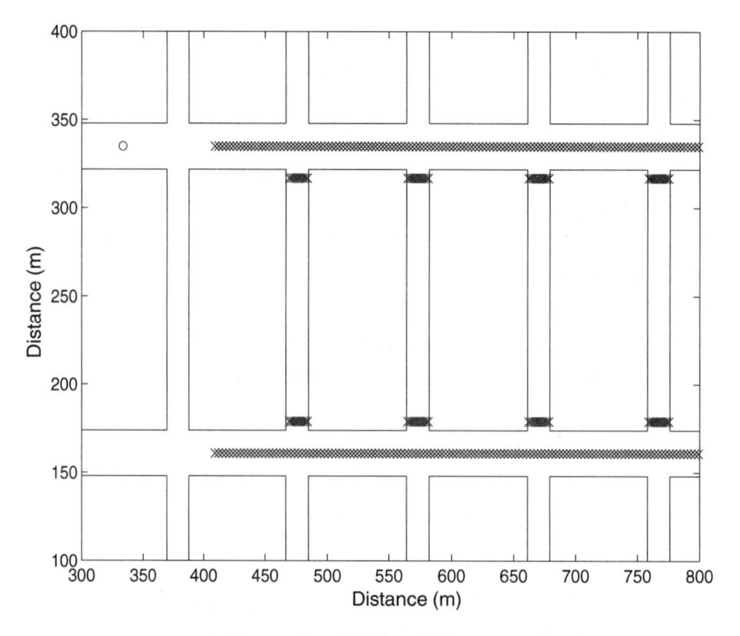

Figure 8-7 Base station and subscriber locations on the LOS, perpendicular, and parallel streets in a Manhattan building environment.

8.2b Predictions made using two-dimensional ray methods

To examine the signal characteristics on various streets, subscriber locations were placed at the points indicated by the ×'s in Figure 8–7, along with a midblock base station shown as a circle. The subscriber locations include points on the LOS street, points across the perpendicular street near the LOS street and the other end of the block, and points on the first parallel street. Computed path gain is shown in Figure 8–8 for a base station antenna height of 9 m, subscriber antenna height of 1.8 m, and a frequency of 894 MHz. It is seen from the ×'s in Figure 8–8 that for locations on the perpendicular street that are 5 m in from the building line on the LOS street, the signal is already on average 15 dB below that in the middle of the LOS street and has about a 10-dB variation from one side of the street to the other. At the corresponding locations at the other end of the block, the signal, plotted as circles in Figure 8–8, is about 40 dB lower than on the LOS street, in agreement with the discussion in Section 8–2a. The signal has little variation from one side of the perpendicular street to the other and is almost the same level as it is in the middle of the intersection on the first parallel street, whose computed signal is indicated by dots in Figure 8–8.

Because of the large loss found when the waves turn a corner, the signal reaching locations on a perpendicular street does so primarily via a single turn off the LOS street. If the base station is located in the middle of a block, signals to midblock subscriber locations on parallel streets must make a second turn. Thus in Figure 8–8, the signal at midblock locations on the first paral-

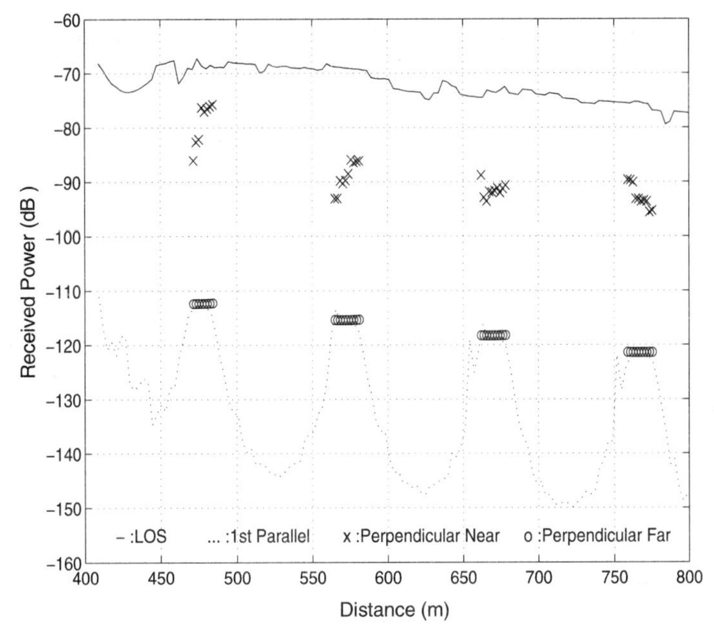

Figure 8-8 Path gain from the base station to the subscriber locations of Figure 8–7.

lel street are seen to be nearly 30 dB lower than at the nearest intersection. The variation shown along the parallel street has been observed in measurements made in London [3]. If the base station is located in an intersection, all locations can be reached via a single turn. In this case it has been shown that the equal path loss contours have a hyperbolic shape [11–13].

8.3 Outdoor predictions using a three-dimensional building data base

Because turning corners involves a large loss, significant contributions to the received signal come from paths that go over the buildings unless the buildings are very tall compared to the base station. To account for rays that can go over the buildings, as well as over and around, requires a three-dimensional building database. The oldest and simplest type of building databases gives the height of the building on a pixel-by-pixel basis. However, many codes make use of a vector database in which a building, or building element, is described as a collection of planar polygons, with the spatial coordinates of the vertices listed in the database. If the building walls are restricted to be vertical, the description simplifies to the footprint of the building, or building element, and the heights of the corners. If the heights of the corners are with reference to a datum plane rather than the height above ground, terrain information will be accounted for in the ray paths that lie over buildings. Terrain may also be taken into account for portions of the ray paths that go around buildings.

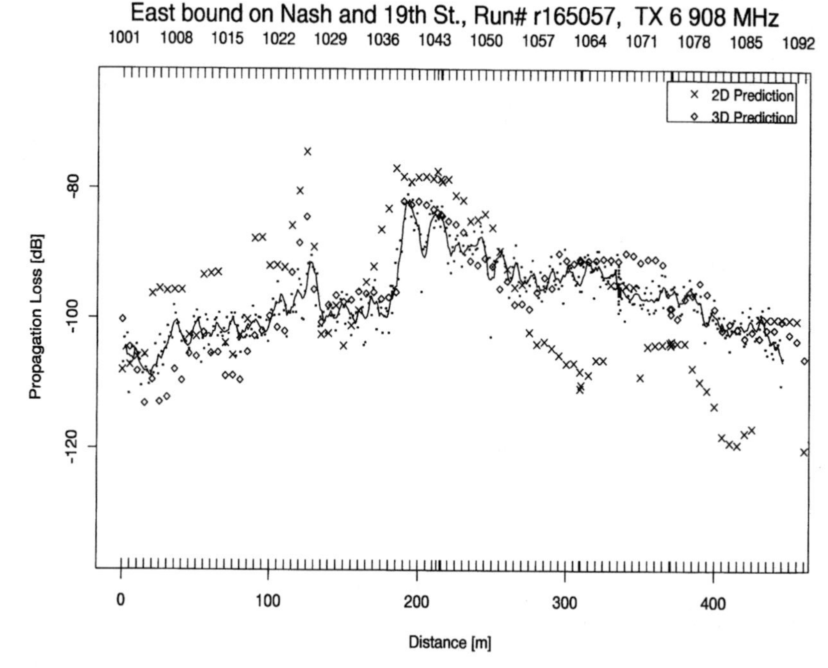

Figure 8-9 Comparison between 908-MHz measurements and predictions made using three- and two-dimensional building databases for Rosslyn, Virginia. The receiver locations are as shown in Figure 8–4, and a directive base station antenna pointing east is located at Tx6 at a height h_{BS} = 44 m [17](©1999 IEEE).

8.3a Three-dimensional pincushion method

Several groups have reported on computer codes that will run with a three-dimensional database, always with some level of approximation or limitation on the rays considered, and with restrictions on the database. For example, one such code involves numerical integrations, and although very accurate, is limited to a few buildings [14]. Another uses UTD approximations but limits the buildings to rectangular parallelepipeds on a rectangular street grid, with individual streets having the same width for their entire length [15].

A rather general code written to work with a three-dimensional building database has been developed at Bell Laboratories [16,17]. It assumes that the building walls and roofs can be described by a collection of vertical and horizontal planes, but otherwise does not restrict building shape. This code makes use of the pincushion method in three dimensions for finding the rays. The rays are launched from the source at small incremental angles over the surface of a unit sphere, and traced in much the same manner as is done in two dimensions. Predictions obtained using the three-dimensional code have been compared to measurements made in Rosslyn, Virginia [17]. Figure 8–9 shows such a comparison at 908 MHz for subscriber locations 1001 through 1092 in the building footprint of Figure 8–4, and for h_m = 2.5 m. A directional antenna

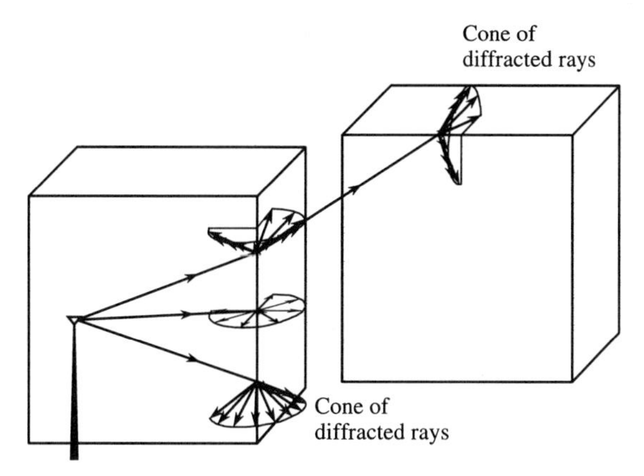

Figure 8-10 Rays incident on vertical and horizontal corners of buildings excite diffracted rays that lie in cones whose half-angle is equal to the angle between the incident ray and the corner.

of 30° beam width and 6° downtilt was used at base station site Tx6, which is situated atop a building at a height $h_{BS} = 44$ m. Predictions made using the three-dimensional database are seen to follow the measurements fairly closely, and are much more accurate than the those made using only a two-dimensional building database, whose predictions are also shown in Figure 8–9. Since the two-dimensional database assumes the buildings to be infinitely tall, it can give high predictions in some locations by including rays that are reflected at heights above the actual building, and give low predictions at other locations by blocking rays that actually pass over low buildings.

The pincushion approach in three dimensions can allow for an adequate number of reflections at building surfaces, but cannot treat rays that undergo more than a total of two diffractions at vertical or horizontal building corners. The reason behind this limitation can be understood with reference to Figure 8–10, where we have shown rays that are diffracted at a vertical and a horizontal corner. Each ray incident on the vertical corner generates a cone of diffracted rays, as shown, and each ray in each cone must be traced to find its subsequent path. Because the cones all have different angles with respect to the corner, for the purpose of finding the subsequent paths, the corner must be divided into discrete segments, each of which is treated as a secondary source of rays. These rays are then traced to find their subsequent paths. Similarly, horizontal corners must be divided into discrete segments to find the subsequent paths of the diffracted rays. For a large number of buildings, the running time needed to do all the indicated ray traces from all the segments of all the corners sets a limit of two on the maximum number of diffraction events that a ray can experience. Approaches to decrease running time and/or increase functionality are discussed below.

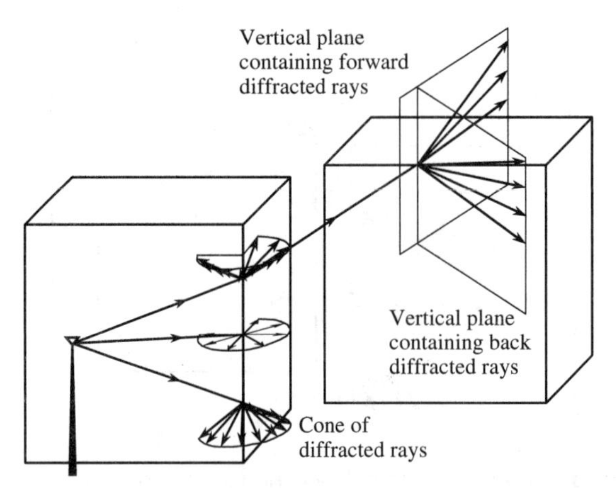

Figure 8-11 In the VPL method, the rays diffracted at vertical edges lie in a cone, but the rays diffracted at the horizontal edges are taken to lie in the vertical plane of the incident ray or the vertical plane of the reflected ray.

8.3b Vertical plane launch method

The assumption that the building walls are vertical vastly simplifies finding the vertical component of the paths of rays reflected from the walls and diffracted at vertical corners. For rays that do not undergo diffraction at horizontal corners, all segments of the ray between the antennas, or between antennas and the ground reflection point, have the same slope in the vertical plane. The sections of vertical planes containing the segments of such rays can be found using the two-dimensional pincushion method, and the actual ray segments within the planes are then found by analytic methods. Although this approach can include rays that travel over buildings, it requires a further approximation to include rays diffracted at horizontal corners. To include these rays, it is necessary to assume that the ray segments leaving the horizontal corner lie in the vertical plane of the incident ray, or in the vertical plane of the reflected ray, as shown in Figure 8–11. In other words, the ray cone generated at the horizontal corner in Figure 8–10 is broken at the top and unrolled into two vertical planes. Restricting the rays diffracted at the horizontal corners to lie in the vertical plane of the incident or reflected ray distorts the subsequent path. This distortion is small when the horizontal corner is close to being perpendicular to the vertical plane containing the incident ray, or the slope of the incident and diffracted rays in the vertical plane is small.

The vertical plane launch (VPL) method is suggested in Figure 8–12, where the base station is treated as launching a series of vertical planes, which are allowed to go over buildings as well as being reflected at vertical walls and diffracted at vertical corners [18]. Unfolding the vertical planes, the vertical trajectory of the rays can be found analytically, thereby giving the ray path in three dimensions, as shown for several rays in Figure 8–12. As viewed from above, the vertical planes launched by the source appear as rays in the two-dimensional pincushion method, as shown in Figure 8–13. However, because the planes can go over the buildings, the rays now

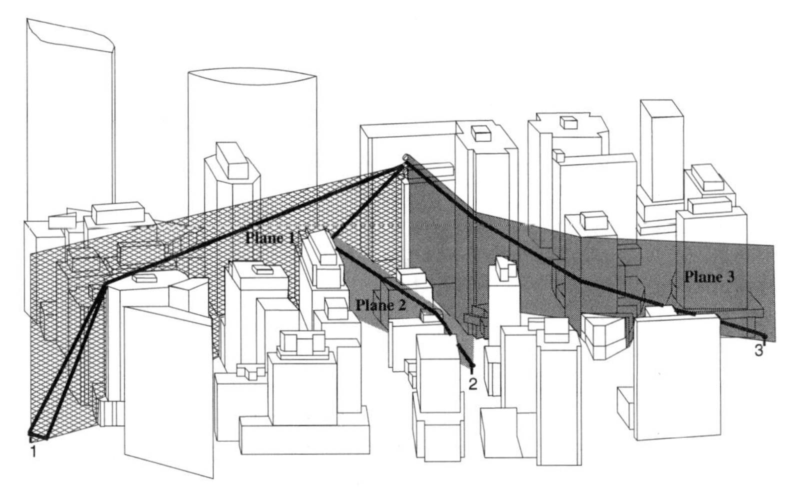

Figure 8-12 Perspective view of the buildings in Rosslyn, Virginia, showing the vertical planes launched from a base station at site Tx5 of Figure 8–4 and the ray paths in those planes [18](©1998 IEEE). The bottom of this figure corresponds to the top of Figure 8–4.

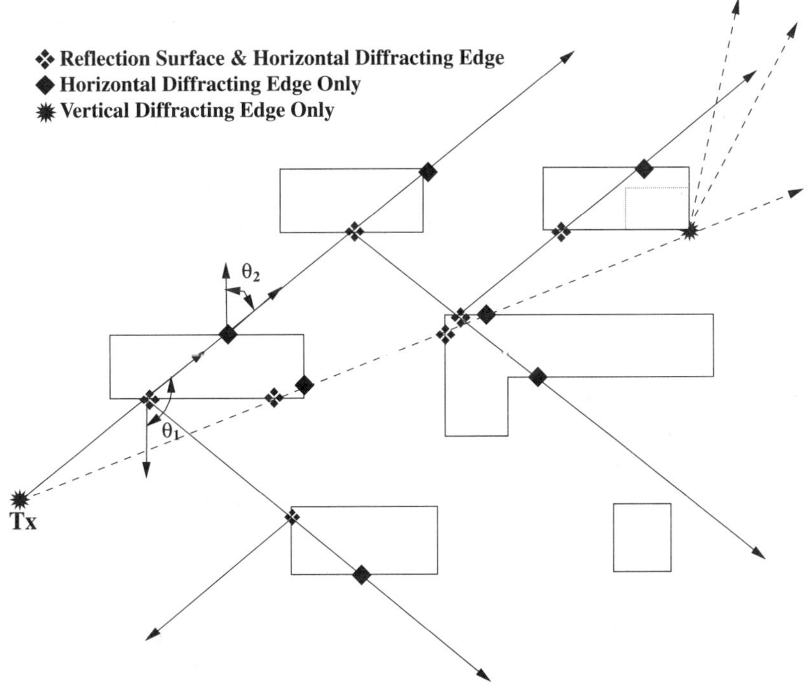

Figure 8-13 Top view of the vertical planes used in the VPL method [18](©1998 IEEE).

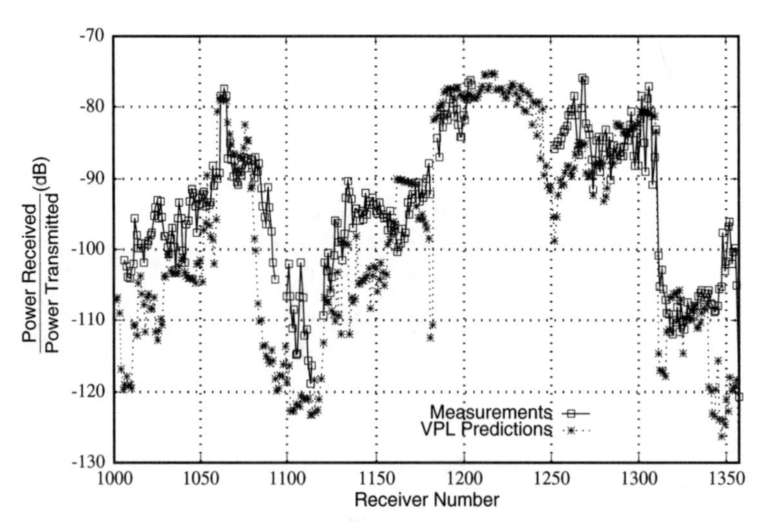

Figure 8-14 Comparison between 908-MHz measurements and predictions made using the VPL method for the three-dimensional building database of Rosslyn, Virginia. The receiver locations are as shown in Figure 8–4, and a directive base station antenna pointing north is located at Tx5 at a height h_{BS} = 42 m [18](©1998 IEEE).

form a binary tree, with branching occurring at the exterior surfaces of the buildings (where the angle θ in Figure 8–13 is greater than 90°). Vertical corners that are illuminated act as secondary sources whose rays are traced by again using the pincushion method. Because of the branching of the rays, this method is not as fast as in the simple two-dimensional case, and various algorithms are employed to prune the branches [18]. As compared to three-dimensional ray tracing, each vertical corner initiates only a single two-dimensional ray trace, while diffraction at horizontal corners is treated analytically without requiring additional ray tracing. As a result, the VPL method can account for double diffraction at vertical edges, in addition to multiple diffractions over buildings, which can be very important when the buildings are of roughly uniform height.

Predictions made using the VPL method are compared in Figure 8–14 with measurements at 900 MHz made in Rosslyn, Virginia, using a directional base station antenna having a 30° beam width and 6° downtilt. The antenna is at site Tx5 in Figure 8–4, which is situated atop a building at a height h_{BS} = 42 m. The buildings shown in Figure 8–12 are a simplified version of the database of Rosslyn, Virginia, for the same area as that shown in Figure 8–4, with the top of Figure 8–4 corresponding to the bottom of Figure 8–12. The base station location in Figure 8–12 corresponds to site Tx5. The predictions and measurements shown in Figure 8–14 are for the subscriber locations in Figure 8–4 [18]. Again the predictions follow the measurements but give greater variations of the received signal. For all base station locations, the prediction errors typically had an average of a few decibels, and a standard deviation of 7 to 10 dB [18].

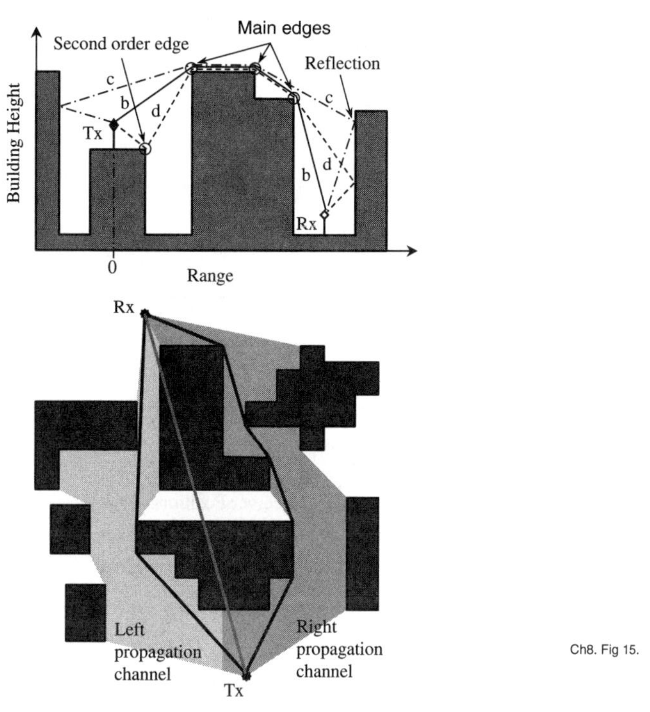

Figure 8-15 Simplified ray paths of the SP-VP method are restricted to lie in either the vertical or slant planes containing the base station and subscriber [19](©1993 IEEE).

8.3c Slant plane–vertical plane method

The slant plane–vertical plane (SP-VP) method makes further restrictions on, and approximations to, the actual rays in three dimensions to gain computational speed [19]. Vertical and slant planes are erected that contain the base station and subscriber antennas, as shown in Figure 8–15, and ray tracing is limited to these two planes. Rays that lie outside these planes are ignored. This approach is sometimes referred to as the *2.5D method.* For low base station antennas, tracing is done in the horizontal plane rather than in slant planes, so that all subscriber locations can be treated at the same time. Rays diffracted over buildings are accounted for by the vertical plane. Forcing the paths of reflected rays, such as those labeled "c" in Figure 8–15 to lie in the vertical plane can be a significant distortion if the wall of the reflecting building is not perpendicular to the vertical plane. This and other forms of out-of-plane reflection and diffraction at vertical corners are sources of error, which is found to be significant for high base station antennas [8,18].

By augmenting the two-dimensional ray method with paths that are diffracted over the buildings, the SP-VP method improves prediction accuracy when the buildings are of nearly uniform height. Such an example is offered by a comparison with measurements in the city of Aal-

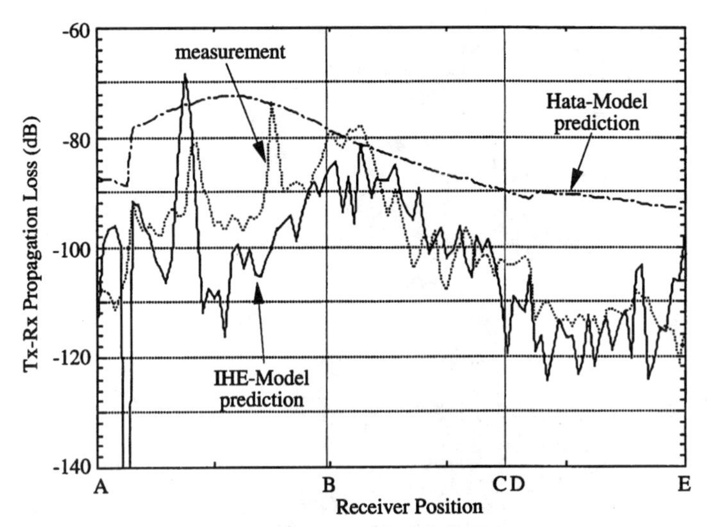

Figure 8-16 Comparison between 955-MHz measurements and predictions made using the SP-VP method and the Hata model for Aalborg, Denmark [19](©1993 IEEE).

borg, Denmark, most of whose buildings are three to five stories high. Figure 8–16 shows a comparison between the SP-VP predictions and measurements at 955 MHz over portions of a 750-m square area in Aalborg [19]. Predictions made by the Hata model are also shown. As in other ray methods, the predictions follow the measurements, except at some locations, where they give much greater variation. However, for this microcellular environment, the SP-VP predictions are clearly more accurate than those found from the Hata model.

8.3d Monte Carlo simulation of higher-order channel statistics

In the foregoing discussion, the use of site-specific codes was illustrated by predictions of the small-area average path gain for narrowband systems, which is important for system coverage. Site-specific predictions are now being used in Monte Carlo simulations to find higher-order statistical parameters for the channel. For example, by studying the variation in small-area averages from one small area to the next, it is possible to find the correlation length of shadow fading. If predictions are compared for the signals from two base stations, it is possible to study the correlation of the shadow fading to different base stations, which has an important effect on CDMA capacity. Although results are not yet available, it is anticipated that the simulations will show how these and other channel parameters depend on variables such as the antenna height and distance to the base station R. Moreover, it is hoped that the simulations will reveal the relationship between the statistical channel parameters and statistical parameters that describe a particular building environment or city.

Because ray codes can be made to output all aspects of the individual rays, measurable channel parameters other than the narrowband signal level can be simulated. For example, in

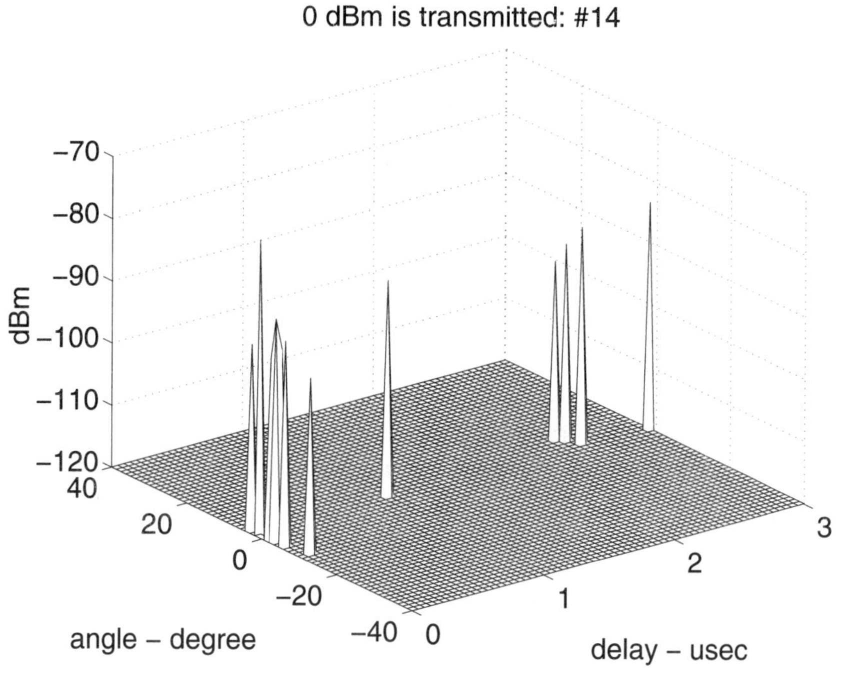

Figure 8-17 VPL simulation of the angle of arrival and time delay of rays reaching an elevated base station in Munich from a 900-MHz street-level subscriber.

studying smart antennas, it is the time delays $t_i = S_i/c$, where c is the speed of light, and the directions of arrival of the individual rays that are of interest. Figure 8–17 is an example of the time delay, angle of arrival, and path gain of the individual rays arriving at an elevated base station from a street-level subscriber. The simulation was made at 900 MHz using the VPL method and a database for Munich [20]. The angle of arrival is measured from the vertical plane defined by the base station and subscriber, and the time delay is referenced to the first arrival. Figure 8–17 is similar to the measurement-based plot of ray arrivals shown in Figure 2–22. In Figure 8–17, the arriving rays are grouped into four clusters in time, with the last cluster delayed by about 3 μs.

When representing the spreading of signals in time and angle, it is common to make use of the RMS delay spread τ_m and RMS angle spread $\Delta\psi_m$ for various subscriber locations $m = 1, 2, \ldots$ distributed within a particular building environment. The average over a small area of the RMS delay spread can be computed from the rays traveling between a particular subscriber and the base station. Let t_{im} be the travel time of the ith ray from the mth subscriber to the base station, and let $P_m^{(i)}$ be the path gain of the ray. Then the RMS delay spread for the subscriber is

$$\tau_m = \left\{ \frac{\sum_i [t_{im} - \langle t_m \rangle]^2 P_m^{(i)}}{\sum_i P_m^{(i)}} \right\}^{1/2} \tag{8–12}$$

where the mean time delay is

$$\langle t_m \rangle = \frac{\sum_i t_{im} P_m^{(i)}}{\sum_i P_m^{(i)}} \tag{8–13}$$

Similarly, let ψ_{im} be the angle of arrival at the base station of the *i*th ray, as measured from the vertical plane defined by the base station and the *m*th subscriber. The RMS angle spread for the subscriber is

$$\Delta\psi_m = \left\{ \frac{\sum_i [\psi_{im} - \langle \psi_m \rangle]^2 P_m^{(i)}}{\sum_i P_m^{(i)}} \right\}^{1/2} \tag{8–14}$$

where the mean angle spread is

$$\langle \psi_m \rangle = \frac{\sum_i \psi_{im} P_m^{(i)}}{\sum_i P_m^{(i)}} \tag{8–15}$$

Using the VPL method, τ_m and $\Delta\psi_m$ have been computed at 900 MHz for various subscriber locations in sections of Seoul, Munich, and Rosslyn. The histograms of building heights in the sections of each city used in the simulations are shown in Figure 8–18. Rosslyn has a large range of building heights, including very tall buildings. Munich has low buildings with large footprints; the buildings in Seoul are taller but have smaller footprints. The three areas therefore provide examples of very different building statistics, which can affect τ_m and $\Delta\psi_m$. One example of the results obtained from simulations is given in Figure 8–19, where the cumulative distribution functions of τ_m and $\Delta\psi_m$ are plotted. For these simulations the base station antenna was assumed to be 5 m above the tallest building, and the mobiles were located at distances R_m ranging from about 300 to 900 m. From Figure 8–19 it is seen that the RMS delay spread for high base station antennas is independent of the building statistics, while the RMS angle spread is significantly different in the three cities. The median delay spread of about 0.2 μs and the angle

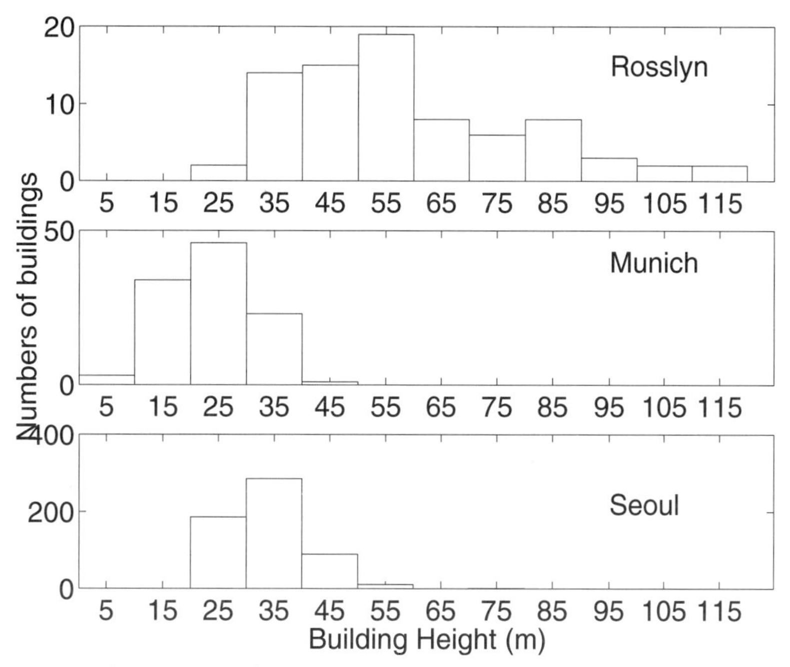

Figure 8-18 Histograms of the building heights in areas of three cities used for simulations.

spreads are consistent with measurements for short ranges [21]. More extensive simulations are expected to give a better understanding of the relation between the statistical channel parameters and the building environment.

8.4 Indoor site-specific predictions

Theoretical studies of propagation in tunnels and inside buildings predate the corresponding outdoor studies. The earliest studies of propagation in tunnels made use of the modal approach, which views the tunnel as an electromagnetic waveguide that supports a spectrum of modes [22]. Other rigorous approaches to solving Maxwell's equations have been used with success to study sharp bends and junctions in tunnels [23] and propagation in corridors [24]. Because the walls of the tunnel are not perfectly reflecting, all the modes exhibit some attenuation with distance, which blurs the distinction between modes that are propagating and modes that are below cut off. Nevertheless, at UHF and microwave frequencies, very many modes must be included to describe the fields.

At UHF frequencies, the radio channel inside tunnels has characteristics that are similar to those of the outdoor channel [25]. Rather than sum many modes to describe these characteristics, it is more convenient to use ray approximations for the fields [26,27]. For tunnels of rectangular cross section, the ray methods are like those in the outdoor environment, except that

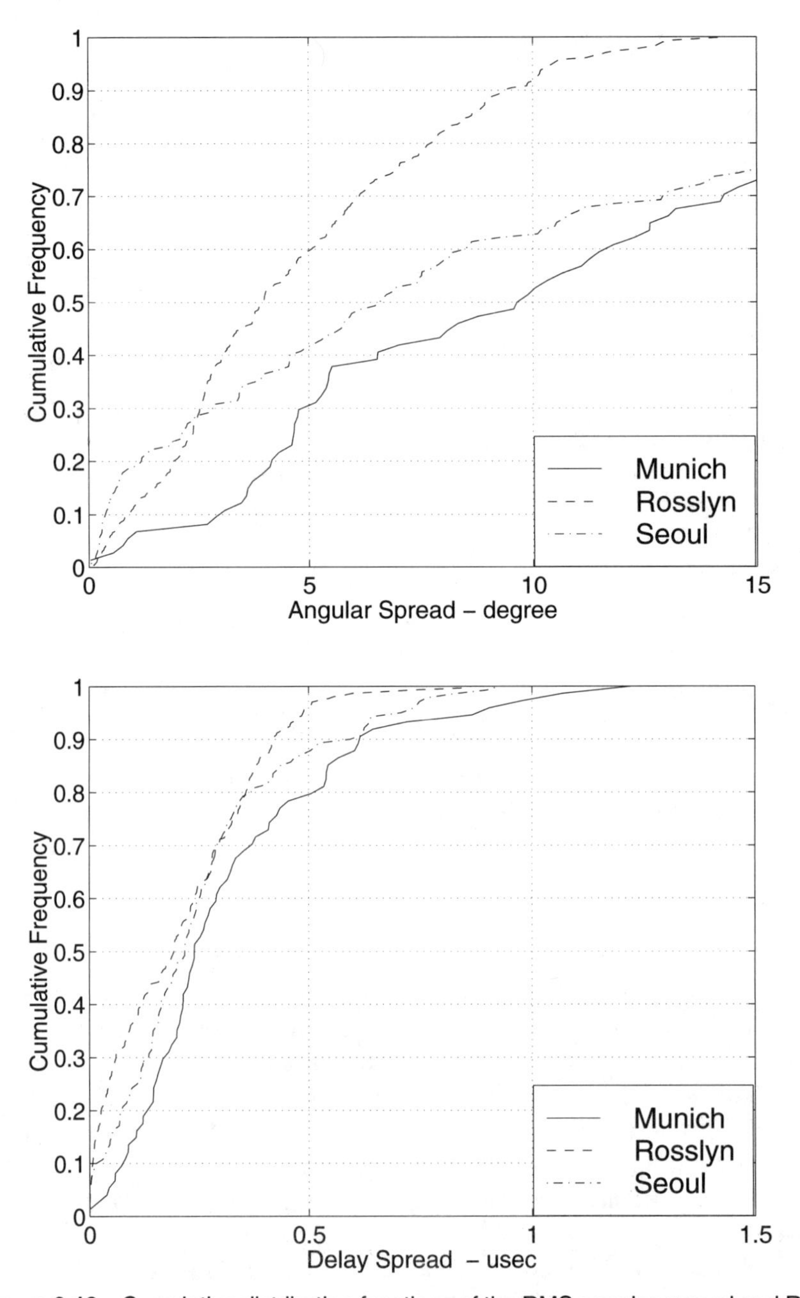

Figure 8-19 Cumulative distribution functions of the RMS angular spread and RMS delay spread found from simulations in three cities.

reflections from the ceiling and floor of the tunnel must be accounted for. This can be done by using multiple images of the transmitting antenna in the ceiling and floor. Transmission around bends and into side tunnels is analogous to propagation in a high-rise building environment and may be studied using the ray techniques developed for use with a two-dimensional building database. When the ceiling is curved rather than flat, more complicated ray techniques must be used to account for the effect of the curvature on the divergence of the ray tubes [28].

Ray approximations have been used to predict the propagation characteristics within buildings. The simplest ray approximations are used to compute the path loss by considering only the direct ray from the base station (called an access point for applications inside buildings) to the subscriber. In this approach, wall and floor loss is added to the free-space loss associated with the direct ray [29,30]. With some adjustment of the wall loss for a particular building, the standard deviation of the prediction error can be as low as 3 dB for links on one floor of a building [29]. More general ray models make use of ray tracing models in two or three dimensions for predictions that include both reflection and transmission at walls, ceilings and floors [31–34]. The two-dimensional codes are used for coverage over a single floor and neglect reflections at ceilings and floors. Because of the reflections at walls, the rays tracing procedure creates a binary tree that must be constructed to find the contributing rays, much as shown in Figure 8–13. Either the image or pincushion method may be used to find the rays. In three dimensions the rays can penetrate the floors and ceiling to reach other floors as well as being reflected from the floors and ceilings. The ray tracing procedures are essentially the same as those discussed for outdoor propagation. Transmission through walls is discussed in Chapter 3. However, some of the physical issues associated with transmission and reflections from floors and ceilings in office buildings are different, as discussed below.

8.4a Transmission through floors

When the access point and the subscriber are on different floors, the direct ray between them passes through the intervening floors. Modern office buildings are usually constructed with concrete floors and drop ceilings of acoustical material supported by a metal frame. The acoustical material has a low dielectric constant and is readily penetrated by an incident wave. However, the space between the drop ceiling and the floor above contains an irregular collection of supporting beams, ventilation ducts, lighting fixtures, pipes, and so on, all of which scatter and attenuate the wave. Although it is difficult to characterize the scattering, an effective floor loss can be inferred from measurements made with antennas on different floors of a building. For reinforced concrete or precast concrete floors, transmission loss has been measured at 10 dB or more [30,35,36]. Floors constructed of concrete poured over corrugated steel panels show much greater loss [35]. In this case the received signal may in fact be associated with other paths involving diffraction outside the building, or in stairwells and elevator shafts, rather than transmission through the floor. The direct ray and two diffraction paths are suggested in Figure 8–20 for a very simple geometry [37]. Here a hotel is shown in cross section with the transmitter

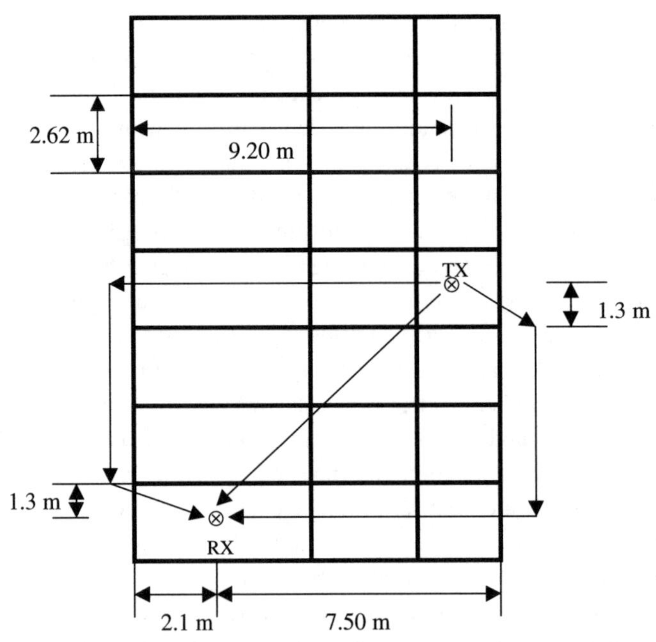

Figure 8-20 Cross section of hotel building showing the direct path and two paths involving double diffraction between subscriber and access point located on different floors.

located in the hallway outside a room on one floor and the receiver on a lower floor in the sleeping area of the room directly below the transmitter.

The direct path from the transmitter to the receiver passes through two walls and the intervening floors. Floors of the hotel are constructed of precast concrete panels, with the underside of the floor panels being the ceiling of the floor below, and without suspended ducts, light fixtures, and so on. In this case, the floor loss will be somewhat smaller, and the direct path signal will be given by the free-space path gain, reduced by the transmission loss through the walls and floors. In addition, rays can exit the building through large windows lining the hallway, diffract down along the building face, and diffract back into the lower floor through windows located there to illuminate the receiver. A second path involves diffraction out of and back in through the bedroom windows. The path gain for these rays is given by the last term in (8–3) for two diffractions and no reflections, which must be further reduced by the transmission loss through the windows (taken as 0.25 dB) and the walls (taken as 2.2 dB). The measured value of the small-area average path gain [37] is plotted in Figure 8–21 as a function of floor separation at 852 MHz. The signal computed for the direct path, assuming 7-dB transmission loss through the floor, and the sum of the signals computed for the two diffraction paths, using the absorbing screen diffraction coefficient, are also shown. The initial rapid decrease of the signal arises from the transmission loss through the floor, while the later slow decrease results from the diffracted signals, which have a large loss for a separation of one floor but vary only slowly with increased

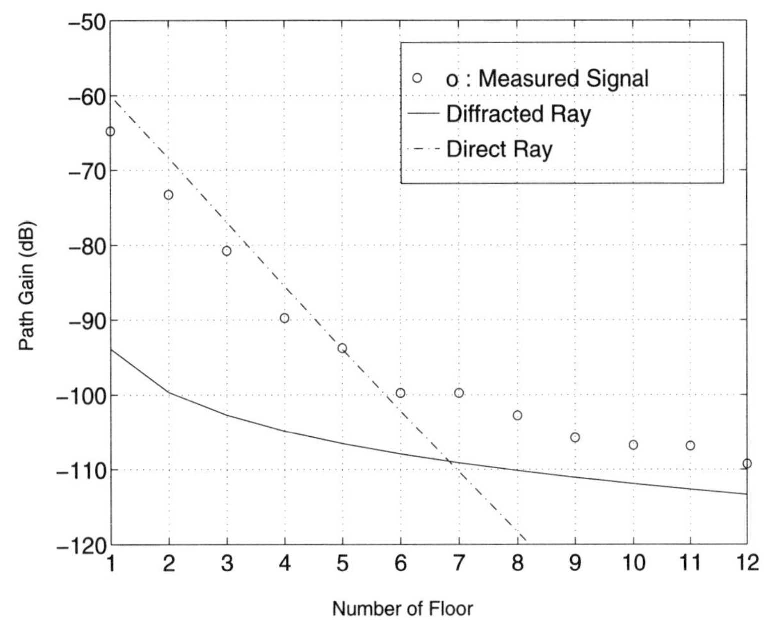

Figure 8-21 Comparison of measured and computed small-area average path gain between subscriber and access point in the hotel as a function of the number of floors separating them. The direct and diffracted ray path gains are shown separately.

floor separation. This example indicates the importance of diffraction paths in propagation between floors.

8.4b Effect of furniture and ceiling structure on propagation over a floor

As noted in the previous section, the region above the acoustical ceiling of a modern office building is electrically very irregular, and hence will strongly scatter incident waves. Similarly, furnishings placed on the floor, such as desks, cubicle partitions, filing cabinets, and work benches, will scatter the waves and prevent them from being specularly reflected from the floor, except in hallways. By allowing for specular reflection from floors and ceilings, three-dimensional ray tracing fails to account for these significant phenomena. These observations suggest that propagation over one floor of such a building must take place in the clear space between the furniture and ceiling fixtures, as shown symbolically in Figure 8–22. When the access point and subscriber antennas are located in the clear space, as in Figure 8–22, the path loss mechanism can be understood in terms of the Fresnel ellipse about the unfolded ray path of length S_j. If the antennas are close enough so that the ellipse lies entirely within the clear space, the fields associated with the ray will not be affected by the presence of the scatterers, and the path loss will have the $1/S_j^2$ dependence of free space. As the separation between transmitter and receiver increases, the Fresnel ellipse will grow in size so that the scatters lie within it, as in Figure 8–22, and at some distance S_j the path loss will be greater than that of free space. The distance at which

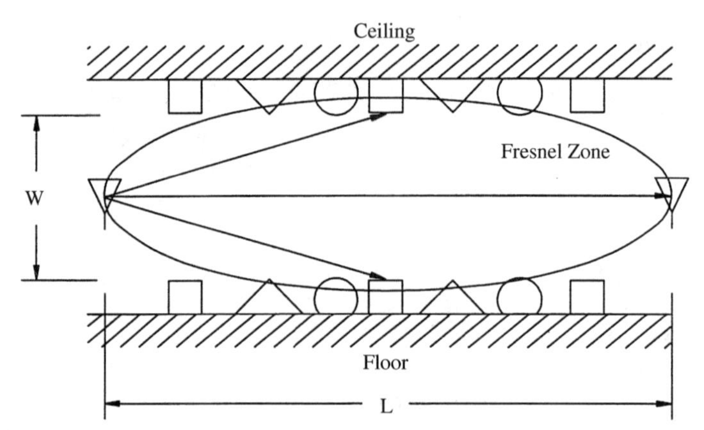

Figure 8-22 Fresnel zone for propagation between subscriber and access point located on the same floor of an office building showing propagation through the clear space between furniture and ceiling fixtures [31](©1992 IEEE).

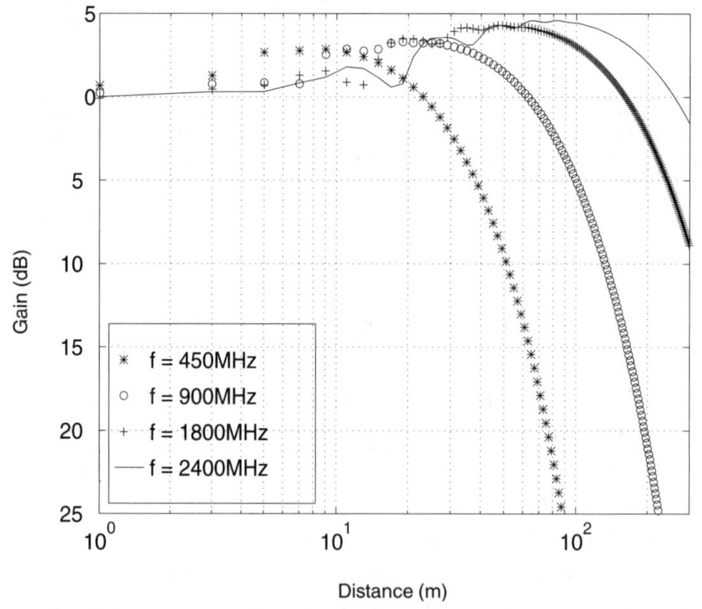

Figure 8-23 Excess path loss versus distance for propagation through the clear space.

the ellipse first encounters the scatters is W^2/λ, where W is the width of the clear space, and is seen to be proportional to frequency.

The effect of the scatters has been simulated using pairs of absorbing screens that extend down from the ceiling and up from the floor [31]. The spacing between successive pairs of screens is assumed to be uniform, but the opening between a pair of screens is taken to be a ran-

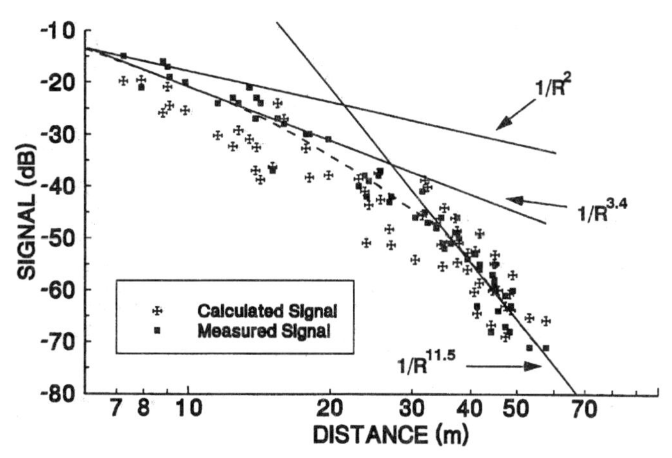

Figure 8-24 Comparison of measured and predicted small-area average signal strength at 852 MHz over one floor of a large office building showing increasing slope index [31](©1992 IEEE).

dom variable. Starting with a line source parallel to the edges of the screens, the field in successive openings is computed by the methods discussed in Chapter 6. In this case the numerical integration is simpler since the integration has a finite upper limit. Recognizing that the exact signal pattern will be dependent on the nearby scattering objects, $|E|^2$ is averaged over the opening between each pair of screens. The excess path gain at this transverse plane is then defined by dividing the average of $|E|^2$ by the corresponding value for free-space propagation. The excess path gain computed in this way for 2-m separation between successive pairs of screens, and for openings that are randomly distributed between 1.5 and 2 m, are plotted as a function of path length S in Figure 8–23 for several frequencies. (The corresponding plot in [31] is for a smaller clear space than is stated there.) It is seen from Figure 8–23 that the average signal initially increases with S, as if the ceiling and floor were giving constructive reflections, but beyond a certain distance the signal decreases dramatically. At the point where the excess path gain falls below 0 dB, the Fresnel width $\sqrt{\lambda S}$ ranges from 3.7 m at 450 MHz to 5.6 m at 2.4 GHz. Thus when the Fresnel zone grows to several times the clear space width W, the average signal will decrease below the free-space value. For a given distance, the excess path loss is seen to decrease with increasing frequency.

For two-dimensional ray tracing over a floor, the rays arriving at the subscriber location directly and via multiple reflections are first found. The total path gain for each ray is the product of (1) the free-space path gain, (2) the excess path gain $\mathrm{EG}(S_j)$ obtained from Figure 8–23 for the total path length S_j of the ray, 3) the power transmission coefficients of the walls the ray passes through, and (4) the power reflection coefficients of the wall the ray is reflected from. Thus for the jth ray, the path gain $P^{(j)}$ is

$$P^{(j)} = \left(\frac{\lambda}{4\pi S_j}\right)^2 \mathrm{EG}(S_j) \prod_n |\Gamma_n(\theta_{nj})|^2 \prod_m |T_m(\theta_{mj})|^2 \tag{8–16}$$

where the products are over the wall reflections, denoted by n, and the transmissions through walls, denoted by m. Because the reflection coefficients for interior walls are small, except for glancing incidence, inclusion of rays that undergo up to three reflections is reported to be sufficient for predictions [32].

Using the foregoing method, predictions have been made for one floor of a modern rectangular office building whose longest side is just over 80 m [31]. The walls are made of gypsum board, and the building has acoustic drop ceilings, giving a clear space that was taken to be 1.5 m. Computed and measured signals at 852 MHz for various locations on the floor are plotted in Figure 8–21 as a function of the straight-line distance to the base station. A third-order polynomial curve obtained by a least-squares fit to the measured data is shown in Figure 8–24, together with slope lines for large and small distances. The change in slope with distance shown in Figure 8–24 is due to a combination of the transmission through walls and the influence of the clear space on the excess path loss. The full change in slope is not obtained when the excess path loss is omitted from the predictions. At the higher frequencies used for PCS and unlicensed operation, the effect of the clear space is less significant and can be neglected except for very large buildings.

Summary

Computer codes based on ray approximations for computing the electromagnetic fields have been developed to predict the propagation characteristics in specific building environments. Some codes are intended to work with a two-dimensional database of the building footprints and are appropriate for low base station antennas in high-rise environments. For base station antennas mounted at or near the rooftops, or in an environment of mixed building heights, significant ray paths lie over the buildings and the code must make use of a three-dimensional building database to make accurate predictions. Ray codes have also been developed for propagation inside tunnels and inside buildings. Because the ray codes treat the building walls as specular reflectors, the outdoor ray codes are valid only over distances for which the Fresnel zone about the ray is smaller than that of the buildings. At 900 MHz this limit is about 1 km, so the codes can be used for microcellular applications.

Running time of the ray codes, which for three-dimensional codes is measured in hours for a single processor, is a limiting factor in their use. For a full three-dimensional ray trace, running time limits the number of diffraction events a ray is allowed to undergo to two. When predictions of received signal strength or path loss are compared with measurements, the distribution of errors is found to have an average of a few decibels, and a standard deviation of 6 to 10 dB, depending on the uniformity of the building heights and the distance from the base station. Various levels of approximation, such as in the VPL and SP-VP methods, have been introduced to improve functionality or reduce running time.

Because the codes can give the directions of the rays at both ends of the links, and the path length as well as amplitude of the ray fields, they can be used for system simulations. They may be used alone to predict characteristics, such as time-delay profile, angle of arrival distribution,

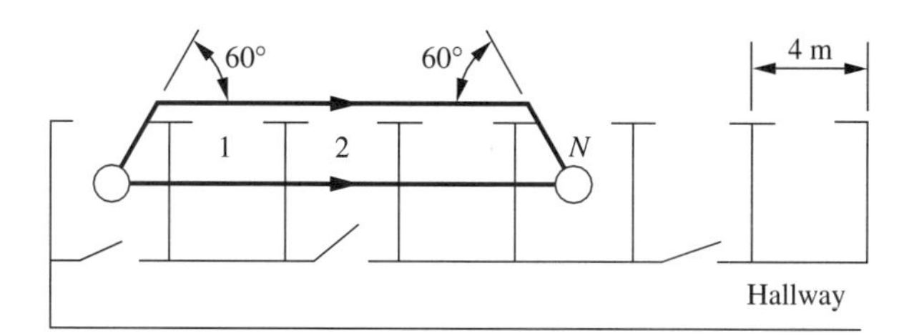

Figure 8-25

correlation of fading to different base stations, or as part of a larger event-driven program that simulates system performance. Propagation characteristics may be sensitive to the building environment, so that measurements in one city may not apply to another. Because the propagation characteristics are time consuming to measure, Monte Carlo simulation of different cities using ray codes may prove to be an important tool for understanding such dependence.

Problems

8.1 Consider rays of the type shown in Figure 8–6a that turn a corner by reflection only and just graze the corner at $(-w_x/2, w_z/2)$. Assume that x_{BS} and the other dimensions are as used in the example discussed in Section 8–2a.

(a) For such a ray making $N_L = 3$ reflections on the LOS street, find the number of reflections N_P on the perpendicular street after which the ray crosses the z axis at the closest point above $z =$ 100 m. Find the value of z at the crossing and compute the path gain of the ray at this point.

(b) Repeat part (a) for a ray that makes $N_L = 4$ reflections on the LOS street.

8.2 Various diffraction coefficients can be used to compute the path gain for the diffracted rays of Figure 8–6b. Compute the path gain for $|x_{BS}| = z_m = 100$ m using:

(a) The diffraction coefficients for a conducting 90° wedge ($\Gamma_0 = \Gamma_n = -1$).

(b) Felsen's diffraction coefficients for wedges with absorbing boundary conditions.

8.3 There is a small fan or wedge of rays that make $N_L = 3$ reflections on the LOS street and N_P reflections on the perpendicular street in Figure 8–6a. One edge of the fan is given by the ray that just grazes the corner at $(-w_x/2, w_z/2)$. The other edge is given by the ray that is first reflected on the perpendicular street at a point that is just next to the corner at $(w_x/2, w_z/2)$. Assume that $w_x = w_z = 24$ m, $L_x = 80$ m, and $L_z = 180$ m, and that $x_{BS} = -145$.

(a) Find the intervals along the $z > 0$ axis that are illuminated by this fan of rays, out to a distance of $z = 200$ m.

(b) Compute and plot as a function of z the contributions to the path gain from the rays in this fan.

8.4 Accounting for the actual diffraction angles, compute and plot the contributions to the path gain from the rays diffracted at the four corners in Figure 8–6b for distances in the range $w_z/2 < z \leq 200$ m. Assume that $f = 1.8$ GHz, $w_x = w_z = 24$ m, $L_x = 80$ m, and $L_z = 180$ m, and that $x_{BS} = -145$ m. For simplicity, use Felsen's diffraction coefficients for absorbing building walls. Compare this result to that of Problem 8–3.

8.5 Identical offices 4 × 5 m in size are arranged along an outside wall with windows, as shown in top view in Figure P8–5. A transmitter at the center of one room communicates with receivers at the center of the other room. For the direct ray, assume that the transmission loss through each wall is 6 dB. The diffracted ray leaves the building through a window, turns through 60°, and reenters the building through another window, again turning through 60°. Neglect transmission loss through the window glass. As a function of room number $N \geq 2$, compute and plot the path gain on a decibel scale for the direct and diffracted rays. Also plot the total path gain of the two rays.

References

1. M. O. Al-Nuaimi and M. S. Ding, Prediction Models and Measurements of Microwave Signals Scattered from Buildings, *IEEE Trans. Antennas Propagat.*, vol. 42, pp. 1126–1137, 1994.
2. U. Dersch and E. Zollinger, Propagation Mechanisms in Microcell and Indoor Environments, *IEEE Trans. Veh. Technol.*, vol. 43, pp. 1058–1066, 1994.
3. S. T. S. Chia, R. Steele, E. Green, and A. Baran, Propagation and Bit Error Ratio Measurements for a Microcellular System, *J. IRE*, vol. 57, pp. S255–S266, 1987.
4. M. C. Lawton and J. P. McGeehan, The Application of a Deterministic Ray Launching Algorithm for the Prediction of Radio Channel Characteristics in Small-Cell Environments, *IEEE Trans. Veh. Technol.*, vol. 43, pp. 955–969, 1994.
5. S. Y. Tan and H. S. Tan, Propagation Model for Microcellular Communications Applied to Path Loss Measurements in Ottawa City Streets, *IEEE Trans. on Veh. Technol.*, vol. 44, pp. 313–317, 1995.
6. S. Y. Tan and H. S. Tan, A Microcellular Communications Propagation Model Based on the Uniform Theory of Diffraction and Multiple Image Theory, *IEEE Trans. Antennas Propagat.*, vol. 44, pp. 1317–1325, 1996.
7. K. Rizik, J.-F. Wagen, and F. Gardiol, Two-Dimensional Ray-Tracing Modeling for Propagation Prediction in Microcellular Environments, *IEEE Trans. Veh. Technol.*, vol. VT-46, pp. 508–517, 1997.
8. G. Liang, Ray Based Models for Site Specific Propagation Prediction, Ph.D. dissertation, Polytechnic University, Brooklyn, N.Y., 1997.
9. O. Landron, M. J. Feuerstein, and T. S. Rappaport, A Comparison of Theoretical and Empirical Reflection Coefficients for Typical Exterior Wall Surfaces in a Mobile Radio Environment, *IEEE Trans. Antennas Propagat.*, vol. 44, pp. 341–351, 1996.
10. L. Piazzi and H. L. Bertoni, Achievable Accuracy of Site-Specific Path-Loss Predictions in Residential Environments, *IEEE Trans. Veh. Technol.*, vol. 48, pp. 922–930, 1999.
11. V. Erceg, A. J. Rustako, Jr., and R. S. Roman, Diffraction Around Corners and Its Effects on the Microcell Coverage Area in Urban and Suburban Environments at 900 MHz, 2 GHz, and 6 GHz, *IEEE Trans. Veh. Technol.*, vol. 43, pp. 762–766, 1994.
12. C. Demetrescu, B. V. Budaev, C. C. Constantinou, and M. J. Mehler, Electromagnetic Scattering by a Resistive Wedge, *IEEE Trans. Antennas Propagat.*, vol. 47, pp. 47–54, 1999.
13. F. Nui and H. L. Bertoni, Path Loss and Cell Coverage of Urban Microcells in High-Rise Building Environments, *Proc. IEEE GLOBECOM*, pp. 266–270, 1993.
14. C. C. Constantinou and L. C. Ong, Urban Radiowave Propagation: A 3-D Path-Integral Wave Analysis, *IEEE Trans. Antennas Propagat.*, vol. 46, pp. 211–217, 1998.
15. A. G. Kanatas, I. D. Kountouris, G. B. Kostraras, and P. Constantinou, A UTD Propagation Model in Urban Microcellular Environments, *IEEE Trans. on Veh. Technol.*, vol. 46, pp. 185–193, 1997.
16. V. Erceg, S. J. Fortune, J. Ling, A. J. Rustako, and R. A. Valenzuela, Comparisons of a Computer-Based Propagation Prediction Tool with Experimental Data Collected in Urban Microcellular Environments, *IEEE J. Sel. Areas Commun.*, vol. 15, pp. 677–684, 1997.
17. S. C. Kim, B. J. Guarino, Jr., T. M. Willis III, V. Erceg, S. J. Fortune, R. Valenzuela, L. W. Thomas, J. Ling, and J. D. Moore, Radio Propagation Measurements and Prediction Using Three Dimensional Ray Tracing in Urban Environments at 908 MHz and 1.9 GHz, *IEEE Trans. Veh. Technol.*, vol. 48, pp. 931–946, 1999.

18. G. Liang and H. L. Bertoni, A New Approach to 3-D Ray Tracing for Propagation Prediction in Cities, *IEEE Trans. Antennas Propagat.*, vol. 46, pp. 853–863, 1998.
19. T. Kurner, D. J. Cichon, and W. Wiesbeck, Concepts and Results for 3D Digital Terrain-Based Wave Propagation Models: An Overview, *IEEE J. Sel. Areas Commun.*, vol. 11, pp. 1002–1012, 1993.
20. H. L. Bertoni, P. Pongsilamanee, C. Cheon, and G. Liang, Sources and Statistics of Multipath Arrival at Elevated Base Station Antenna, *Proc. IEEE Vehicular Technology Conference*, 1999.
21. M. Nilsson et al., Measurements of the Spatio-Temporal Polarization Characteristics of a Radio Channel at 1800 MHz, *Proc. IEEE Vehicular Technology Conference*, 1999.
22. A. G. Emslie, R. L. Lagace, and P. F. Strong, Theory of the Propagation of UHF Radio Waves in Coal Mine Tunnels, *IEEE Trans. Antennas Propagat.*, vol. AP-23, pp. 192–205, 1975.
23. K. Sakai and M. Koshiba, Analysis of Electromagnetic Field Distribution in Tunnels by the Boundary-Element Method, *IEE Proc.*, vol. 137, Pt. H, pp. 202–208, 1990.
24. G. M. Whitman, K.-S. Kim, and E. Niver, A Theoretical Model for Radio Signal Attenuation inside Buildings, *IEEE Trans. Veh. Technol.*, vol. 44, pp. 621–629, 1995.
25. Y. P. Zhang and Y. Hwang, Characterization of UHF Radio Propagation Channels in Tunnel Environments for Microcellular and Personal Communications, *IEEE Trans. Veh. Technol.*, vol. 47, pp. 283–296, 1998.
26. S. Chen and S. Jeng, SBR Image Approach for Radio Wave Propagation in Tunnels with and without Traffic, *IEEE Trans. Veh. Technol.*, vol. 45, pp. 570–578, 1996.
27. Y. P. Zhang, Y. Hwang, and R. G. Kouyoumjian, UTD Prediction of Radio Wave Propagation Characteristics in Tunnel Microcellular Environments, Part 2: Analysis and Measurements, *IEEE Trans. Antennas Propagat.*, vol. 46, pp. 1337–1345, 1998.
28. D. Cichon, T. Zwick, and W. Wiesbeck, Ray Optical Modeling of Wireless Communications in High-Speed Railway Tunnels, *Proc. IEEE Vehicular Technology Conference*, pp. 546–550, 1996.
29. J. F. Lafortune and M. Lecours, Measurement and Modeling of Propagation Losses in a Building at 900 MHz, *IEEE Trans. Veh. Technol.*, vol. 39, pp. 101–108, 1990.
30. S. Y. Seidel and T. S. Rappaport, 914 MHz Path Loss Predictions Models for Indoor Wireless Communications in Multifloored Buildings, *IEEE Trans. Antennas Propagat.*, vol. 40, pp. 207–217, 1992.
31. W. Honcharenko, H. L. Bertoni, J. Dailing, J. Qian, and H. D. Yee, Mechanisms Governing Propagation on Single Floors in Modern Office Buildings, *IEEE Trans. Veh. Technol.*, vol. 41, pp. 496–504, 1992.
32. R. A. Valenzuela, A Ray Tracing Approach to Predicting Indoor Wireless Transmission, *Proc. IEEE Vehicular Technology Conference*, pp. 214–218, 1993.
33. S. Y. Seidel and T. S. Rappaport, Site-Specific Propagation Prediction for Wireless In building Personal Communication System Design, *IEEE Trans. Veh. Technol.*, vol. 43, pp. 879–891, 1994.
34. J. H. Tarng, W. R. Chang, and B. J. Hsu, Three-Dimensional Modeling of 900-MHz and 2.44-GHz Radio Propagation in Corridors, *IEEE Trans. Veh. Technol.*, vol. 46, pp. 519–526, 1997.
35. H. W. Arnold, R. R. Murray, and D. C. Cox, 815 MHz Radio Attenuation Measured within Two Commercial Buildings, *IEEE Trans. Antennas Propagat.*, vol. 37, pp. 1335–1339, 1989.
36. T. B. Gibson and D. C. Jenn, Prediction and Measurement of Wall Insertion Loss, *IEEE Trans. Antennas Propagat.*, vol. 47, pp. 55–57, 1999.
37. W. Honcharenko, H. L. Bertoni, and J. Dailing, Mechanisms Governing Propagation between Different Floors in Buildings, *IEEE Trans. Veh. Technol.*, vol. 41, pp. 787–790, 1993.

INDEX

G

H

I

K

L

R

S

T

U

W